114 Advances in Polymer Science

Polymer Analysis and Characterization

With contributions by
V. A. Bershtein, G. C. Berry, N. Ise
J. Lesec, H. Matsuoka,
I. S. Osad'ko, V. A. Ryzhov, J.-L. Viovy

With 151 Figures and 7 Tables

Springer-Verlag
Berlin Heidelberg New York
London Paris Tokyo
Hong Kong Barcelona Budapest

ISBN 3-540-57238-4 Springer-Verlag Berlin Heidelberg New York
ISBN 0-387-57238-4 Springer-Verlag New York Berlin Heidelberg

© Springer-Verlag Berlin Heidelberg 1994
Library of Congress Catalog Card Number 61-642
Printed in Germany

Typesetting: Macmillan India Ltd., Bangalore-25
Printing: Saladruck, Berlin; Bookbinding: Lüderitz & Bauer, Berlin
SPIN: 10126175   02/3020 - 5 4 3 2 1 0- Printed on acid-free paper

Editors

Table of Contents

Separation of Macromolecules in Gels: Permeation Chromatography and Electrophoresis

J. L. Viovy[1] and J. Lesec[2]

[1] Laboratoire de PhysicoChimie Théorique (U.A. CNRS 1382), E.S.P.C.I. 10 rue Vauquelin, 75231 Paris, Cedex 05, France

[2] Laboratoire de Physicochimie Macromoléculaire (U.A. CNRS 278), E.S.P.C.I. 10 rue Vauquelin, 75231 Paris Cedex 05, France

The principal methods for separating macromolecules in gels, permeation chromatography and electrophoresis, are reviewed. The emphasis of the review is put on the steric exclusion between the macromolecules to separate and the gel, and on the migration mechanisms. In gel permeation chromatography, a heterogeneous assembly of gel particles leads to a mechanism of steric exclusion in which the largest molecules migrate the fastest. In gel electrophoresis, the steric exclusion occurs in a volume homogeneously filled with gel, and the reverse, i.e. the smallest molecules migrating fastest, generally occurs. These general rules offer exceptions, however, and several mechanisms of migration used for practical applications are described in the review. The present experimental state-of-the-art is also discussed in some detail, and the different aspects and limitations of the two techniques are discussed (gel media, instruments, detection and analysis). The conclusion points to the probable convergence of these two complementary techniques in the near future.

Advances in Polymer Science, Vol. 114
© Springer-Verlag Berlin Heidelberg 1994

List of Abbreviations and Symbols

GPC	Gel Permeation Chromatography
GE	Gel Electrophoresis
GFC	Gel Filtration Chromatography
HPC	High Performance Chromatography
SEC	Size Exclusion Chromatography
V_e	Elution volume
K_a	Adsorption constant
S	Specific surface area
V_o	Volume of interstititial solvent
V_{gel}	Gel Volume
V_{mat}	Matrix Volume
V_p	Pores volume
K_{gpc}	Partition coefficient
K_{excl}	Exclusion coefficient
N	Chain size
Z_{pore}	Partition function of a chain in a pore
Z_{free}	Partition function of a chain in free solvent
k	Boltzmann constant
ΔF	Excess free energy
T	Absolute temperature
d_p	Particle diameter
w	Peak width
M	Molecular weight
η	Viscosity
V_h	Hydrodynamic volume
N_p	Number of theoretical plates
HETP	Height Equivalent of a Theoretical Plate
w	Peak width
F(V)	Experimental chromatogram
W(V)	Theoretical chromatogram
G(V)	Instrumental function
ISF	Instrumental Spreading Function

OD	Optical Density
ID	Internal Diameter
HPLC	High Performance Liquid Chromatography
UV	Ultra Violet
LALLS	Low-Angle Laser Light Scattering
MALLS	Multi-Angle Laser Light Scattering
THF	Tetrahydrofuran
ΔP	Pressure drop
r_c	Capillary radius
l_c	Capillary length
Q	Solvent flow rate
M_n	Number-average molecular weight
M_w	Weight-average molecular weight
M_v	Viscosity-average molecular weight
M_z	z-average molecular weight
I_p	Polydispersity index
Da	Dalton
DNA	DeoxyriboNucleic Acid
bp, kbp	Base-pair, kilobase-pair
T%	Total acrylamide content
C%	Bisacrylamide/total acrylamide content
μ	Electrophoretic mobility
μ_o	Electrophoretic mobility in free solvent
μ_1	Limiting mobility
f	fractional volume
s, 1, n	Surface, line and number density of the matrix
L, S, V	Mean length, excluded area and excluded volume of a particle
K_r	Retardation coefficient
R, r	Particle radius, fiber radius
R_s	Polymer Stokes radius
a	Average pore diameter
ζ	Friction coeffcient
t_{rep}	Reptation time
t_A	Equilibration time of a blob
q	Blob charge
E	Electric field
$\varepsilon = \dfrac{Eqa}{kT}$	Reduced electric field
$\dot{S}$	Curvilinear velocity
F	Effective force
t_{dis}	Disengagement time
ρ	Reduced end-to-end projection
$Q(\theta, \varepsilon)$	Distribution function of blob orientations
N*	Critical chain size for mobility saturation
t_{or}	Orientation time of a chain
τ	Pulse time

1 Introduction

Macromolecules are generally difficult to purify and to characterize. In particular, the crystalization processes widely used for small molecules are often inapplicable to the separation and/or the characterization of macromolecules with a large molecular weight. As an alternative, very sophisticated and efficient separation methods using the migration properties of macromolecules in gels were developed.

The typical scale of the microscopic structure of gels can be varied in a large range, typically from a few Å to several μm. Because of this rather unique property, gels are very extensively used in industry and research for the separation of molecules, macromolecules and microparticles. The simplest of such applications is gel filtration. The gel there acts as a membrane in which only particles smaller than its average pore size can penetrate, leading to the separation of a population of particles or molecules in two "families". The natural weakness of swollen gels, however, generally make them less suitable than solid organic or inorganic membranes for this purpose (indeed, the distinction between gels and membranes is often rather arbitrary, since the latter are often prepared by a sol-gel process, and may be partly "reswollen" by vector fluids). In this review, we focus on more specific transport properties in gels, which have provided separation technology in the last few decades with most spectacular and powerful tools, i.e. Gel Permeation Chromatography (GPC) and Gel Electrophoresis (GE). These techniques are very important tools for the characterization of polymers and biomolecules, and they are also extensively used on a preparative scale. Gel filtration, gel electrophoresis and GPC present numerous common features. They use an external field (hydrodynamic for filtration and GPC, electric for electrophoresis) to "push" solutes (particles or molecules) in the gel, which in turn hinders motion by restricting the space available to them. In spite of this common "excluded volume" origin, the migration mechanisms in gels may be rather subtle and very size-or-shape-specific. Describing these mechanisms, their use and their limits, is the aim of the following sections. There is a very strong interconnection of research and applications in this field, and it is probably fair to say that our understanding of transport in gels would be much poorer, had not progress been constantly promoted by technological needs and questions. Therefore, we also strongly refer to methodological aspects in the following presentation. A brief history of GPC is given in Sect. 2.1, and the different mechanisms of solute retention and transport are described in Sect. 2.2. The applications and methodology are discussed in Sect. 2.3. Electrophoretic transport is considered in Sect. 3. A brief history, and a description of the most extensively used gels are given in Sect. 3.1. The mechanisms of separation in electrophoresis are recalled and discussed in Sect. 3.2. Finally, we describe in Sect. 3.3 recent methodological and fundamental progress in electrophoresis, based on the introduction of pulsed fields and spectroscopic techniques of investigation. A general conclusion is given in Sect. 4.

2 Gel Permeation Chromatography (GPC)

2.1 History of GPC

The fractionation of macromolecules according to molecular size by liquid chromatography has its origin in the University of Uppsala in 1959. Using cross-linked dextran gels swollen in aqueous media, J. Porath and P. Flodin [1] obtained the size-separation of various water-soluble polymers, introducing the term of Gel Filtration Chromatography (GFC) [2] and the Sephadex family [3] still widely used today by biochemists. These "soft gel" packings only operate under low pressure which involves very long analysis times.

Some years later, J.C. Moore [4] in the Dow Chemical Company, achieved the separation of organo-soluble polymers using packings based on cross-linked polystyrene gels, immediately commercialized by Waters Associates under the name of Styragel [5]. These semi-rigid packings, partially swollen with organic solvents, could operate under moderate pressure, leading to a shorter analysis time. A new instrument, the Waters GPC 100 liquid chromatograph was simultaneously introduced, giving birth to Gel Permeation Chromatography (GPC) which came rapidly into extensive use for polymer analysis in industry as well as in the university laboratories.

The first totally rigid porous packing was introduced in 1966 by De Vries et al. [6–7] in Pechiney-St-Gobain. It consisted of porous silica gel commercialized under the name of Spherosil or Porasil. These packings were fully compatible with both aqueous and organic solvents and very pressure-resistant, but their use was restricted by the strong interactions between the silica surface and a number of solutes, leading to abnormal retention in GPC experiments.

In 1974, the development of new packings consisting of small porous particles with a typical diameter of around 10 μm instead of 50–100 μm initiated technological improvement. High Performance (HP) chromatography was born. High pressure technology and the reduction of column volumes, as a consequence of the smaller particle diameter and of the high efficiency of the columns, decreased analysis times from a few hours to some ten minutes. The first small particles introduced commercially for HPGPC were μ-Styragel by Waters Associates [5, 8] with the same chemical structure as Styragel but a diameter of 10 μm. Other similar packings [9, 10] were introduced at once. At the same time, Kirkland [11, 12] and Unger [13, 14] described porous silica micro-beads capable of withstanding high pressure and compatible with the majority of solvents. In order to prevent interactions with solutes, other packings were developed, involving silica surface modifications by organic grafting (μ-Bondagel [5], TSK-Gel SW [15], Lichrosphere-diol [16], etc.) or by organic coating [17] as SynChropak Catsec [18].

The most recent developments led to the commercialization of smaller cross-linked polystyrene beads with a diameter of around 7 μm and a very narrow particle size distribution, providing a very high column efficiency: ultra-Styragel [5], TSK-gel [15], Shodex-A gel [19], PL-gel [20]. At the same time, manufacturers have developed new families of porous hydrophylic gels for aqueous GPC with the same characteristics as organic GPC packings. They are fully compatible with aqueous solvents and withstand relatively high pressure (TSK-Gel PW [15], Shodex OH-pak and Ion-pak [19], PL-aquagel [20], . . .).

Modern high performance Size Exclusion Chromatography (HPSEC) or simply Size Exclusion Chromatography (SEC) is now the most commonly used term for GPC when using small porous particles, since, as well as for Gel Filtration Chromatography (GFC), the size exclusion by the gel is the dominant fractionation mechanism. As we shall see, however, several other mechanisms (often unwanted) may be responsible of chromatographic retention of macromolecules on a cross-linked gel.

2.2 Solute Retention in GPC

2.2.1 The Chromatographic System

Gel permeation chromatography is a particular liquid chromatography process in which the stationary phase is a porous cross-linked gel carefully packed in a chromatographic column and swollen by a mobile phase. For chromatographic reasons, the gel generally consists of spherical beads with the smallest possible diameter and a narrow distribution in size. For retention mechanism reasons, the size of the pores should have the same order of magnitude as that of the macromolecules to be separated in solution. Therefore, it is generally not possible to use gel beads with a diameter much smaller than 10 μm, in contrast to modern conventional Liquid Chromatography (HPLC).

A GPC column is represented in Fig. 1. The total column volume V_{col} can be considered as the addition of two volumes: the dead volume V_o corresponding to the interstitial solvent and the gel volume V_{gel}. V_{gel} is in turn divided into the gel matrix volume V_{mat} and the porous volume V_p filled up with the solvent:

$$V_{col} = V_o + V_{mat} + V_p \tag{1}$$

When considering GPC solute retention, only the total solvent volume V_m in the column must be taken into account: if K_{gpc} is the partition coefficient between the mobile phase V_o and the stationary phase V_p, the macromolecule elution volume V_e can be written:

$$V_e = V_o + K_{gpc}V_p \tag{2}$$

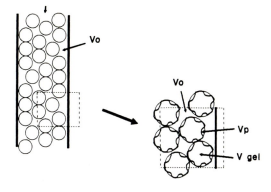

Fig. 1. Representation of the GPC column; V_o is the dead volume, V_p is the porous volume and V_{gel} is the gel volume

where K_{gpc} varies from 0 for excluded macromolecules (no access to the stationary phase) to 1 for macromolecules in total permeation (total access to the stationary phase). As we shall see later, K_{gpc} depends on the size of macromolecules, which means that the biggest macromolecules, excluded from the gel, elute first at V_o, then intermediate ones, the smallest ones being eluted last at $V_o + V_p$.

In fact, in order to cover the whole range of molecular weights of usual polymers, packings should have pore sizes from a few Angstroms to a few thousands of Angstroms. For technical reasons, it is only possible to synthesize packings with a limited range of pore sizes, and the GPC column is generally an assembly of several columns in series, packed with several gels of different porosities (Fig. 2). Another possibility is to use only one column packed with a mixture of several different gels with various porosities (mixed beds).

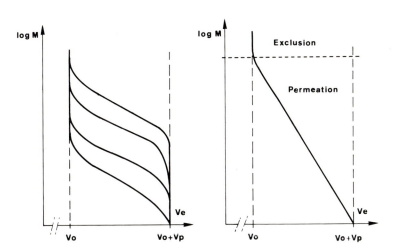

Fig. 2. Influence of pore size packing on the response in size of a GPC column: *on the left*, responses corresponding to various pore sizes are represented. The response corresponding to a combination of these pore sizes is represented *on the right*

2.2.2 Retention Mechanisms

Many attempts have been made to explain polymer fractionation in GPC (see the reviews of Audebert [22] and Hagnauer [23]), and the size exclusion mechanism is now widely accepted and demonstrated [24–28]. In actual experiments, however, several additional mechanisms based on interactions between the solute and the stationary phase may occur in the column, such as adsorption and liquid-liquid partition.

Adsorption Mechanism. Adsorption mechanism mainly occurs with mineral porous gel particles well-known for their surface activity. This is the classical mechanism of solid-liquid adsorption chromatography. The elution volume takes the form:

$$V_e = V_o + K_a S \tag{3}$$

where K_a is a constant representing adsorption forces and S the specific surface area. In the first approximation, the adsorption energy per molecule linearly increases with the number of adsorption sites it occupies on the surface. This generally leads to a retention time which exponentially increases with molecular weight, and it can produce an irreversible adsorption of the polymer onto the stationary phase. This phenomenon is extremely undesirable in classical GPC experiments and the trend is to modify the surface by grafting or coating organic compounds onto silanol groups, mainly responsible of these effects, to prevent adsorption when using those kinds of particles as the GPC stationary phase [15–18].

Liquid-Gel Partition Mechanism. Liquid-gel partition mechanism is the classical liquid chromatography process where the mobile phase is the solvent and the stationary phase is the swollen gel. In modern GPC where gels are highly cross-linked, only very small molecules are capable to enter the gel matrix, but with "soft gels" or "semi-rigid gels", the characteristic dimensions of gels are greater and macromolecules can penetrate the gel. In this case, the elution volume can be written according to the partition equilibrium:

$$V_e = V_m + K_{par} V_{mat} \tag{4}$$

where V_m is the volume of mobile phase ($V_m = V_o + V_p$), V_{mat} is the volume of the gel matrix and K_{par} is the partition coefficient. This equation represents the equilibrium theory of conventional liquid chromatography. The partition coefficient K_{par} depends on thermodynamic interactions between solutes and the swollen gel but also on the solutes molecular weight.

Figure 3 represents the liquid-liquid partition equilibrium. When the solute enters the pore volume V_p (by a size exclusion mechanism which will be discussed later), it has access to the gel matrix swollen by the solvent and a thermodynamic equilibrium may occur between these two phases. This equilibrium depends mainly on the relative solute affinity for the gel. Therefore, the elution volume will depend on the solute size, since the gel volume accessible to the solute is controlled by steric exclusion, but also on interactions between the

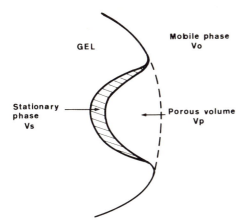

Fig. 3. Liquid-liquid partition equilibrium: V_o is the volume of the mobile phase, V_p is the porous volume and V_s is the volume of the stationary phase

solute and the gel (the latter depending on chemical affinity and on molecular weight).

This phenomenon mainly occurs in the low molecular weight range but it is not reproducible from one macromolecule to the other. As shown later, the GPC column has to be calibrated in molecular weight for general use with many different polymers, and the liquid partition mechanism appears as a disturbing phenomenon which should be minimized by a suitable selection of stationary and mobile phases. A general rule is to run GPC experiments with a thermo-dynamically good solvent of the polymer as mobile phase, the stronger the affinity of the solute for the solvent, the lower the affinity for the gel, and conversely. When using a poor solvent of the polymer, abnormal retentions may occur leading to a wrong interpretation of GPC chromatograms.

Size Exclusion Mechanism. Steric exclusion (or size exclusion) is the main process of polymer fractionation by GPC. This mechanism is based on a thermodynamic equilibrium between two phases: the interstitial solvent in the dead volume V_o and the solvent filling the porous volume V_p. If K_{gpc} is the partition coefficient, the elution volume V_e of a macromolecule is defined by:

$$V_e = V_o + K_{gpc}V_p \tag{5}$$

The evolution of the partition coefficient as a function of chain size and topology on the one hand, and of pore size distributions on the other, has a purely entropic origin, and it is well predicted by the theory of Casassa [29–35]: The partition function for a chain of a given size N trapped in a pore with volume V_p and inpenetrable walls, Z_{pore}, can be obtained by an enumeration of all possible random paths of N steps which start in the pore volume and never cross the wall (e.g. such as the full line and unlike the dotted line in Fig. 4).

The partition function of the same chain in an equivalent element of the dead volume V_o, Z_{free}, is given by the same enumeration without the condition of not crossing a wall: It is larger than the Z_{pore} by an amount representing all conformations such as the dotted line in Fig. 4. Therefore, entering the pore

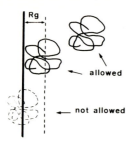

Fig. 4. Representation of different conformations of a randomly coiled polymer near a wall. *Solid line*: authorized conformation; *Dashed line*: forbidden conformation

implies for the chain an excess free energy of order

$$\Delta F = -kT \log(Z_{pore}/Z_{free})$$

By application of the Boltzmann equipartition law, this leads to a depletion of chains in the porous volume by a factor.

$$K_{excl} = \exp(-\Delta F/kT) = Z_{pore}/Z_{free}, \quad 0 < K_{excl} < 1 \tag{6}$$

An equivalent although mathematically less straightforward way of considering size exclusion is to say that the steric repulsion of the wall creates in its vicinity a depletion layer of order R_g (Fig. 5), so that only a fraction $V'_p = K_{excl} V_p$ of the pore volume is really available to the chain in the pore (Fig. 6).

The topology in actual gels is more complicated, but gel strands are uncrossable to macromolecular solutes, and they play the role of the "walls" pictured above. The K_{excl} coefficient depends on the macromolecule size and varies from 0 for a very high molecular weight polymer unable to penetrate the gel to 1 for a very small macromolecule capable of entering the total gel volume. Consequently, the elution volume of a macromolecule through a GPC column takes the following form:

$$V_e = V_o + K_{excl} V_p \tag{7}$$

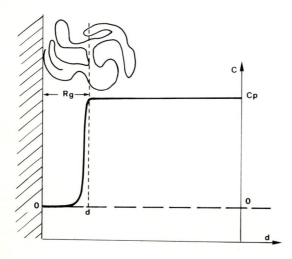

Fig. 5. Concentration profile of a randomly coiled macromolecule near a wall. R_g is the radius of gyration, C_p is the polymer concentration and d is the limiting distance for exclusion

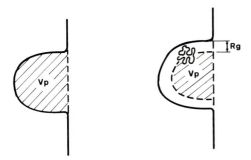

Fig. 6. Representation of the size exclusion of a macromolecule by the wall inside a pore

Accordingly, for a pure GPC experiment only controlled by the size exclusion mechanism, we have $K_{gpc} = K_{excl}$.

It is interesting to remark that, in this mechanism, gels with a monodisperse distribution of pore sizes are capable of performing the fractionation of molecules over a certain range of molecular weights (around two decades for random-coil polymers, around the half for rod-like polymers [21]). In practice, these ranges are greater since GPC packings generally have a substantial pore-size distribution which increases their fractionation capacity (but decreases the maximum resolving power). In general, as this distribution is not very wide for synthesis reasons, it is usual to mix together several gels of different porosities or to assemble several columns with different fractionation ranges in series in order to form a "custom-made column" appropriate for a specific problem. Another interesting approach is the bimodal distribution [36] used by Du Pont [37] with their porous silica microspheres (PSM). These packings have only two kinds of pores of different sizes. Nevertheless, they are characterized by a good linearity of the calibration curve over a wide range of molecular weights.

Another particularity of GPC is that the partition coefficient K_{gpc} only varies from 0 to 1, which means that all the solutes are eluted between two limits V_o and $V_o + V_p$ (volume V_p being experimentally of the same order of magnitude as volume V_o). This restricted volume of fractionation contrasts with other chromatographic methods. Figure 7 schematically represents the variations of elution volume in a GPC column with a single pore size distribution as a function of polymer size.

The mechanism of size exclusion in GPC is only controlled by entropy considerations, contrary to all other mechanisms encountered in chromatography, including the previously described adsorption and liquid-partition mechanisms based upon enthalpy differences. This unique feature makes the generality of size exclusion chromatography. To preserve this generality, however, the polymer analyst must be very careful with adsorption and liquid liquid partition, since they are very difficult to detect yet nevertheless strongly disturb the GPC experiment. In particular, the stationary and mobile phases must be chosen according to solute properties in order to minimize such effects.

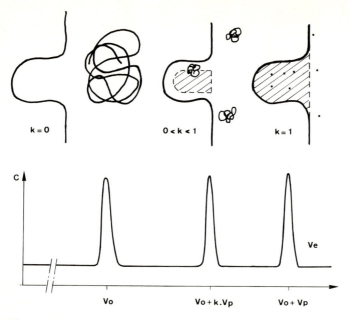

Fig. 7. Representation of the size exclusion of a macromolecule in a GPC column. k = 0: excluded macromolecule; 0 < k < 1: selective permeation

2.2.3 The GPC Column

Chromatographic Characteristics of the Column. Leaving aside the spurious non-exclusion effects discussed previously, the main characteristic of a GPC column is its efficiency defined by the number N_p of theoretical plates, or its reciprocal HETP (height equivalent to a theoretical plate). These parameters have been widely discussed in classical chromatography. They have their origin in a non-uniform migration of the solute across the column. This effect, also called band broadening, is represented in Fig. 8: As the solute migrates through the column, diffusion effects and differences in paths arising from packing heterogeneities makes the band containing the solute spread out progressively, leading to a Gaussian concentration profile at the outlet of the column. This effect depends on the particle diameter d_p (it approximately varies as d_p^{-2}), on the shape of the gel beads, and on their size distribution. Spherical particles with the narrowest possible distribution represent the best geometry. Good gel packing (Fig. 8a) provides a sharp peak while poor gel packing (Fig. 8b) produces a broad peak. Another possible cause of band broadening is a non-uniform packing: even with good gel particles, when the column has not been packed homogeneously, global band broadening may occur, as shown in Fig. 8c.

The plate number N_p of a column (Fig. 9) is defined by:

$$N_p = 16V_e^2/w^2 \qquad (8)$$

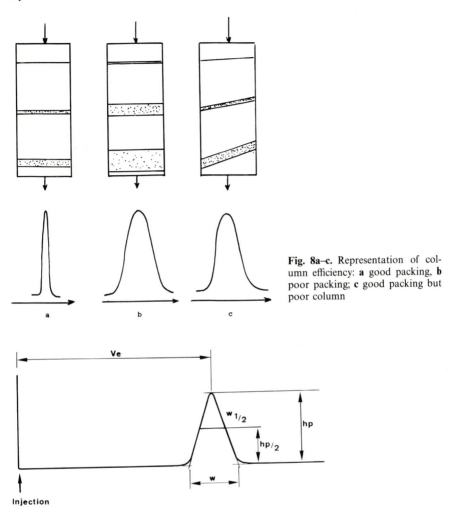

Fig. 8a–c. Representation of column efficiency: **a** good packing, **b** poor packing; **c** good packing but poor column

Fig. 9. Method for evaluating the plate count of a GPC column: V_e is the elution volume, w is the peak width and h_p is the peak height

or other similar relationships [21] where V_e represents the elution volume and w the peak width. This plate number is easy to determine experimentally as shown in Fig. 9. Modern GPC columns are generally made of stainless steel tubing, 3/8″ O.D. and 9/32″ I.D., each end being equipped with low porosity end-fittings to keep the gel inside the column. Their plate count is usually more than 10 000 plates/foot which means that, for a classical set of 4 columns, a peak corresponding to a monodisperse low molecular weight compound is eluted around 40 ml and has a width smaller than 800 µl , which gives a minimum of error in data interpretation.

The band-broadening effect is a very significant phenomenon in GPC since it strongly disturbs data treatment and it distorts the representation of polymer molecular weight distributions. It may also lead to a wrong interpretation of chromatograms and to an inaccurate molecular weight determination, especially for narrow distribution polymers. When the distribution is broad, the relative influence of peak spreading decreases and becomes negligible when polydispersity is very important. Many attempts have been made to evaluate this effect in order to correct it during data treatment [38]. Basically, the problem is to solve Tung's integral equation [39, 40] which describes the band-broadening effect on the GPC chromatogram:

$$F(V) = \int_{-\infty}^{+\infty} G(V - y)\, W(y)\, dy \tag{9}$$

where W(y) is the true chromatogram, F(V) is the experimental chromatogram and G(V) is a function representing instrumental peak spreading. For infinitely accurate data, the determination of W is a rather straightforward problem which can be solved, for instance, by a polynomial series. For actual imperfect and noisy data, however, Eq. (9) enters the vast class of "inverse problems", well known for their technical difficulty. Several methods have been proposed [41–43], but they do not seem to give a definitive answer to this issue. More recently, an other approach based on Instrumental Spreading Shape Function (ISF) has been proposed [44–46] introducing a polyplatykurtic coefficient to describe a function similar to the one of Provder and Rosen [47].

Column Calibration. The response to molecular weight is the main function of a GPC column. This property is represented by a calibration curve in which the logarithm of the molecular weight is approximately a linear function of the elution volume (Fig. 10). Obviously, this function mainly depends on the pore size distribution of the gel, which is generally unknown. Consequently, the calibration curve has to be determined experimentally. Several methods have been established but the most common uses a set of very narrow standards with various molecular weights in order to determine their elution volume. However, this procedure is only valid for a given polymer/solvent system, each new one needs its own calibration curve (Fig. 10).

In 1966, Benoit et al. [48] introduced the universal calibration method utilizing the hydrodynamic volume of macromolecules. This volume V_h can be expressed by the theory [49] as the product of the molecular weight M by the intrinsic viscosity [η] :

$$V_h = [\eta]M \tag{10}$$

This method of calibration is totally consistent with Casassa's theory [29] based on the polymer chain radius. It provides a unique curve when Log ([η]M) is plotted against elution volume, whatever the nature of the polymer (Fig. 11). Since the intrinsic viscosity is a function of the molecular weight through the empirical Mark-Houwink relationship:

$$[\eta] = KM^a \tag{11}$$

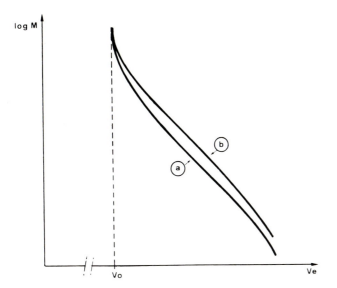

Fig. 10. Classical calibration curves Log(M) = f(V$_e$) for different polymers A and B

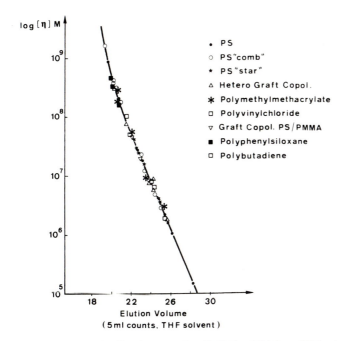

Fig. 11. Universal calibration curve Log{[η]M} = f(V$_e$) for a GPC column

the only knowledge of the two coefficients K and a is necessary to apply the universal calibration method. Unfortunately, these parameters are only known accurately for a few polymers. The only route to universal calibration, as we shall see later, is to couple GPC with viscometric detection, which provides continuous viscometric information and allows the determination of these coefficients.

Universal calibration has been widely verified in various conditions, even in polar solvents like water. Most often, the discrepancies reported in some particular cases could be traced to mechanisms, other than steric exclusion, taking place in the system and disturbing sample elution.

2.3 Applications of GPC

2.3.1 Instrumentation

The chromatograph. GPC equipments are very similar to HPLC ones, now widely available from several manufacturers. They consist of a liquid chromatograph composed of a solvent reservoir, a pumping system, an injector, a column set and a concentration detector eventually coupled to additional detectors (Fig. 12). The solvent is generally filtered by an in-line filter to ensure proper operation. The pumping system has to be accurate and capable of running under high pressure (more than 100 bar with modern columns). It must deliver the solvent at an extremely constant and reproducible flow rate, so that elution

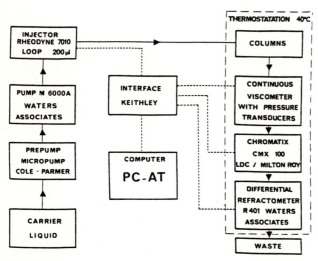

Fig. 12. Schematic diagram of a GPC instrument with triple detection (light scattering, viscometer, refractometer)

times can be accurately converted into elution volumes for the use of the calibration relationship between molecular weights and elution volumes. The typical flow rate is 1 ml/min.

The pumping system is followed by an injector for sample introduction. The most widely used device is a loop injector which permits the sample to be introduced with great accuracy into pressurized solvents. For classical columns (3/8″ O.D., 1 foot long), the injection volume must not be greater than 100 µl per column. In classical GPC, an accurate knowledge of injection volumes and sample concentrations was not necessary. In modern GPC using coupled mass detectors, however, these two parameters must be determined accurately since they are used in molecular weight calculations. The trend is to use a programmable automatic injector. All the components of the GPC instrument are connected by low dead volume fittings and stainless steel capillary tubings (1/16″ O.D., 9/1000″ I.D.) in order to minimize instrumental band spreading.

Temperature is not an important parameter in GPC since it does not affect the fractionation mechanism. It must be maintained constant (generally at room temperature), however, to obtain a good reproducibility and stability, especially for the detector response. It is sometimes necessary to operate at elevated temperature in order to decrease the viscosity of the solvent and to facilitate the steric exclusion equilibrium. This increases column efficiency and reduces pressure drop in the system. Moreover, some polymers require high temperature conditions to dissolve and special equipment able to withstand high temperature running must be used in these particular cases. The Waters Associates GPC 150C and GPC 150CV [5] are completely integrated and automatic chromatographs capable of running at up to 150 °C – this is particularly useful for polyolefine analysis.

Detectors. The detector is an important part of the GPC. In contrast to HPLC, the most common GPC detector is the differential refractometer which provides a signal proportional to polymer concentration. This detector is not very sensitive, but it has the advantage of being non-specific and quasi-universal. On the other hand, the refractometer is extremely sensitive to temperature variations since a sensitivity around 10^{-7} refractive index units is necessary, so that the temperature must be carefully stabilized. The cell usually has a volume of around 10 µl to minimize remixing and band spreading. The trend in modern GPC is to couple one or several other detectors to the concentration detector providing complementary information on the polymer.

The spectrometer is sometimes used alone as a concentration detector but it is more commonly used as a secondary detector for composition analysis, especially with copolymers [50–52]. Obviously, solutes must absorb at a particular wavelength to which the spectrometer has to be set. The solvent must be transparent at this wavelength, which may cause problems of selection with regard to polymer solubility. An ultraviolet (UV) spectrometer is the most commonly used, and it is very concentration-sensitive when the sample has a good absorption coefficient. Classical UV detectors operate at 254 nm but many multi-wavelength spectrometers are available. Other spectrometers like infrared spectrometers (IR) are also used, for example in polyolefine analysis [53]. Detector cells usually have a volume around 10 µl.

In modern GPC, there is a trend towards coupling the concentration detector with detectors sensitive to molecular weight. Such detectors provide very useful specific information complementary to the concentration information. Two techniques drawn from polymer physics are presently available, namely viscometric and light scattering detection. These two techniques run in a similar way to static measurements but they provide values at every point of the polymer distribution. They allow an on-line study of macromolecular properties as a function of elution volume, the role of the GPC column being to fractionate the polymer and to deliver fractions to the detectors for measurement.

Viscometric detection is a tool for universal calibration since it provides information on the intrinsic viscosity (required for the use of the $[\eta]M$ parameter). In early developments, a classical Ubbelhode viscometer [54, 55] was simply added after the refractometer. This assembly was no longer usable with the small volume of modern GPC columns. In 1972, Ouano [56] proposed an other type of viscometer based on the measurement of the pressure drop in a capillary tube using pressure transducers. A similar device, shown in Fig. 13 was then extensively studied [57–61, 63]. It has been recently improved by the use of a differential transducer and a low dead volume capillary. It is now commercially available from Waters Associates [5] with the GPC 150 CV high temperature chromatograph [62].

Viscometric detection is based on the Poiseuille relationship:

$$\Delta P = \frac{8\, l_c}{\pi\, r_c^4}\eta Q \tag{12}$$

where ΔP is the pressure drop, l_c and r_c are the length and the radius of the capillary, respectively, η is the viscosity of the eluent and Q is the solvent flow rate. At constant flow rate Q, ΔP is directly proportional to the viscosity η. At

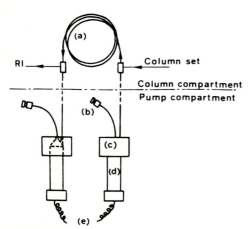

Fig. 13. Schematic diagram of the single capillary viscometer: (*a*) capillary; (*b*) purge; (*c*) transducer assay; (*d*) pressure transducer; (*e*) electronic unit

constant viscosity, ΔP is proportional to the flow rate Q, which provides a very sensitive flow-meter. Consequently, the single capillary viscometer [56–63], represented in Fig. 13, is very sensitive to flow fluctuations and it requires an enhanced pumping system with an extremely stable flow rate of solvent. For this reason, it can be considered as a very efficient diagnostic tool for trouble-shooting in the pumping system, in addition to its more fundamental use in universal calibration. Also, viscometric detection is particularly useful for long-chain branching determination [59, 67, 68]. Other similar detectors [63–65] were developed at the same time and are now commercially available. These other models [64, 65] are more complicated and were designed to be insensitive to flow variations. They are available from Viscotek [66]. Both have several capillaries and several transducers: They make use of a delay volume for flow rate compensation by baseline measurement.

The second type of molecular-weight sensitive detector is the light scattering photometer. Firstly described by Ouano and Kaye [69], it is now available from Chromatix [70]. It uses the Low Angle Laser Light Scattering technology (LALLS) which consists of measuring scattered light at a very low angle (around 5 °). Therefore, one can assume that the form factor is equal to unity and that the signal is directly proportional to the weight average molecular weight. The major interest of this detector is that calibration is no longer necessary, since absolute molecular weights are provided as a function of elution volume. When the light scattering detector is used in conjunction with a calibration curve, it provides information on polymer retention or on long-chain branching [71]. More recently, another detector using Multi Angle technology (MALLS) was introduced [72, 73]. Measurements are carried out at 16 simultaneous angles and signals are extrapolated to zero angle by a software program. This device provides absolute information on the radius of gyration.

Another type of detector was recently described by W. Yau et al. [74]. It is a continuous osmometer based on the pressure drop in a capillary column filled up with small beads of a swollen gel. The variation of osmometric pressure produces gel swelling variations when the polymer is present in the column and, accordingly, a variation of the pressure drop. The measurement is based on the same technology as the Yau viscometer [65, 68]. This detector is sensitive to the number average molecular weight and, consequently, very efficient in the low molecular weight region. The relationship between detector response and absolute molecular weight, however, is not clearly established at this time. This device is not yet commercially available.

Data Acquisition. The first purpose of a GPC analysis is to calculate average molecular weights and molecular weight distributions. The mathematical procedure is relatively simple and it can be achieved from the chromatogram on a strip-chart recorder. When complicated calculations are needed, especially in multidetection GPC, highly sophisticated computer methods are required and the previous procedure is no longer possible. Due to the tremendous expansion of computers during the past ten years, it is now common to include a micro-computer in the GPC instrumentation. Detectors are connected to the computer through an interface which performs analog to digital conversion of signals. Data are logged on mass storage devices for further computations with appropriate GPC software.

2.3.2 Experimental Conditions

Gel/Solvent Selection. Selection of the gel/solvent system is mainly directed by sample solubility. The mobile phase must be a "good" solvent of the polymer to minimize non-exclusion effects, but it must also be a good swelling agent of the gel to avoid gel collapse and column shrinkage. For organic polymers, polystyrene gels [5, 15, 19, 20] are very appropriate and the most commonly used solvent is tetrahydrofuran (THF). It dissolves many polymers and it is very compatible with polystyrene gel packings. It can be replaced by toluene, chloroform or methylene chloride. Unfortunately, some important industrial polymers are not soluble in THF and their analysis must be carried out in other solvents, mainly at high temperature [21]. Polyolefins are analyzed in *o*-dichlorobenzene or 1,2,4-trichlorobenzene at 135–145 °C, polyamides in metacresol or benzyl alcohol at 135 °C [75], polyesters in *N*-methylpyrrolidone and polyurethanes in dimethylformamide.

Water soluble polymers are obviously analyzed with water as the mobile phase. Salts should be added to avoid polyelectrolyte effects when the polymer is charged. Porous silica gels with a grafted surface [5, 15, 16] are often used, but very efficient porous hydrophylic gels [5, 15, 19, 20] are now available.

Polymer Analysis. The first level of data analysis is the survey of the chromatogram on the strip-chart recorder. Several qualitative pieces of information can be obtained by looking at the polymer peak which is an image of the molecular weight distribution. For control operations. chromatograms are considered as finger-prints that can be superposed for comparison. The shape of the chromatogram (symmetrical, non-symmetrical, monomodal, bimodal, etc.) characterizes roughly the distribution function and the peak width(s) give an idea of the polymer polydispersity. Elution volume at the peak apex can be easily measured, providing an evaluation of the average molecular weight using a calibration curve established with narrow standards. When quantitative information is needed, a second level of data analysis is required involving peak summation methods. Digitized peak heights must be manually measured and introduced into a calculator, that is a tedious procedure and the third level of data analysis with automatic acquisition and data reduction by an in-line microcomputer is generally preferred nowadays.

Low Molecular Weight Compounds. GPC may be used for low molecular weight compounds since efficient gels with small pore sizes are available. Since these gels are less cross-linked than the regular ones, they are more sensitive to swelling. This property restricts the choice of solvents. In this case, elution volumes are less controlled by steric exclusion mechanism and the relationship between molecular size and molecular weight is not well-defined and depends on molecular structure. In addition, with regard to the particular behavior of the swollen gel the partition between the mobile phase and the swollen gel may occur and disturb elution volumes. Consequently, molecular weight determinations should be regarded with caution even when conditions seem to be controlled well. Qualitative separation in the low molecular weight range is particularly useful, however, for instance for the characterization of additives in plastics. In some cases, it can be used even for quantitative determination since such peaks are well separated from the polymer peak (Fig. 14).

When the column set is efficient exclusively in the low molecular weight range, the main purpose of a GPC experiment is essentially to enumerate the number of components in a mixture or to quantify one or more of these. This kind of experiment is closer to HPLC than GPC, except that all the components are eluted between two limits (V_o, $V_o + V_p$). This specific behavior is used to obtain a fast estimation of the composition of a complex mixture without any optimization of the mobile phase. When a very high resolution is needed for the separation of two similar molecules, for example isomers [76–77], the resolution can be greatly increased by the "recycling" technique which consists of re-introducing eluates into the column by switching eluates towards the pump when they come out of detector. This procedure is equivalent to increasing the number of columns and it allows to reach an extremely high efficiency [78] as shown in Fig. 15.

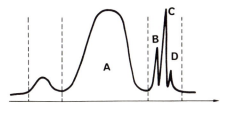

Fig. 14. Evidence of additives in plastics by GPC. (A): polymer peak; (B): plasticizer; (C): stabilizer; (D): dye

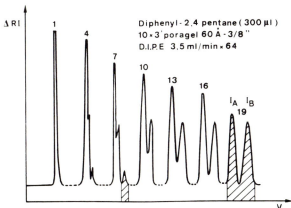

Fig. 15. Separation of diphenyl-2,4-pentane stereoisomers by recycle-GPC in di-isopropyl ether (DIPE), [78]. 19 cycles were necessary to separate the two isomers I_A and I_B

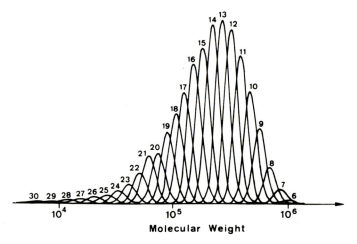

Fig. 16. Preparative fractionation of standard polystyrene NBS 706 into narrow fractions [81]

Preparative GPC. Since GPC is a non-destructive technique, it is possible to collect eluates as they come out of the detector. Evaporation of solvent or polymer precipitation allows recuperation of fractions for preparative purposes [79, 80]. However, the amount of sample is extremely small in analytical GPC, and a special apparatus with a large-diameter column must be used to obtain a sufficient amount of material. One purpose of preparative GPC is to purify a polymer by the "heart-cut" technique which consists in collecting only the heart of a broad polymer distribution and rejecting the wings containing the lowest and highest molecular weight components. This procedure provides a better defined material with a reduced polydispersity, but the fractionation can be carried out more accurately in order to prepare several narrow standards by cutting the distribution into several narrow fractions [81] as shown in Fig. 16. Another possibility is the purification of low molecular weight compounds [78] using low porosity gels, a procedure by which large amounts of purified materials can be obtained by increasing sample load. Generally, industrial fractionations are rather commonly performed especially with biological substances.

2.3.3 Molecular Weight Calculations

Classical Calculations. As the chromatogram provides continuous information on concentration, data reduction amounts to summing up the properties of "slices" of the chromatogram, each slice being characterized by its elution volume V_i and its intensity h_i proportional to the concentration C_i. Molecular weights M_i are determined from V_i through the calibration curve

$Log(M_i) = f(V_i)$. The differential molecular weight distribution is represented by C_i versus $Log(M_i)$ and the different average molecular weights are the moments of this distribution. The number average $\bar{M}_n$, the viscosity average $\bar{M}_v$, the weight average $\bar{M}_w$ and the z-average $\bar{M}_z$ molecular weights can thus be calculated using the following relationships:

$$\bar{M}_n = \frac{\sum C_i}{\sum C_i / M_i}$$

$$\bar{M}_v = \left[\frac{\sum C_i M_i^a}{\sum C_i}\right]^{1/a}$$

$$\bar{M}_w = \frac{\sum C_i M_i}{\sum C_i}$$

$$\bar{M}_z = \frac{\sum C_i M_i^2}{\sum C_i M_i} \tag{13}$$

The polydispersity index is defined as: $Ip = \bar{M}_w / \bar{M}_n$. It characterizes the degree-of-polymerization distribution function.

Calculations with Multidetection. The coupling with spectrometry (mainly UV and IR spectrometry) does not bring any improvement of molecular weight calculations. The ratio between the spectrometer signal and the one from the refractometer, however, may provide the distribution in composition, for instance for copolymers or complex biopolymers.

The viscometry coupling is mainly used in conjunction with universal calibration. The viscometer data provides the average intrinsic viscosity but also the variations of instantaneous intrinsic viscosity versus molecular weight, allowing the determination of the Mark-Houwink coefficients and, consequently, the calculation of real molecular weights in universal units. In addition, by comparing the experimental viscosity law with this of the corresponding linear polymer, the g' distribution of long-chain branching and the λ distribution of branching frequency can be generated.

The light scattering coupling is characteristic in providing the absolute molecular weight along the concentration profile. This unique property eliminates the use of a preliminary calibration. Such data may also be used for the determination of the long-chain branching distribution.

The trend to-day is multidetection using several detectors connected in series. Many combinations are possible providing more information on the sample and, for example, the triple coupling refractometry-viscometry-light scattering provides several cross-checks on the behavior of polymers during elution, especially in complicated systems such as water-soluble polymers [82, 83]. The comparison between molecular weights calculated using a light-scattering detector independent of elution volumes, and those obtained by the universal calibration and viscometric detection dependent on elution volumes allows the study of the retention mechanisms.

3 Gel Electrophoresis

3.1 History and Current Gels for Electrophoresis

3.1.1 Introduction

Gel electrophoresis is nowadays one of the most common analytical tools for proteins, amino acids and, to a lesser extent, polysaccharides. The idea of separating charged particles or ions in solution using electric fields is rather old, and separation of proteins by zone boundary electrophoresis was performed as early as 1937 [84]. The first difficulty encountered is convection, which leads to transport processes faster than electrophoresis itself if rather drastic precautions are not taken. In the last few years, rotative zone stabilization [85], laminar flows [86] and most notably capillary electrophoresis [87], have led to several breakthroughs in liquid electrophoresis. The first practical developments and the large majority of present applications of electrophoresis, however, rely on porous matrices. Various matrices such as paper [88, 89], starch [90], agar gum [91] or pectin [92] were used in early applications. Among those, only paper has survived for routine non-demanding applications, and most electrophoresis work is now performed on synthetic or modified natural gels (reviewed in Ref. [93]).

3.1.2 Cellulose Acetate

Cellulose acetate is prepared from natural cellulose, by the acetate esterification of the two hydroxyl groups in each hexose of the polysaccharide chain. Developed in the 1950s [94], cellulose acetate and in particular the commercial gelatinized film "Cellogel" are still widely used for clinical applications thanks to their easy handling as ready-made dry membranes which can be equilibrated with buffer and dried again for storage. It has rather large and non-uniform pores, however, and its action in separation processes is essentially anticonvective. This medium is efficient for separating particles with different charge/friction ratios (which includes many biologically important proteins and polysaccharides) but it offers little discrimination by size.

3.1.3 Agarose

Agarose is also a natural polysaccharide (galactan), extracted from seeweed [95]. The basic repeating unit (agarobiose) is given in Fig. 17.
    This formula, however, is somewhat idealized, and natural agaroses always have some degree of substitution by sulfate, pyruvate and methoxyl groups.

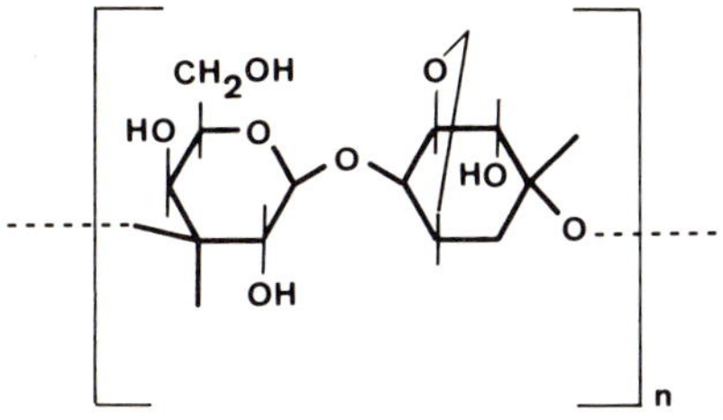

Fig. 17. Agarobiose unit

These substitutions depend on the particular species of seeweed used, and on the preparation procedure. They significantly influence physical properties such as gelling and melting temperatures, pore size and electroendosmosis (the global move of solvent induced by charges on the gel in the presence of an electric field). There is no "generic" agarose, and the commercially available products often present minor differences in properties (although different batches in a given brand are now highly reproducible thanks to optimized preparation procedures). Typical agarose has a molecular weight of about 120 kDa [96]. On cooling, molecules are believed to organize themselves into double helices, which in turn aggregate to form thick, rigid and interconnected "pillars" [97]. This high degree of aggregation is responsible of the high strength of agarose gels at low concentration: 0.5% gels can be manipulated with moderate care in routine applications. Gelification still occurs at concentrations as low as 0.1% but such gels must be supported by special attachments. This aggregation also results in large pores (500–5000 Å) which make agarose suitable for the separation of large molecules (typically, from 10^5 to 10^7 Da, and even much more for the special case of linear DNA, as described in Sect. 3.2.2). Specialized agaroses are now prepared by controlled substitutions (see e.g. Ref. 93). For instance, the controlled introduction of hydroxyethyl groups seems to reduce the aggregation of agarose helixes, leading to matrices with lower melting and gelling temperatures, smaller pore sizes, and lower gel strength. The FMC corporation (formerly "Marine Colloids") is famous for its leadership in agarose, but many companies (Biorad, GenApex, IBF, Serva, etc.) now offer agaroses for various applications, which mostly differ by their melting and gelling temperatures. "Very low-melt" agaroses (15–45 °C), such as SeaPrep FMC or NuSieve GTG have the smallest pores: They are used for proteins in the range 50–500 kDa, or for preparative processes in which high temperatures should be avoided. "Low-melt" agaroses (25–65 °C, SeaPlaque FMC, Gen–Apex DNA) are easier to handle, and they also have many preparative applications. The most widely used agaroses, however, are of the "high-melt" type (35–95 °C, SeaKem FMC, Indubiose IBF, HSB Gen–Apex, etc.). Recently, agaroses with a still higher gelling temperature and gel strength have been proposed for pulsed electrophoresis of very large DNA (Fastlane GTG). The degree of electroendosmosis can also be controlled by selected substitution (LE, ME HEEO SeaKem, IsoGel).

3.1.4 Polyacrylamide

Polyacrylamide gels were applied to electrophoresis in the early sixties [98–102], and they are now the most widely used matrices. Their popularity stems from various properties, such as:

— A very good mechanical resistance, which allows the handling of gels up to 50×20 cm in size and down to a few tens of millimeters in thickness, for instance in high resolution DNA sequencing applications.

— Good optical clarity

— Electric neutrality, which prevents electroendosmosis

— A large range of pore sizes (typically 100–1000 Å) available with good homogeneity and good reproducibility, which allow the separation of molecules from typically 1000 to 10^6 Da.

Gels for electrophoresis are generally obtained by the polymerization of acrylamide and N,N'-methylenebisacrylamide ("Bis"), catalyzed by persulfate and/or N,N,N',N'-tetramethylenediamine ("TEMED"). The convention is to call T the total acrylamide + Bis weight percent, and C the percentage of Bis in the total weight of acrylamide. Gels with good mechanical properties are obtained in a rather narrow range of C (typically 3–6%), but T can be varied from a few percent to typically 30%. The porosity then decreases with increasing T in a roughly linear manner [93]. Since all components of polyacrylamide gels are well defined chemical compounds, very reproducible properties can be obtained whatever the commercial origin of the components. Ready-made dried membranes also become more and more popular, in particular for gels with concentration gradients which are more difficult to prepare with a good reproductibility in the laboratory (Phastgel Pharmacia, MiniGels DPC, etc.).

Chemical modifications of "generic" acrylamide gels have also been proposed. For instance, the modification of the coordination of the crosslinker can affect the structure of pores [103], and various substitutions can provide the gel with special chemical properties or affinities with a given electrolyte. The most famous of such modified acrylamide are "immobiline" gels [104–105], which provide accurate Ph gradients along the gel, or the more recent Hydrolink gels, which offers sieving properties intermediate between acrylamide and agarose, and amphiphilic properties.

3.2 Separation Mechanisms in Gel Electrophoresis

3.2.1 Steric Exclusion

In the most straightforward approach to gel electrophoresis, the gel essentially acts as a sieve with a distribution of pore sizes. The connection between this mechanism and this of gel filtration was first made by Morris [106]. This author assumed that the ratio of the electrophoretic mobility in the gel μ, relative to the mobility in free solvent μ_0 is f, the fractional volume available to the particle in the gel:

$$\mu = f\mu_0 \tag{14}$$

The distribution of pore sizes in a suspension of long fibers, and the corresponding probability of penetration for spherical particles, was derived from purely geometrical considerations by Ogston [107]. A similar model, in which the obstacles can be approximated as two-dimensional objects (like thin sheets) was treated by Giddings [108]. Finally, these models were generalized by Rodbard and Chrambach [109–111] to particles and obstacles of arbitrary size. The fractional free volume is expressed as $f_2 = \exp(-sL)$ for a particle of mean length L in a suspension of planes with surface area s per unit volume, $f_1 = \exp(-lS/4)$ for a particle with excluded area S in a suspension of long fibers with total length 1 per unit volume, and $f_0 = \exp(-nV)$ for a particle with excluded volume V in a suspension of point objects with number density n. Ignoring coupling terms, the results are generalized as

$$f = \exp(-(sL + lS + nV)) \tag{15}$$

for a particle in a more complicated suspension. Relation [15] leads to the following expression for the mobility:

$$\log \mu = \log \mu_0 - K_r T \tag{16}$$

where T is the gel concentration (proportional to s, l or n). K_r is the retardation coefficient, which depends on the size and shape of the particle. Relation [16] has been widely verified in starch, polyacrylamide and agarose, and it is still currently used for molecular weight determination by means of the Ferguson plot [112] (Fig. 18).

Once the linear plot of K_r versus R has been determined using standards of known radius and charge, the measure of the retardation coefficient for an unknown particle gives access to its effective radius and to its effective charge.

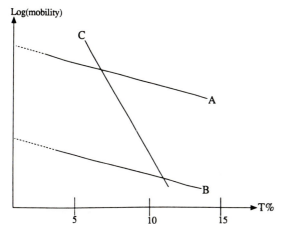

Fig. 18. "Ferguson plot" of mobility versus gel concentration. According to steric exclusion theories, [107–112], Log(mobility) should decay linearly with gel concentration T%. The intercept at T% = 0 is the free mobility (proportional to charge ratio), whereas the slope absolute value increases with particle size, i.e. with the importance of steric exclusion. For instance, *A* corresponds to particles with a high charge ratio (high free flow mobility) and a small size (weak slope), *B* to a particle with low charge ratio and small size, *C* to a particle with a high charge ratio and a large size (steep slope)

For a spherical particle, $S = 4\Pi(R + r)^2$ and $V = (4\Pi/3)(R + r)^3$ where R is the effective particle radius, and r the fiber radius of the gel. This suggests that the retardation coefficient K_r should scale as

$$K_r = c_1(R + r)^2 \qquad (17)$$

for 1-D gels (fiber-like obstacle) and as

$$K_r = c_2(R + r)^3 \qquad (18)$$

for O-D gels (globular obstacles). Relation (17) seems to prevail for both agarose and acrylamide, suggesting that these gels are valuably modeled by a suspension of rods. For highly crosslinked polyacrylamide with $T > 20\%$, however, the slope of R versus K_r decreases to about 0.3, suggesting a crossover to a regime in which the gel is better represented by pointlike obstacles.

Note that, in contrast with GPC, migration occurs in a uniform gel phase (there is no equivalent of the "interstitial solvent" here) so that smaller particles migrate faster.

3.2.2 Biased Reptation

Flexible macromolecules such as denatured proteins or single strand and duplex DNA, are also separated in gel electrophoresis. When the radius of gyration is comparable with the pore size of the gel, the Ogston approach described in the previous section is applicable [113], provided the "particle radius" is replaced by the macromolecule "Stokes radius" R_s, as determined for instance in sedimentation experiments. According to this model, however, particles with a radius of gyration much larger than the pores, such a chromosomal DNA, should not penetrate the gel at all. Actually, this is not observed, and very large DNA, for instance, migrates in gels at a velocity which saturates at a limiting finite value when the size of the chain increases [114–116]. The clue to this striking behavior was discovered in 1982, when Lumpkin and Zimm [117] and Lerman and Frisch [118] independently remarked that thin flexible chains can thread their way in the gel (Fig. 19) by a reptation process similar to the one proposed by De Gennes for entangled synthetic polymers [119].

These ideas are the basis of biased reptation theories [120–150] and computer simulations [122–127, 141–147], which have in the last years encountered considerable success in the description of gel electrophoresis of DNA.

The sequence of pores followed by a polyelectrolyte chain (DNA, denaturated protein, etc.) is called the "tube" [148], whereas the section of chain contained in one pore (of typical size a), is called a "blob" [119, 149]. Escape from the "tube" along the chain contour can occur only by loops or "hernias" (Fig. 19b), in which each pore of the gel is crossed by the chain twice instead of once. In the absence of electric field, De Gennes has shown that such events are exponentially unprobable due to an associated entropy loss of order kT per blob escaped [149] : Macroscopic diffusion occurs by a reptation motion in which

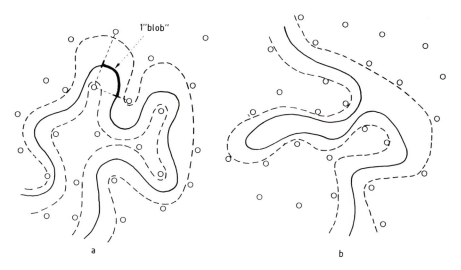

Fig. 19a. Schematic representation of the chain (*full line*) and its "tube" (*dotted lines*) in the gel (*bold circles*) **b** the same with a loop or "hernia"

only chain ends create new tube sections. The conformation of the chain is then a random sequence of N connected, non-overlapping blobs. In the absence of external field, the square averaged end-to-end distance of the chain is $\langle R^2 \rangle = Na^2$, as in a free theta solvent, but the time taken by the chain to diffuse on a distance R is the reptation time.

$$t_{rept} = \zeta a^2 N^3 / 2kT = N^3 t_A \tag{19}$$

where t_A is the equilibrium time of a blob, and ζ is the blob friction coefficient. With an external field along, say, z, the tube remains relevant (i.e. there is no significant formation of "hernias") as long as the electric gain in free energy obtained by "pulling out" a blob is smaller than the entropic loss, i.e.

$$\varepsilon = Eqa/2kT < c \tag{20}$$

where q is the blob charge, and c a number between, say, 1 and 10. In such conditions, components of the electric forces perpendicular to the local tube axis are cancelled by the tube reaction, and the longitudinal components induce a curvilinear electrophoretic drift, or "biased reptation" [120–122]. It obeys the equation:

$$\dot{s} = F_{eff}/N\zeta = qE_zR_z/Na\zeta \tag{21}$$

F_{eff} is the total effective electric force on the chain (effective means projected onto the primitive path) and R_z is the projection of the end-to-end vector onto the field direction. For long chains and practically useful fields, this drift rapidly dominates over reptation, so that the disengagement time t_{dis} is obtained as:

$$t_{dis} = N a/\dot{s} \tag{22}$$

The electrophoretic mobility is then obtained as:

$$\mu = V_z/E_z = \langle R_z \rangle / E_z \langle t_{dis} \rangle = \mu_1 \langle \rho^2 \rangle \tag{23}$$

where we introduced the limiting mobility $\mu_1 = q/\zeta$ and the reduced end-to-end projection $\langle \rho^2 \rangle = \langle (R_z/Na)^2 \rangle$.

Since R_z appears in Eqs. (21) and (23), it is crucial to know whether it is altered by the electric field, and how. When the chain migrates, its "head" has to choose a new "gate" in the gel for each step of size a. In the presence of the field, the simplest assumption relates the orientation probability for the head segment, $Q(\theta, \varepsilon)$ to the Boltzmann distribution [120–122]

$$Q(\theta, \varepsilon) = \cos \theta \exp(\varepsilon \cos \theta)/2 \, sh(\cos \theta) \tag{24}$$

Then, the probability distribution function (pdf) for the end-to-end projection R_z is this of a sequence of N segments with orientational pdf Q, and its square average can be derived as:

$$\langle \rho^2 \rangle = \langle (R_z/Na)^2 \rangle = \langle \cos^2\theta \rangle / N + \langle \cos \theta \rangle^2 (N - 1)/N \tag{25}$$

with $\langle \cos \theta \rangle = 1/(th \, \varepsilon) - 1/\varepsilon$, which tends to $\varepsilon/3$ for $\varepsilon \ll 1$, and $\langle \cos^2\theta \rangle = 1 + 2/(\varepsilon^2) - 2/(\varepsilon \, th \, \varepsilon)$, which tends to 1/3 at small ε.

The E dependence of R_z implied by Eqs. (24) and (25) introduces into the dynamic equations retarded field-dependent features which are at the heart of the complexity of conventional and pulsed electrophoresis of flexible chains. The model predicts two regimes which crossover for

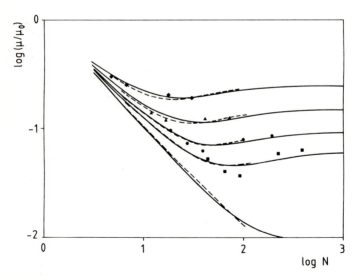

Fig. 20. Evolution of the mobility versus molecular weight for different scaled electric fields ($\varepsilon = 0.2$, 0.435, 0.54, 0.69 and 0.9, from the *lower* to the *upper curve*). The *full lines* are analytical results from Ref. [128], the *dotted lines* are the computer simulations from Ref. [125]. The *points* are fitted experimental data from Ref. [125]

$$N = N^* = 1/(3\langle\cos\theta\rangle^2) \cong 3/\varepsilon^2 \qquad (26)$$

i) $N < N^*$. In this regime, the first term in Eq. (25) dominates, and the mobility scales as $N^{-1}E^0$.

ii) $N > N^*$. This regime is dominated by the second term in Eq. (25). It asymptotically leads to a mobility scaling as N^0E^2.

As quoted by Slater et al. [125], the description sketched above is over-simplified. A more rigorous treatment [125, 128, 135, 150] predicts a non-monotonous M_w dependence of the mobility around N^*, which has been confirmed experimentally [125, 128] (Fig. 20). The asymptotic behavior in regimes i and ii, however, is not modified. The saturation of the mobility at large N is a dramatic nuisance for the electrophoretic separation of large biomolecules such as chromosomal DNA with sizes ranging from a few to 2×10^5 kilobase-pairs (kbp) (a base pair has a molecular weight around 360 Da). Until the recent introduction of pulsed electrophoresis, the practical range of separation hardly reached 100 kbp.

3.3 Recent Trends and Progress

3.3.1 Pulsed Electrophoresis

In its simplest version, this method [151–154] introduced by Schwartz and Cantor in 1984 [151] amounts to alternately applying electric pulses to the gel with different directions and/or orientations. Since then, numerous variations upon this idea have been proposed [155–166], and very puzzling and compli-cated behavior has been observed in some cases. However, the most common applications, which use two alternating field pulses of equal duration and equal amplitude at an angle close to 120°, can be understood rather simply on the basis of the reptation model [129]. According to Eqs. (23) and (25), the mobility of a long chain (larger than N^*) is an increasing function of the electric field, as a consequence of the orientation of the tube along E (Fig. 21a).

This orientation, however, is only progressively built up by reptation occurring after the onset of the field. The time necessary to reach the steady state velocity (orientation time t_{or}), scales as NE^{-2}. If the field is alternated between two orientations faster than this time, the tube presents several sections with different orientations (Fig. 21c), as a "memory" of earlier pulses, and the velocity is smaller than the steady-state one (Fig. 21a). In practice, when two fields are alternately applied during a time τ to a mixture of chains, the smaller chains with $t_{or} \ll \tau$ rapidly reach the steady state mobility at the beginning of each pulse (Fig. 21d), whereas chains with $t_{or} \ll \tau$ adopt an average (and weaker) orientation and velocity (Fig. 21b, c). An approximate analytical solution of the reptation model in this situation is possible [131, 167], leading to the general evolution shown in Fig. 22.

The mobility presents a steep (i.e. very selective in terms of size) section, which can be tuned by varying the pulse time. Presently, this method remains

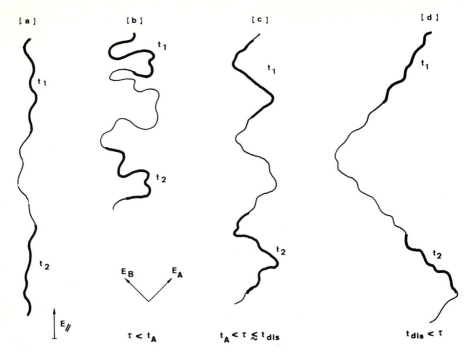

Fig. 21a–d. Different regimes of migration in crossed-field PFG. *Bold lines* represent chain conformations at different times t_1, and t_2, while the *thin lines* represent the tube. **a** permanent field; **b** "short pulses" regime ($\tau \ll t_A$) showing the reduction of overall oreintation; **c** crossover regime ($\tau \simeq t_{or}$), showing the chain in a "zigzag" conformation; **d** "macroscopic" regime, ($\tau > t_{or}$), showing the path as a sequence of fully oriented sections

limited to sizes smaller than 6 to 10 Mbp, probably because of trapping effects which are presently poorly understood [126].

Another puzzle is provided by field inversion electrophoresis, a method in which pulses of equal amplitude and unequal durations are alternately applied in the forward and backward directions [155, 168–169]. A very dramatic non-monotonous behavior of the mobility as a function of pulse times (at fixed size), or as a function of size (at fixed pulse times), in which long chains can migrate faster than smaller ones by orders of magnitude, may be obtained in some cases. Several theoretical models based on rather different ideas [126, 132, 139, 142, 147] qualitatively predict non-monotonous behaviors, but no analytical theory presently describes field inversion electrophoresis quantitatively. A novel and promising approach to gel electrophoresis of flexible chains, in which a bulky protein is attached at one end of the chain, has been proposed recently [170]. For a well-chosen porosity of the gel, a mobility decaying exponentially with the length of the chain and leading to very efficient separation, is observed. It is associated with the trapping of the protein in the smaller pores of the gel (Fig. 23).

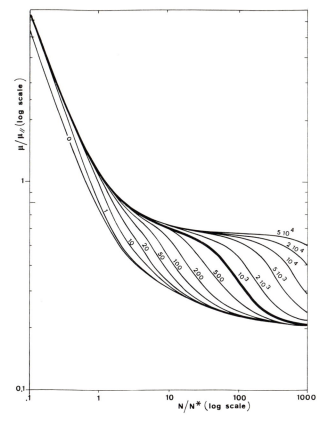

Fig. 22. Universal mobility versus molecular weight curves for different reduced pulse times (field angle = 110°). Values of the reduced pulse time, τ/t_A, are indicated on the curves

A recent theory based on this idea seems to account qualitatively for experimental results [133, 171]. In particular, both experiments and theory suggest that the exponentially selective zone can be displaced along the scale of chain sizes by periodic field inversion, inducing field-activated detrapping (Fig. 24).

Detrapping by field-inversion has also been successfully applied to the migration of linear duplex DNA [166], and of μm-sized latex spheres [158] in agarose.

3.3.2 Computer Simulations

Numerical simulations have been very helpful in the understanding of electrophoretic transport mechanisms in gels. As a first contribution, numerical simulations of the biased reptation model have permitted progress beyond the

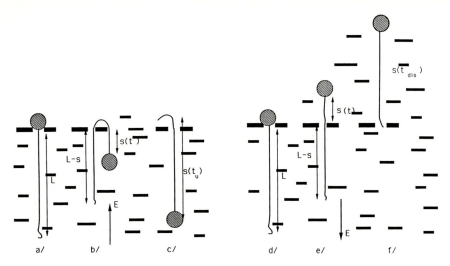

Fig. 23. Schematic representation of the trapping of protein-labeled DNA in a gel: *a/* Chain in the trap, with field pulling down *b/* Chain in the course of crossing the potential barrier (thermal detrapping), with field pulling down *c/* Chain detrapped *d/* same as a/ *e/* Chain in the course of detrapping activated by a reverse field-pulse (field pulling up)*f/* Chain detrapped by the reverse pulse

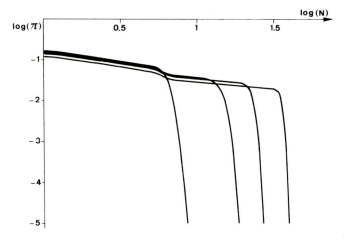

Fig. 24. Mobility versus size for protein-labeled DNA in field inversion electrophoresis, with a trap density of 10^{-3} and a reduced field $\varepsilon = 0.5$. The curves correspond, from left to right, to field-inversion pulses of 0, 200, 400 and 800 reduced units. (The forward pulse is 5000 reduced units in all cases)

limitations of analytical calculations, which are difficult in the general case. In particular, several simulations taking full account of internal modes [127, 134, 146, 147], have shown that they are responsible of strong stretching and collapsing motions, which are believed to participate in the spectacular non-monotonous behavior of the mobility as a function of chain size observed in field-inversion electrophoresis.

A more drastic criticism of biased reptation ideas was put forward by simulations of flexible chains on a 2D or 3D lattice of obstacles [141–145], which do not a priori postulate any tube: For rather strong but still realistic fields, these lattice chains present an intermittent motion which partly contradicts reptation ideas: It is characterized by a collapse of the "head" which immobilizes the chain, followed by a progressive development of one or several "hernias" in the direction of the field, ultimately defining a new "tube" in which the chain starts moving again. When the field is reversed with a period in the same range as the spontaneous period of this intermittent dynamics, chains remain "pinned" around a cluster of hernias for a long time, and the mobility is strongly reduced: this effect is responsible for the spectacular mobility behavior observed in field-inversion electrophoresis [172]. Computer simulations of chains with internal modes and hernias were also used recently to explain why the separating power of crossed-field electrophoresis discontinuously increases just above an angle of 90 degrees [147].

The models described above, biased reptation and lattice chains, assume a homogeneous gel. An opposite approach, in which gel inhomogeneities play a dominant role, was recently proposed by Zimm [139]. In this model, the "tube" is represented by a series of "lakes" in which the chain can expand, and "straits" in which it is strongly confined. Finally, simulations of electophoretic drift in 3D random percolation clusters of obstacles, have started [173].

3.3.3 Spectroscopy and Microscopy

Experimentally, efforts have also been aimed at the investigation of migration mechanisms in gel electrophoresis. Fluorescence polarization [174], linear dichroism [175–178] and electric birefringence [179–181] confirmed the orientation of chains by the electric field and its connection with the enhanced mobility of long chains. They also revealed a strong transient overshoot of the orientation at the onset of the field, associated with field-inversion spectacular dynamic effects, and permitted the measure of relaxation times which confirmed the relevance of reptative motions [179] for some electrophoretic conditions. The most spectacular experimental achievement in the field, however, is the onset of fluorescence videomicroscopy techniques, which allow direct observation of the shape and motion of individual labeled DNA molecules in real time and real space [182–184]. These observations unambiguously confirmed that long DNA molecules adopt, in agarose gels, very elongated conformations, and reptative-like motions [118–132]. The dynamics, however, are much more complicated than in the biased reptation model, and seems to combine several features suggested earlier in the literature: Chains of intermediate size (typically a few hundreds of kbp) undergo strong collapse and stretching motions closely resembling 2D lattice simulations by Deutsch et al. [141–143]. The connection of these motions with field-inversion dynamics was also confirmed. Longer chains (Mbp range), however, do not seem to collapse. Instead, they present several denser regions along their path, from which hernias often protrude without developing as in the case of smaller chains. This latter behavior presents some resemblance with the "lakes and straits" model of Zimm [56], although the model does not allow for real "hernias". Finally, even chains too small to adopt really reptative motions seem to have a rather non-uniform migration, suggesting that gel inhomogeneities also plays a significant role. Videomicroscopy, which is still limited in its quantitative interpretation by technical difficulties in obtaining well-controlled and homogeneous fields and matrices on a miniaturized scale, will, in the next few years, certainly provide a great deal of original information on the migration of macromolecules in gels.

4 Conclusions

We have reviewed various mechanisms for the transport of macromolecules in gels, which are the basis of important separation technologies and represent a major application of gels nowadays. This review is not exhaustive. In particular, we deliberately omitted the very rich and industrially important processes based on specific affinities between the gel and the solutes. In these affinity chromatography [185] and affinity electrophoresis [186] methods, the interactions at the origin of the separation are very local, and they do not make use of the gel structure per se: The gel there only acts as a convenient porous and immobile support, and its structure is not really relevant to the separation. We restricted ourselves to processes based on the gel being a "physically active" (as opposed to a "chemically active") object, i.e. to the geometry of gels and solutes in the general sense (topology, shape, conformations), and to entropic (as opposed to enthalpic) solute-gel interactions. Even this apparently simple type of interactions leads to nontrivial and spectacular transport properties.

Besides this unifying feature of using size exclusion as the basic solute-gel interaction, current chromatographic and electrophoretic separation methods present strong differences. The first trivial one is the driving force. The use of pressure or electric fields obviously requires very different technologies, and are applied to different solutes. This difference, however, is not so profound as it may appear: Neutral molecules can be separated in electrophoresis, using suitable electroendosmosis effects. In this case, the electric field is used instead of pressure to promote solvent flow. Also, as seen in Sect. 3.2.1 the mechanism of separation in regular gel electrophoresis is very similar to that of gel filtration, regardless of the force "pushing" the molecules. Finally, the development of gel electrophoresis in capillaries [87] makes use of many components and methods of chromatography, in particular for detection, and tends to reduce the difference between these methods from the technological point of view. From the point of view of solute-gel interactions, however, a major difference remains: GPC uses two phases, and the separation occurs because macromolecules can be either within the gel or outside the gel. The transport in the gel is only there to allow for thermodynamic equilibrium between the gel and the interstitial phase, but it does not (ideally) intervene in the net migration of solutes. The situation in gel electrophoresis is the exact opposite, since the net transport arises solely from migration inside the gel: there is no interstitial phase. The qualitative consequence of this fundamental difference is that, as a general rule, large chains migrate faster in GPC, and slower in electrophoresis and in gel filtration (from the point of view of solute-gel interactions, gel filtration is closer to electrophoresis than GPC).

As discussed in Sect. 2.2.2, size exclusion is rather well understood, and quantitative theories are available for molecules of various shapes and topologies. Its effectiveness, however, is determined by actual thermodynamic equilibrium, i.e. by transport processes inside the gel. The sieving mechanism which dominates electrophoresis is also rather well understood for small macromolecules (Sect. 2.2.1), but the situation of large macromolecules is much more complex, and a lot of progress remains to be made in the conceptual side. In particular, recent videomicroscopy experiments and computer simulations suggest that, even in the so-called "homogeneous" gels used in electrophoresis (as opposed to the bead packings used in GPC), gel inhomogeneities play a considerable role in the complex deformation processes undergone by very large flexible chains. It is also possible that chains do not distribute themselves evenly in the gel, and preferentially follow "channels" of larger pore sizes.

Indeed, all gels are inhomogeneous in nature, and the "two phases" and "one phase" pictures respectively used to represent GPC and electrophoresis matrices are idealized limiting cases. They can be continuously linked by various inhomogeneous systems, and all kinds of mixed mechanisms can be imagined, and probably occur in real systems to some extent. A theoretical unification of size exclusion and sieving, in inhomgeneous gel systems, would probably be an interesting aim for future research, although inhomogeneities in gels are still very poorly understood. From the technological point of view, deliberately hybridizing these two processes may also lead to interesting separation properties.

5 References

1. Porath J, Flodin P (1959) Nature 183: 1657
2. Porath J (1967) Lab Pract 16: 838
3. Pharmacia Fine Chemicals - Uppsala (Sweden)
4. Moore JC (1964) J Polym Sci A2: 835
5. Waters Associates, Milford, MA (USA)
6. De Vries AJ, Le Page M, Beau R, Guillemin CL (1967) Anal Chem 39: 935
7. Le Page M, Beau R, De Vries AJ (1968) J Polym Sci C 21: 119
8. Limpert RJ, Cotter RL, Dark WA (1974) Am lab May 63
9. Kato Y, Kido S, Hashimoto T (1973) J Polym Sci Polym Phys Ed 11: 2329

10. Kato Y, Kido S, Yamamoto M, Hashimoto T (1973) J Polym Sci Polym Phys Ed 11: 2329
11. Kirkland JJ (1972) J Chromatogr Sci 10: 593
12. Kirkland J (1976) J Chromatogr 125: 231
13. Probst J, Unger K, Cantow HJ (1974) Angew Makromol Chem 35: 177
14. Unger K, Kern R, Ninou MC, Krebs KF (1974) J Chromatogr 99: 435
15. Toyo Soda Manufacturing Ltd, Tokyo (Japan)
16. Merck E, Darmstadt (Germany)
17. Letot L, Lesec J, Quivoron C (1974) J Liq Chromatogr 5: 217
18. Synchrom Inc., Linden, IN (USA)
19. Showa Denko KK, Tokyo (Japan)
20. Polymer Laboratories Ltd, Shropshire (England)
21. Yau WW, Kirkland JJ, Bly DD (1979) Modern size exclusion chromatography, Wiley Interscience, New York
22. Audebert R (1979) Polymer 20: 1561
23. Hagnauer GL (1982) Anal Chem 54: 265R
24. Van Kreveld M (1975) J Polym Sci (Phys) 13: 2253
25. Kubin M (1975) J Chromatogr 108: 1
26. Doi M (1975) J Chem Soc Faraday Trans II: 71: 1720
27. Ambler MR, Mc Intyre D (1975) J Polym Sci B 13: 589
28. Basedow AM, Ebert KH, Ederer HJ, Fosshag E (1980) J Chromatogr 192: 259
29. Casassa EF (1971) J Phys Chem 75: 3929
30. Casassa EF (1972) J Polym Sci Part A2: 10: 381
31. Casassa EF (1967) J Polym Sci Part B 5: 773
32. Casassa EF, Tagami Y (1969) Macromolecules 2: 14
33. Casassa EF (1971) Separation Sci 6: 305
34. Yau WW, Malone CP (1971) Polym Prep 12: 797
35. Casassa EF (1976) Macromolecules 9: 182
36. Yau WW, Ginnard CR, Kirkland JJ (1978) J Chromatogr 149: 465
37. Du Pont de Nemours, Wilmington, DE (USA)
38. Hamielec AE, In: Janca J (ed) Steric exclusion liquid chromatography of polymers, vol 25, Marcel Dekker, New York, p 117
39. Tung LH, Runyan JR (1969) J Appl Polym Sci 13: 2397
40. Tung LH (1969) J Appl Polym Sci 13: 775
41. Yau WW, Stoklosa HJ, Bly DD (1977) J Appl Polym Sci 21: 1911
42. Balke ST, Hamielec AE (1969) J Appl Polym Sci 13: 1381
43. Hamielec AE (1982) Pure & Applied Chem 54: 293
44. Omorodion SNE, Hamielec AE (1989) J Liq Chrom 12: 1131
45. Omorodion SNE, Hamielec AE (1989) J Liq Chrom 12: 1155
46. Omorodion SNE, Hamielec AE (1989) J Liq Chrom 12: 2635
47. Provder T, Rosen EM (1970) Sep Sci 5: 437
48. Grubisic Z, Rempp P, Benoit H (1967) J Polym Sci Part B 5: 753
49. Tanford C (1961) Physical Chemistry of Macromolecules, Wiley, New York
50. Adams H (1971) Sep Sci 6: 259
51. Albaugh EW, Talarico PC (1972) J Chromatogr 74: 233
52. Runyon JR, Barnes DE, Rudd JF, Tung LH (1969) J Appl Polym Sci 13: 2359
53. Ross JH, Casto ME (1968) J Polym Sci C 21: 143
54. Goedhart D, Opschoor A (1970) J Polym Sci A2 8: 1227
55. Grubisic-Gallot Z, Picot M, Gramain P, Benoit H (1972) J Appl Polym Sci 16: 2931
56. Ouano AC (1972) J Polym Sci A1 10: 2169
57. Lesec J, Quivoron C (1976) Analusis 4: 399
58. Letot L, Lesec J, Quivoron C (1980) J Liq Chromatogr 3: 427
59. Lecacheux D, Lesec J, Quivoron C (1982) J Appl Polym Sci 27: 4867
60. Lesec J, Lecacheux D, Marot G (1988) J Liq Chromatogr 11: 2571
61. Lecacheux D, Lesec J, Prechner R (1981) Fr Pat 82402324.6 (1984) US Pat 4478071
62. Ekmanis JL (1989) paper # 1657, Pittsburg Conf, Atlanta, GA; Ekmanis JL (1989) Int GPC Symposium Newton MA (USA)
63. Kuo CY, Provder T, Kohler ME, Kah AF (1987) In: Provder T (ed) Detection and data analysis in SEC, ACS Symp Ser, p 130
64. Haney MA (1984) US Pat 4463598; Haney MA (1985) J Appl Polym Sci 30: 3037
65. Abbot SD, Yau WW (1986) US Pat 4578990 and 4627271

66. Viscotek Corp, Porter, TX (USA)
67. Lecacheux D, Lesec J, Quivoron C, Prechner R, Panaras R, Benoit H (1984) J Appl Polym Sci 29: 1569
68. Yau WW, Remeter SW (1990) J Liq Chromatogr 13: 627
69. Ouano AC, Kaye W (1974) J Polym Sci A1 12: 1151
70. Chromatix-LDC division, Riviera beach, FL (USA)
71. Yu LP, Rollings JE (1988) J Appl Polym Sci 35: 1085
72. Wyatt Technology Corp, Santa Barbara, CA (USA)
73. Wyatt PJ, Jackson C, Wyatt GK (1988) Am Lab 20: 108
74. Yau WW (1989) Int GPC Symposium, Newton, MA (USA)
75. Marot G, Lesec J (1988) J Liq Chromatogr 11: 3305
76. Lesec J, Lafuma F, Quivoron C (1974) J Chrom Sci 12: 683
77. Lesec J, Lafuma F, Quivoron C (1977) J of Chromatogr 138: 89
78. Lesec J, Quivoron C (1979) J Liq Chromatogr 2: 467
79. Ekmanis JL (1987) In: Provder T (ed) Detection and Data analysis in SEC ACS symp ser 47
80. Lesec J (1985) J Liq Chromatogr 8: 875
81. Kato Y, Kametani T, Furukawa K, Hashimoto T (1975) J Polym Sci, Part A2 13: 1695
82. Lesec J, Volet G (1990) J Polym Sci, Appl Polym Symp 45: 177
83. Lesec J, Volet G (1990) J Liq Chromatogr 13: 831
84. Tiselius A (1937) Trans Faraday Soc 33: 524
85. Hjerten S (1967) J Chromatogr 9: 122
86. Foret F, Bocek P (1990) Electrophoresis 11: 661
87. Hjerten S (1990) Electrophoresis 11: 664
88. Von Klobusitsky D, Köning P (1939) Arch Exp Pathol Pharmacol 192: 271
89. Wieland T, Fischer E (1948) Naturwissenschaften 35: 29
90 Smithies O (1955) J Biochem 61: 629
91. Grabar P, Williams CA (1953) Biochim Biophys Acta 10: 193
92. Grabar P, Nowinsky WW, Generaux BD (1956) Nature 178: 430
93. Righetti PG (1989) J Biochem Biophys Meth 19: 1
94. Kohn J (1957) Nature 180: 986
95. Grabar P (1957) Meth Biochem Anal 7: 1
96. Hickson TGL, Polson A (1965) Biochim Biophys Acta 53: 514
97. Atwood TK, Nelmes BJ, Sellen DB (1988) Biopolymers 27: 201
98. Raymond S, Weintraub L (1959) Science 130: 711
99. Davis BJ (1964) Ann NY Acad Sci 121: 404
100. Ornstein L (1964) Ann NY Acad Sci 121: 321
101. Hjerten S (1963) J Chromatogr 11: 66
102. Hjerten S (1962) Anal Biochem 3: 109
103. Righetti PG, Gelfi C, Bossi ML, Boschetti E (1987) Electrophoresis 8: 62
104. Chiari M, Casale E, Santaniello E, Righetti PG (1989) Appl Theor Electrophoresis 1: 99
105. Chiari M, Casale E, Santaniello E, Righetti PG (1989) Appl Theor Electrophoresis 1: 103
106. Morris CJ (1967) In: Peeters (ed) H Protides of the biological fluids vol 14, Elsevier, New York
107. Ogston AG (1958) Trans Faraday Soc 54: 1754
108. Giddings JC, Kucera E, Russel CP, Myers MN (1968) J Phys Chem 72: 4397
109. Rodbard D, Chrambach A (1970) Proc Natl Acad Sci USA 4: 970
110. Rodbard D, Chrambach A (1971) Anal Biochem 40: 95
111. Lunney J, Rodbard D, Chrambach A (1971) Anal Biochem 40: 158
112. Ferguson KA (1964) Metabolism 13: 985
113. Maniatis T, Jeffrey A, Vand de Sande H (1975) Biochemistry 14: 3787
114. Flint D, Harrington RE (1972) Biochemistry 11: 4858
115. Mc Donnel MW, Simon MN, Studier FW (1977) J Mol Biol 110: 119
116. Hervet H, Bean CP (1987) Biopolymers 26: 727
117. Lumpkin OJ, Zimm BH (1982) Biopolymers 21: 2315
118. Lerman LS, Frisch HL (1982) Biopolymers 21: 995
119. de Gennes PG (1971) J Chem Phys 55: 572
120. Lumpkin OJ, Dejardin P, Zimm BH (1985) Biopolymers 24: 1573
121. Slater GW, Noolandi J (1985) Phys Rev Lett 55: 1579
122. Slater GW, Noolandi J (1986) Biopolymers 25: 431
123. Slater GW, Rousseau J, Noolandi J, Turmel C, Lalande M (1988) Biopolymers 27: 509
124. Slater GW, Rousseau J, Noolandi J (1987) Biopolymers 26: 863

125. Noolandi J, Rousseau J, Slater GW, Turmel C, Lalande M (1987) Phys Rev Lett 58: 2428
126. Noolandi J, Slater GW, Lim HA, Viovy JL (1989) Science 243: 1456
127. Slater GW, Noolandi J (1989) Proc of the International Symposium on new trends in physics and physical chemistry of polymers honoring Prof PG de Gennes, Toronto, Plenum Press, New York
128. Viovy JL (1988) Europhys Lett 7: 657
129. Viovy JL (1987) biopolymers 26: 1929
130. Viovy JL (1987) C R Acad Sci Paris 305: 181
131. Viovy JL (1989) Electrophoresis 10: 429
132. Viovy JL (1988) Phys rev Lett 60: 855
133. Défontaines AD, Viovy JL (1991) Proc "First international conference on electrophoresis, supercomputing and the human genome, Tallahassee), Cantor CR, Lim HA (eds) World Scientific, Singapore
134. Lim HA, Slater GW, Noolandi J (1990) J Chem Phys 92: 709
135. Déjardin P (1989) Phys Rev A 40: 4752
136. Levene SD, Zimm BH (1989) Science 245: 396
137. Lumpkin O (1989) Phys Rev A 40: 2634
138. Lumpkin O, Levene SD, Zimm BH (1989) Phys Rev A 39: 6557
139. Zimm BH (1988) Phys Rev Lett 61: 2965
140. Jamil T, Frisch HL, Lerman LS (1989) Biopolymers 28: 1413
141. Deutsch JM (1987) Phys Rev Lett 59: 1255
142. Deutsch JM (1989) J Chem Phys 90: 7436
143. a: Deutsch JM, Madden TL (1989) J Chem Phys 90: 2476
     b: Deutsch JM (1988) Science 240: 922
144. Lalande M, Noolandi J, Turmel C, Rousseau J, Slater GW (1987) Proc Natl Acad Sci USA 84: 8011
145. Kremer K (1988) Poymer Comm 29: 292
146. Duke TAJ (1989) Phys Rev Lett 62: 2877
147. Duke TAJ, Viovy JL (1992) Phys Rev Lett 68: 542
148. Edwards SF (1967) Proc Phys Soc 92: 9
149. De Gennes PG (1979) Scaling concepts in polymer physics, Cornell University Press, Ithaca, New York
150. Doi M, Kobayashi T, Makino Y, Ogawa M, Slater GW, Noolandi J (1988) Phys Rev Lett 61: 1893
151. Schwartz DC, Cantor CR (1984) Cell 37: 67
152. Carle GF, Olson MV (1984) Nucleic Acids Res 12: 5647
153. Vollrath D, Davies RW (1987) Nucleic Acids Res 15: 7865
154. Southern E, Anand R, Brown WRA, Fletcher DS (1987) Nucleic Acids Res 15: 5925
155. Carle GF, Frank M, Olson MV (1986) Science 232: 65
156. Chu G (1989) Electrophoresis 10: 290
157. Gardiner K, Laas W, Patterson D (1989) Electrophoresis 10: 296
158. Griess GA, Serwer P (1990) Biopolymers 29: 1863
159. Lai E, Davi NA, Hood LE (1989) Electrophoresis 10: 65
160. Lai E, Birren BW, Clark SM, Hood LE (1988) Nuclei Acids Res 16: 10376
161. Louie D, Serwer P (1989) Applied and Th Electrophoresis 1: 169
162. Serwer P (1987) Electrophoresis 8: 301
163. Serwer P, Hayes JH (1989) Applied and Th Electrophoresis, 1: 95
164. Serwer P, Hayes JH (1987) Electrophoresis 8: 244
165. Sor F (1987) Nucleic Acids Res 15: 4853
166. Turmel C, Brassard E, Slater GW, Noolandi J (1990) Nucleic Acids Res 18: 569
167. Slater GW, Noolandi J (1989) Electrophoresis 10: 413
168. Crater GD, Gregg MC, Holtzwarth G (1989) Electrophoresis 10: 310
169. Heller C, Polhl FM (1989) Nucleic Acids Res 17: 5989
170. Ulanowsky L, Drouin G, Gilbert W (1990) Nature 343: 190
171. Viovy JL, Défontaines AD (1991) In: Ulanowsky L Burmeister M (eds) Pulsed-Field Electrophoresis Methods in Molecular Biology series, Humana Press
172. Duke TAJ, Viovy JL, J Chem Phys 96: 8552
173. Melenkevitz J, Muthukumar M (1989) In Science at the Von Neumann National Supercomputer Center Annual report p 145–151
174. Hurley I (1986) Biopolymers 25: 539

175. Akerman B, Jonssohn M, Nordén B (1985) J Chem Soc Chem Comm 422
176. Holtzwarth G, Mc Kee CB, Steiger S, Crater G (1987) Nucleic Acids Res 15: 10031
177. Jonssohn M, Akerman B, Nordén B (1988) Biopolymers 27381
178. Akerman B, Jonssohn M, Nordén B, Lalande M, to appear
179. Sturm J, Weill G (1989) Phys Rev Lett 62: 1484
180. Chu B, Wang Z, Wu C (1988) Biopolymers 27: 2005
181. Stellwagen NC, Stellwagen J (1989) Electrophoresis 10: 332
182. Smith SB, Aldridge PK, Callis JB (1989) Science 243: 203
183. Schwartz DC, Koval M (1989) Nature 338: 520
184. Gurierri S, Rizzarelli E, Beach D, Bustamante C (1990) Biochemistry 29: 3396
185. see e.g. Chaiken J, Chiancone E, Fontana A, Neri P (1987) Macromolecular recognition: principles and methods Humana Press, p 309
186. see e.g. Andrews AT (1986) Electrophoresis: Theory techniques and biochemical and clinical applications, 2nd edn, Clarendon, Oxford

Editor: Prof. K. Duśek
Received 12 August 1992

Far Infrared Spectroscopy of Polymers

V.A. Bershtein, V.A. Ryzhov
A.F. Ioffe Physico-Technical Institute of the Russian Academy of Sciences,
194021 St. Petersburg, Russia

A comprehensive review of polymer studies using spectroscopy in the low-frequency infrared region (far infrared, FIR) presented here shows that contemporary spectroscopic studies in the far infrared region have become an important tool for characterizing the molecular dynamics and inter-molecular interactions in polymers which determine to a considerable extent their physical properties. FIR spectroscopy also provides the most informative criteria for the presence of hydrogen bonds in biological matter, multiplets and clusters in ion-containing polymers, polymer crystallinity, etc.

The FIR spectra appeared fruitful in revealing the molecular nature of β, γ and δ-relaxations in glassy polymers. Success has been achieved after the three following steps: (i) finding of empirical correlations between spectral parameters and molecular characteristics of polymers; (ii) estimation of potential barriers and molecular units of motion from the FIR spectra; (iii) comparison of results obtained with the activation energies of relaxation transitions.

Advances in Polymer Science, Vol. 114
© Springer-Verlag Berlin Heidelberg 1994

List of Abbreviations

FIR	far infrared
PE	polyethylene
PMMA	poly(methyl methacrylate)
PVC	poly(vinyl chloride)
PS	polystyrene
PTFS	poly(trifluorostyrene)
PPFS	poly(pentafluorostyrene)
PMS	poly(α-methylstyrene)
PP	polypropylene
PTFE	poly(tetrafluoroethylene)
POE	poly(oxyethylene)
POM	poly(oxymethylene)
PVA	poly(vinyl acetate)
PAN	polyacrylonitrile
PCS	polychlorostyrene
PVF	poly(vinyl fluoride)
PVF_2	poly(vinylidene fluoride)
PBMA	poly(n-butyl methacrylate)
POMA	poly(octyl methacrylate)
PDMA	poly(decyl methacrylate)
PCHMA	poly(cyclohexyl methacrylate)
PC	polycarbonate
PDMS	poly(dimethylsiloxane)
PMAA	poly(methacrylic acid)
PVAl	poly(vinyl alcohol)
PSSA	poly(styrene sulfonic acid)
PET	poly(ethylene terephthalate)

List of Symbols

A	Total intensity of the absorption
A_{sk}	Intensity of the skeletal vibration band
$\alpha(v)$	Absorption coefficient
$\alpha_{sk}(v)$	Absorption coefficient of the skeletal vibration band
v	Wave number in cm^{-1}
$\bar{v}$	Frequency in Hz
$v_1, v_2, \ldots, v_{12},$	
$v_a, v_s, v_\delta, v_\gamma,$	
$\delta_a, \delta_s, \delta_w, \delta_r, \gamma_t$	Nomenclature of normal vibrations
v_{sk}	Wave number, cm^{-1} for skeletal vibrations
v_{loc}	Wave number, cm^{-1} for local torsional vibrations
v_{libr}	Wave number, cm^{-1} for librational vibrations
ω	Angular frequency in Hz
c	Light velocity
k	Boltzmann constant
k	Wave number vector in Eq. (1)
k_1, k_2, k_s, k_b, k_t	Force constants of vibrations
f	Force constant in Eq. (1)
θ	Phase difference
φ	Angle of libration
ϕ	Angle of torsion vibration in backbone chain
m	Atomic mass
M	Repeated unit mass
μ_{eff}	Effective dipole moment
R_{eq}	Equivalent radius of the librational unit
I	Moment of inertia
C^{n+}	Cation with charge n^+
n	Number of monomer units in polymer chains
n_1, n_2	Number of molecules in the ground (l_0) and the first vibrational (l_1) levels, respectively
N	Number of monomer units in oligomers
S	Statistical (Kuhn) segment size of macromolecule in a number of monomer units
$\tan\delta$	Dielectric loss tangent
ε	Unelastic deformation
ε''	Dielectric loss
$\alpha, \beta, \gamma, \delta$	Relaxation processes (transitions)
$T_g, T_\beta, T_\gamma, T_\delta$	Temperatures of α, β, γ and δ relaxation processes
T^*	A temperature of onset (on heating) or stopping (on cooling) of conformational rearrangements in chain
T_2	Characteristic temperature in Gibbs–DiMarzio thermodynamic glass transition theory

T_0	Constant in the Fulcher–Fogel–Tamman and Williams–Landel–Ferry equations
t_d	Debye relaxation time
ΔH_s	Heat of sublimation
α, β, γ	Conformations in crystalline phase of polymers
E_{coh}	Cohesion energy
U or V	A depth of potential well for librational motion in Eq. (8)
v_a	Activation volume
$Q_\alpha, Q_\beta, Q_\gamma, Q_\delta$	Activation energy for α, β, γ and δ relaxations
$Q(\varphi)$	Potential barrier to hindered rotation through the angle φ
Q_{loc}	Potential barrier of local torsional vibrations
Q_{sk}	Potential barrier of skeletal vibrations
Q_{libr}	Potential barrier of librational motion
Q_0 or E_0	Potential barriers for isomeric transitions in Eq. (17)
$\Delta E = E_g - E_t$	Energy difference between isomeric (*gauche-trans*) states

1 Introduction

The subject of this review is the IR spectra of polymers in the frequency range situated between the radio and the mid-infrared regions, usually called by molecular spectroscopist the far infrared region.

An FIR spectrum of a polymer is expected to be complex, including both intra-and intermolecular modes. The latter are external vibrations of the chains, both rotatory and translatory type. The former are likely to include both localized torsions of functional groups and modes which are more delocalized along the chain. Besides, the absorption in this region can arise from such phenomena as "disorder-induced absorption", absorption by relaxational or non-resonant phenomena, and also due to impurities accidentally or deliberately present in the material.

The complexity of the spectrum interpretation is one of the drawbacks of FIR spectroscopy, the other one was for a long time the absence of any appropriate experimental instruments. However, recently rapid progress in the FIR technique, viz. commercial FT-IR spectrometers as well as tunable lasers for the far infrared region, have ensured a successful solution to this problem.

Polymer spectroscopy in the low-frequency infrared region developed slowly also because of a lack of a well-worked-out theory of FIR spectra of polymers and because the main body of experimental evidence was guided by the concept of group frequencies. However, in the far infrared region this simple concept has to take into account intermolecular as well as intramolecular coupling because intermolecular interactions lead to collective excitations in polymers.

As a result, most of the work on far infrared spectra of polymers is limited to mere illustration of the possibilities of using this spectroscopic method.

We are unaware of any attempts to bring the experimental data on FIR spectra of polymers together. The reviews [1, 2, 3] cover only the results obtained until 1970. In a short survey [4] including only 30 references, the authors do not claim completeness, but do give some instructive examples to demonstrate the basic principles of FIR spectroscopy of polymers.

The present review involves the main results concerning the FIR spectra of polymers obtained mainly during the last 10–15 years. In this paper, we point out the usefulness of studying the low-frequency vibrational spectra of polymers. Although we present almost entirely the FIR data, it should be noted that Raman spectroscopy is the necessary complementary method in some cases.

Because the spectra in the region considered are mainly due to the collective excitations in polymers, we first give a sketch of the structure of polymers relevant to vibrational spectra and then briefly explain the theoretical methods necessary for interpretation of the spectra.

In the general part of review, we present the FIR spectra for a large group of the technically and biologically important polymers; the main focus of attention is paid to molecular dynamics and intermolecular interactions of various type;

the direct estimates of the conformational state of chains, and to molecular interpretation of relaxational processes in glassy polymers.

2 Specificity of Low-Frequency Spectra of Polymers

2.1 Theoretical Approach to Interpretation of Spectra

The feature common to high polymers is that a single chemical unit (a monomer, or repeating unit) is repeated many times in a macromolecule. The simplest macromolecule is usually a linear construction of monomers, but macromolecules also may be made up of small number of different chemical units and may have branches or even a network instead of a simple linear chain. Nevertheless, a polymer molecule is a rather linear system with strong covalent bonds in the chain direction and only weak coupling laterally. Its constitution and degree of polymerization strongly influence the conformation of macromolecules – the set of all arrangements of a chain that can be formed by rotation around chemical bonds.

The actual conformations determine the packing of material in condensed state, which will be either amorphous, melt or glass, or partially crystalline. The latter is usually described in terms of a two-phase model in which crystallites are embedded in the amorphous material.

These features of polymers that influence the spectra have demanded specific theoretical methods for assignment of absorption bands in the spectra. The most important method for correlating spectra and polymers structure is the group theory, which made it possible to analyse, to a good approximation, the vibrational modes of portions of a molecule in terms of the local symmetry. In this concept, single atoms bound in a molecule undergo resonance vibrations that are governed solely by the masses involved and by the corresponding force constants. The influence of other chemical bonds are assumed to be negligible or it can be easily calculated [5, 6]. In the mid-infrared, the methods were developed for calculation of the spectral distribution of the absorption coefficient of accuracy quite satisfactory for practical purposes [7].

However, in the low-frequency far-infrared region, this simple concept needs expanding to take into account intramolecular as well as intermolecular coupling because intermolecular interactions lead to collective excitations, similar to phonon in crystals. Description of the low-frequency oscillations would be more adequate if we proceed from spectroscopy of crystals rather than separate molecules.

Now, we illustrate the solid-state physics method by the example of calculation of skeletal modes of a macromolecule. In an approach originating from the spectroscopy of crystals [5, 8, 9], a macromolecule is considered an infinite chain of point masses M spaced an equal distance d apart and held together by chemical bonds that define the force constants f for stretching, deformation or torsional vibration.

The problem can be reduced to calculation of normal vibrational modes of a linear chain. The vibration frequency of such a chain is a periodic function of the wave number vector k:

$$v = \frac{1}{\pi}\,(f/M)^{1/2}\sin(\pi k d) \tag{1}$$

A plot of frequency v versus k is known as a dispersion curve. The first period of this function represents the first Brillouin zone. For a linear diatomic chain in which all bonds are identical and the atoms are equally spaced, but have different masses of M_1 and M_2, the result is:

$$v^2 = \frac{f}{4\pi^2}\left\{(1/M_1 + 1/M_2) \pm \left[(1/M_1 + 1/M_2)^2 - \frac{4}{M_1 M_2}\sin^2(\pi k d)\right]^{1/2}\right\} \tag{2}$$

and thus there are two dispersion curves. The one that passes through the origin is called the acoustical branch, because the frequencies fall in the region of sonic and ultrasonic waves. The second curve is the optical branch and falls in the IR spectral region. The mode of the optical branch corresponding to $k = 0$ describes the stretching of an interatomic bond and thus it is optically active. The acoustical vibration at $k = 0$ does not result in a change in the dipole moment and is optically inactive.

When the repeating unit contains n atoms, there are $3n$ separate branches. Three of these modes correspond to acoustical modes, so $3n-3$ optical branches exist. Of course, some of these branches can be degenerated. Many of the modes of the isolated repeat unit are severely modified by the insertion in a chain, since they are internal frequencies. Others are coupled in the chain; they are known as the skeletal or lattice modes.

2.2 Dispersion Relations for Polyethylene

Consider a simple case of an infinite chain of PE. The repeating unit consists of 6 atoms: CH_2–CH_2, hence, there are 18 normal modes. Figure 1 shows the 9 respective dispersion curves. The branches v_1 and v_6 correspond to the symmetrical and asymmetrical stretching modes of CH_2 group, respectively, v_2 to deformation modes of the same group, v_3 to wagging modes, v_4 to C–C stretching modes, v_5 to deformation skeletal modes, v_7 and v_8 to combination of rocking and torsional modes of CH_2 group and, finally, v_9 to skeletal torsional vibrations. Intersections of these branches with the ordinate at $k = 0$ (the respective phase difference $\theta = 0$ and π in the approach used by a polymer spectroscopist) give frequencies active in the infrared and Raman spectra.

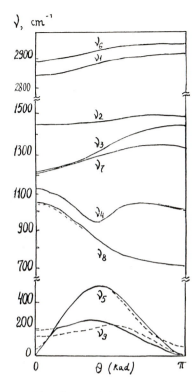

Fig. 1. Dispersion curves of an infinite polymethylene chain (*solid lines*) and an orthorhombic crystal of polyethylene (the *dashed lines* show vibrations perpendicular to the α-axis). From Tasumi and Krimm, Ref. [6]

The shape of the curves characterizes the sensitivity of the respective modes to phase difference between adjacent oscillators and an influence of the chain length on these modes.

As can be seen from Fig. 1, the full agreement between the theoretical and experimental values of frequencies should be expected for v_1, v_2 and v_6 curves whose modes are strongly localized. The skeletal vibrations (curves v_4, v_5 and v_9) are less characteristic. The frequencies of v_5 and v_9 at $\theta = 0$ and π appear zeroth at all, hence the identification of these modes in the spectra requires the calculations of frequencies potentially active for a finite polymer chain.

These calculations are rather difficult, because when chemical units are repeated on a chain in a regular fashion, all of the units have the same energy, so that they are potentially capable of resonating or coupling their vibrational motions. This intermolecular vibrational coupling can lead to development of a series of resolvable vibrational modes characteristic of the length of the coupled units. There are, however, a number of factors, namely, the restrictions imposed by the selection rules, an abrupt decrease in intensity of modes located at considerable distance apart from the edges of the dispersion curve, the defects in real chains, the resolution limitations, etc. All these factors result in such non-characteristic modes giving rise just to series of absorption bands merging into a single asymmetrical absorption band characterizing a finite segment of the coupled units [10, 11], or vibrational segment of the chain [12].

Comparison of frequencies calculated for skeletal vibrations with those observed experimentally in the PE spectrum reveals a satisfactory agreement [5]. In particular, branch v_4 has $k = 0$ modes at $v_0 = 1065$ cm^{-1} and $v_\pi = 1135$ cm^{-1}. The respective bands in the Raman spectrum of PE observed at 1061 and 1131 cm^{-1}.

On increasing the identity period to four CH$_2$ groups, which is equivalent to taking into account the interaction between the neighbouring units, the other modes become potentially active: at $v_{+\pi/2} = 983$ cm^{-1} and $v_{-\pi/2} = 430$ cm^{-1}. In addition, an out-of-plane skeletal mode can be active whose the frequency is given by:

$$v_{\pi/2}^2 = \frac{k_t}{4\pi^2 m} (4 - 2\cos\theta - 4\cos 2\theta + 2\cos 3\theta) \qquad (3)$$

where m is the mass of the CH$_2$-group and k_t is the force constant for rotation about the C–C bond. The value of the latter is not well known, but on the assumption that $k_t = 20$ N/m [5], we find from Eq. (3) that $v_{\pi/2} = 140$ cm^{-1}. It is assumed that the broad absorption band centered at 200 cm^{-1} in the FIR spectrum of PE (Fig. 2a) may be assigned to this mode.

We now consider the polyethylene crystal and look for modifications of the single chain spectrum, which occur in the low-frequency region. Because the main valence bonds in the single chain are very strong compared with the intermolecular interaction in the crystal, it should be expected that the type of dispersion relation to be similar to that for the single chain. However, the single chain "feels" different force constants in the α-axis and β-axis directions. This causes a frequency splitting of all nine branches shown in Fig. 1 (the dashed lines), giving 18 branches. Again, the modes at $\theta = 0$ and π are optically active, so that we obtain 36 frequencies. Only the low-frequency range is affected, essentially because of lattice interaction. Intersection of the branches v_5 and v_9 with the ordinate, at $\theta = 0$ and π gives 8 modes; three of them are acoustical modes and give zero frequencies and five form the lattice modes.

The IR active mode v_5^a corresponds to the well known sharp band in the FIR spectrum of PE with maximum at 73 cm^{-1}. The mode originates from translational vibrations of the lattice, this agrees with the IR dichroism [13] and deuteration studies [14]. Increasing the temperature shifts the band toward lower wave numbers, from 81 to 68 cm^{-1} between 12 and about 400 K [15]. The application of external pressure leads to a frequency shift in the opposite direction. This result makes it possible to discuss lattice anharmonicity and Grueneisen parameters [16].

Another IR active mode v_5^b predicted at 108 cm^{-1} was unobservable at room temperature, but seen in the spectrum at low temperatures (Figs. 2f and 37a, Refs. [4, 17]). It is subtle and, therefore, insufficiently studied.

Comparing the theoretical prediction with the absorption spectrum of semicrystalline PE in the range of 50–500 cm^{-1}, one should admit that the former does not reflect the whole picture observed experimentally. The real spectrum closer reflects the density of phonon states (Fig. 2d) calculated by taking into consideration the nonzero phase difference and "quasi-lattice

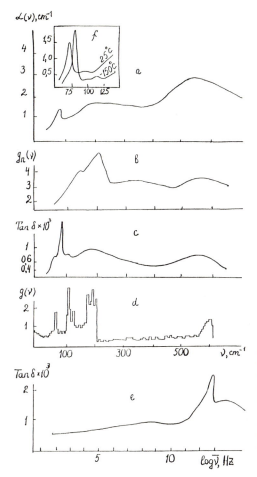

Fig. 2a–e. Polyethylene: absorption spectra in the far infrared region (**a**) from Chantry and Chamberlain, Ref. [2]; (**f**) from Fleming et al., Ref. [17]; (**b**) neutron inelastic scattering data; (**c**) dielectric loss tangent vs. wave number; (**d**) calculated density of phonon states; (**e**) dielectric loss angle tangent vs. log $\bar{v}$. (Reproduced from Bur, Ref. [18])

modes" characterizing the short-range order in amorphous regions of the polymer. These modes of vibrations or density of phonon states have been observed in neutron scattering experiments for PE (Fig. 2b) [18].

Figures 2c and 2e show the low-frequency spectrum of PE within a wide frequency range plotted as tan δ vs. v and vs. $\bar{v}$ dependences where tan δ is the dielectric loss angle tangent, v is the wave number in cm^{-1}, $\bar{v}$ is the frequency in Hz [18, 19]. In addition to the sharp lattice mode at 73 cm^{-1} and the broad band at 200 cm^{-1} discussed above, the two wide regions of the dielectric losses can be seen at 10^8–10^{10} and 10^{12}–10^{13} Hz. These are discussed below.

2.3 Other High-Crystalline Polymers

The theoretical approach used for PE is relatively widely applied to the assignment in FIR spectra of the other high-crystalline polymers, namely PP [2, 20], PTFE [21, 24, 26], POE and POM [22].

Calculations performed for isotactic and syndiotactic PP [20] show that the deformation and torsional vibrations bands should be mainly expected in their low-frequency spectra. Since the relevant modes are not localized within one monomer unit, these absorption bands are satisfactorily identified provided that PP has a relatively high content of crystalline phase. In addition, the number of bands observed experimentally is much greater than that predicted for an isolated macromolecule, most of them are believed to arise from coupled vibrations.

For instance, according to [20, 21], the absorption band of isotactic PP with maximum of 398 cm^{-1} (Fig. 3a) corresponds to the calculated frequency of 392 cm^{-1} of deformational vibrations of the $C-\overset{\underset{|}{C}}{C}H-C$ groups which are symmetric (δ_s) by 35% and asymmetric (δ_a) by 40%. The band at 321 cm^{-1} (calculated frequency: 314 cm^{-1}) is due to δ_a vibrations by 60%, and the band at

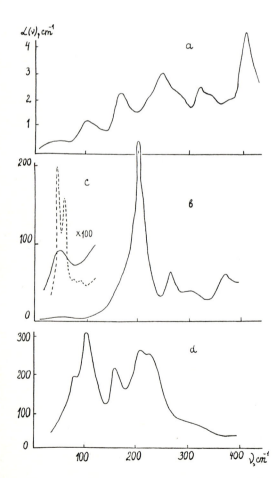

Fig. 3a–d. Far infrared spectra of high-crystalline polymers: (a) isotactic PP, Ref. [20]; (b, c) PTFE (*dashed line* at $-150\,°C$), Refs. [24, 62]; (d) POE, Ref. [153]

251 cm^{-1} (calculated frequency: 272 cm^{-1}) is due to δ_s by 40% and δ_a by 20%. A low-intensity band at 210 cm^{-1} (calculated frequency: 207 cm^{-1}) is due to torsional vibrations of the CH$_3$ group by 95% and the band at 169 cm^{-1} (calculated frequency: 162 cm^{-1}) is due to asymmetric by 40%, symmetric by

30% vibrations of the C–CH–C $\overset{\displaystyle C}{\underset{\displaystyle |}{}}$ group and torsional vibrations around the C–C bond (τ_{CC}) by 15%. Finally, the band at 106 cm^{-1} (calculated frequency: 90 cm^{-1}) is due to skeletal torsional vibrations by 70%.

The above approach has only been partially applied and this is reflected in the uncertainty which exists about the attribution broad band at 55 cm^{-1} in the FIR spectrum PP, assigned tentatively in Ref. [20] to a lattice mode. It would be more reasonable, however, to assign this absorption to "disorder-induced absorption" or quasi-lattice mode [23]. The quasi-lattice absorption is inherent to the FIR spectra of all the disordered condensed media. We shall discuss it later.

As an example, we will now discuss the FIR absorption spectrum of another high-crystalline polymer – PTFE (Fig. 3b). PTFE can be considered as a PE in which all hydrogens are replaced by fluorine atoms. The solid-state phases of PTFE differ from those of PE in so far as the molecule is no longer a planar zig-zag chain but is twisted into a helix and also the intermolecular forces are undoubtedly significantly different in these two cases.

The evaluation of the absorption spectrum of this single chain in terms of the dispersion relations has been carried out by Hannon et al. [24]. The main results are as follows:

The band at 385 cm^{-1} (calculated frequency: 380 cm^{-1}) is due to deformational mode of the CF$_2$ group (v_δ) by 70% and its wagging mode γ_w (CF$_2$) by 25%; the band at 308 cm^{-1} is mostly due to rocking mode γ_r (CF$_2$); the band at 277 cm^{-1} (calculated frequency: 280 cm^{-1}) is due to rocking mode γ_r (CF$_2$) by 60% and the skeletal mode v_5 of the C–C bond by 20%. The most intensive band centered at 203 cm^{-1} (calculated frequency: 192 cm^{-1}) is due to totally rocking oscillations of the CF$_2$ group (γ_t mode). According to the calculations, the skeletal torsional vibrations (modes v_9) must be represented by a single band in the spectrum. There are, however, a few sharp bands at $v < 100$ cm^{-1} due to translational and torsional lattice vibrations at low temperatures (Fig. 3c). At $T \geq 20\,°C$, i.e. above the solid-phase transition in PTFE these bands merge into a broad band with the maximum near 50 cm^{-1}. In Ref. [25], the latter has been assigned to liquid–lattice absorption by analogy with the similar absorption in spectra of PE and PP in this range. Its splitting into a series of bands at lowering the temperature occurs because the oscillations in the crystalline regions of polymer become lattice modes.

PTFE also provides an excellent example for the fact that FIR spectroscopy, especially of polymers, opens the possibility of determining the complete set of dispersion curves for a substance, at least in the low-wave-number region. The basis for this procedure is the breakdown of the selection rules, not only when the crystalline order disappears but also when the chain length becomes so short that boundary conditions are no longer acceptable. As an example, we refer to the study of Chantry et al. [26] on PTFE oligomers.

The theoretical predictions based on the normal vibration analysis also give valuable help in assignments of bands in the FIR spectra of high-crystalline POE [22]. FIR spectrum of POE is shown in Fig. 3d. There is the 215 cm^{-1} (prediction: 211 cm^{-1}) band that should be assigned to skeletal deformation vibrations. The band at 165 cm^{-1} (prediction: 162 cm^{-1}) is related to torsional vibrations around the C–C bond, whereas the band at 106 cm^{-1} (prediction: 108 cm^{-1}) – mostly to those around the C–O bond.

Analogously, the calculations and assignments performed for POM [22] allowed the 428 cm^{-1} (prediction: 444 cm^{-1}) band to be assigned to deformation vibration of the C–O–C angle. The bands at 304 cm^{-1} (calculated frequency: 253 cm^{-1}), 130 cm^{-1} (calculated frequency: 135 cm^{-1}) and 90 cm^{-1} (calculated frequency: 86 cm^{-1}) are due to skeletal torsional vibrations. The assignments revealed an important role of conformational state of the chain, because skeletal modes are highly coupled modes and any change in the chain conformation varies the coupling.

3 Skeletal Deformation and Torsional Vibrations in Amorphous Polymers. Theoretical Predictions and Experiments

3.1 Poly(methyl methacrylate), Poly(vinyl chloride), and Polystyrene

The normal vibration calculations based on a correct structure and correct potential field permit a good correlation to be made between predicted and observed absorption bands in the FIR spectra of high-crystalline polymers in spite of the disordered regions existing in polymer crystals. The size and defects of these crystals influence band shape and position because of finite boundary conditions. It also may give rise to additional absorption bands not predicted by the calculation (because the selection rule cannot be applied in this case). The additional bands are observed, indeed, in the FIR spectra at the frequencies corresponding to the maxima in the spectrum of density of phonon states in the low-frequency region [19, 23].

The successful use of the theoretical methods for assignment of skeletal absorption bands in the FIR spectra of high-crystalline polymers has stimulated their further application to semicrystalline and amorphous polymers, e.g. syndiotactic PMMA [27]. The calculations were performed for an infinite planar zig-zag chain of point masses M_1 and M_2 where M_1 is the mass of a (CH$_2$) group and M_2 is the mass of the (CH$_3$–$\overset{|}{\underset{|}{C}}$–COOCH$_3$) group. Due to the alternate nature of the syndiotactic chain, there are four "masses" in the repeat unit and hence there will be $3N = 12$ skeletal normal vibrations of which four will be non-genuine, i.e. three translations and one rotation about the chain axis. Of the remaining eight genuine vibrations, six will be infrared active and all are Raman active.

The frequencies of the plane skeletal vibrations were evaluated from an equation obtained by Zbinden [8] which is reduced to

$$
\begin{bmatrix}
(M_2 w^2 - G) & G & 0 & 0 \\
G & (M_1 w^2 - G) & 0 & 0 \\
0 & 0 & (M_2 w^2 - R - 4Q) & (4Q + R) \\
0 & 0 & (4Q + R) & (M_1 w^2 - R - 4Q)
\end{bmatrix} = 0 \quad (4)
$$

for $\theta = 0$ and π. (θ is the phase difference for potentially active vibration) and

$$
\begin{bmatrix}
(M_2 w^2 - G - 2F) & 0 & 0 & i(N - 2P) \\
0 & (M_1 w^2 - G - 2F) & i(2P - N) & 0 \\
0 & i(N - 2P) & (M_2 w^2 - R - 2Q) & 0 \\
i(2P - N) & 0 & 0 & (M_1 w^2 - R - 2Q)
\end{bmatrix} = 0 \quad (5)
$$

when $\theta = \pi/2$ and $3\pi/2$. Here, $\omega = 2\pi cv$, where c is the velocity of light; the coefficients are: $G = 2k_s \cos^2 \alpha$, $R = 2k_s \sin^2 \alpha$, $Q = 2k_b \cos^2 \alpha$, $F = 2k_b \sin^2 \alpha$, $P = 2k_b \cos \alpha \sin \alpha$, $N = 2k_s \cos \alpha \sin \alpha$, where α is the angle between any C–C bond, and the chain direction, k_s is the stretching force constant of the C–C bond, k_b is the bending force constant of the C–C–C angle.

Solution of these equations has yielded $v_1 = 0$; $v_2 = 803$ cm^{-1}; $v_3 = 0$; $v_4 = 761$ cm^{-1}; $v_5 = 796$ cm^{-1}; $v_6 = 370$ cm^{-1}; $v_7 = 591$ cm^{-1} and $v_8 = 201$ cm^{-1}.

The solution for the out-of-plane normal vibrations of the zig-zag chain also is given by Zbinden [8] as follows

$$
\omega^2 = \frac{4k_t \sin^2 \theta}{M_1 M_2} [M_1 + M_2 \pm (M_1^2 + M_2^2 + 2M_1 M_2 \cos 2\theta)^{1/2}] \quad (6)
$$

where k_t is the torsional force constant for the C–C–C–C chain. It is seen that for $\theta = 0$ and π the potentially active vibrations are zero vibrations. However, for $\theta = \pi/2$ and $3\pi/2$:

$$
\omega_+^2 = 8k_t/M_2 \quad \text{and} \quad \omega_-^2 = 8k_t/M_1 \quad (7)
$$

Therefore, for the out-of-plane vibrations we obtain $v_9 = 0$; $v_{10} = 182$ cm^{-1}; $v_{11} = 0$; $v_{12} = 73$ cm^{-1}. The calculated frequencies are compared with the observed frequencies and their assignments for syndiotactic PMMA are given in Table 1. For comparison, the respective data for PE were also included in the Table.

Table 1. Comparison of calculated and observed frequencies (cm^{-1}) of skeletal vibrations and assignments of absorption bands in polyethylene, s-PVC and s-PMMA

Polyethyene[a] calculated	s-PVC[b]		s-PMMA[c]		Assignment
	Observed	Calculated	Observed	Calculated	
1135	1125	1054	807	803	v (C–C) $\parallel$ or $[v_+(\pi)]$[d]
1065	1096	1076	749	761	v (C–C) $\perp$ or $[v_+(0)]$ v (C–C)
983	963	993	786	796	$[v_+(\pi/2)]$ and $[v_+(3\pi/2)]$ δ (C–C–C) in-plane
430	487	476	320	370	$[v_-(\pi/2)]$ and $[v_-(3\pi/2)]$ δ (C–C–C) out-of-
140	182	–	225	182	plane or $v(\pi/2)$ and $v(3\pi/2)$

[a] Ref. [5], [b] Ref. [28], [c] Ref. [27], [d] notation Ref. [5]

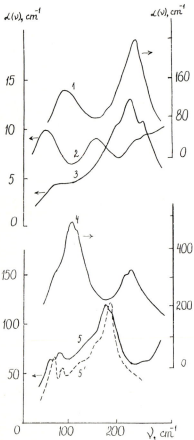

Fig. 4. Far infrared spectra of glassy polymers: (*1*) PVA, Ref. [40]; (*2*) PCS, Ref. [31]; (*3*) PS, Ref. [31]; (*4*) PAN, Ref. [31]; PVC at room (*5*) and liquid nitrogen temperatures (*5'*), Ref. [33]

The FIR spectrum of PMMA is shown in Fig. 21, curve 1. The discrepancy between the theory and the experiment are mainly observed in the region of lowest frequencies and can be attributed to a fault in the approach used in treatment of skeletal modes and as well as to the uncertainty in the choice of force constants and to neglecting interchain interactions.

The values for PE also have been used in the assignment of the skeletal modes of PVC [5, 28]. The FIR spectrum of PVC is shown in Fig. 4, curves 5 and 5′. In addition to the frequencies given in Table 1, this spectrum exhibits bands at 363 cm^{-1} (the calculated mode δ(C–Cl) is 349 cm^{-1}) and 340 cm^{-1} (the calculated mode γ(C–Cl) is 347 cm^{-1}); they are not shown in Fig. 4. There are also expected bands at 86 cm^{-1} and 64 cm^{-1} which are discussed below.

In the FIR spectrum of PS (Fig. 4, curve 3) [5, 29], similar calculations allowed the absorption band at 410 cm^{-1} (the calculated value is 421 cm^{-1}) to be assigned to skeletal vibrations coupled with the 9b mode of the benzene ring. The low-intensity absorption band at 325 cm^{-1} (the calculated value is 358 cm^{-1}) is mainly accounted for by a C–C–C deformation of the backbone chain. A more intensive complicated band at 216 cm^{-1} (the calculated value is 243 cm^{-1}) is assigned to the 9b mode of benzene rings overlapped by chain deformation mode. Absorption bands at 245 and 80 cm^{-1} remained unrecognized in the scope of the approach considered. They will be discussed below.

3.2 Experimental Methods for Assignment of Skeletal Absorption Bands in the Far Infrared Spectra of Polymers

Other attempts to assign absorption bands in the FIR spectra of amorphous polymers using the theoretical normal vibration analysis have revealed its inadequacy in the case of low-frequency skeletal vibrations [5, 7, 8, 21].

It should be noted that also in the mid-infrared region, there are some problems that require a more detailed analysis of spectra than the approach of group frequencies is able to provide. The way out of the situation consists of finding the experimental correlations needed for spectral assignments and then making clear the relations of type "spectrum-structure-property".

In the far infrared where the theory is much less developed, the experimental methods and the correlative approach seem to be the sole promising way for detailed interpretation of the vibrational spectra. The possibilities of the correlative approach for better understanding of the FIR spectra of amorphous polymers have been shown in Refs. [30, 31].

The following linear polymers differing in chemical structure and type of intermolecular interactions were used in these works: polystyrene (PS), polychlorostyrene (PCS), poly(vinyl chloride) (PVC), poly(vinyl fluoride) (PVF), polyacrylonitrile (PAN), poly(methyl methacrylate) (PMMA), poly(butyl methacrylate) (PBMA), poly(cyclohexyl methacrylate) (PCHMA), poly(octyl methacrylate) (POMA), poly(decyl methacrylate) (PDMA), poly(α-methyl styrene) (PMS), and polycarbonate (PC).

Measurements were also carried out for three series of copolymers: copolymers of styrene with methacrylic acid (S-MAA) and crosslinked copolymers of the styrene and divinylbenzene (S-DVB), those of methyl methacrylate and dimethacrylate ethylene glycol (MMA-DMEG) as well as for the set of oligomethacrylates, oligocarbonates and oligo-α-methyl styrenes. The preparation of samples and experimental technique were described in [31–33].

The skeletal deformation and torsional vibration bands were expected in the range of 150–330 cm^{-1}. It was supposed that in some cases, these vibrations may be coupled with internal modes of side groups and the spectral parameters of the absorption bands in this region would be dependent on conformational state of macromolecules. In the analysis of the spectra and assignment of bands, various theoretical and experimental data were taken into consideration.

As noted above, the 216 cm^{-1} band in the FIR spectrum of PS was assigned to internal vibrations of the benzene ring [5] or to 9b mode of ring overlapped by chain deformation mode [7, 34]. As seen in Fig. 4, curve 3, the band has a complicated contour consisting of at least two subbands at 216 and 245 cm^{-1}. The corresponding band pair, shifted towards high frequencies by 15 cm^{-1} (230 ± 5 cm^{-1} and 260 ± 5 cm^{-1}) has been observed for model PS compounds [29], and, according to calculation, the first band (as the 216 cm^{-1} band in PS) was assigned to deformation vibrations of benzene rings, and the second was attributed to the skeletal vibration of the backbone chain. A peak at 250 cm^{-1}, characterizing the motion of the main PS chain, has also been observed in a neutron scattering spectrum [35].

The absorption band at ca 150 cm^{-1} in the PCS spectrum (Fig. 4, curve 2) is due to the out-of-plane vibration of the chlorobenzene rings, according to the assignment in the Raman spectra of polychlorostyrenes [36], whereas the absorption in the range 240–265 cm^{-1} is evidently due to skeletal vibrations of the backbone chain. The bands at 226 and 245 cm^{-1} in PMS spectrum (Fig. 12a) can be identified again by analogy with PS. Most of the low-frequency absorptions in PS, PMS and PCS will be discussed later.

Figure 12b shows the spectrum of PC and its oligomers. By analogy with the just-considered spectra of benzene-ring-containing polymers, the bands at 220 and 260 cm^{-1} are due to the skeletal torsional vibrations in chains as well. Of course, it cannot be considered as an unambiguous assignment but at least it tells us more than a simple assertion without proof that the absorption in the FIR spectrum of PC below 400 cm^{-1} is due to multiphonon processes [37].

As we discussed above on the basis of analysis and calculations of vibrations carried out in Ref. [27], the band at 225 cm^{-1} in the FIR spectrum of PMMA was assigned to the skeletal out-of-plane vibration (Fig. 21, curve 1). The doublet structure of this band (220 and 230 cm^{-1}) is due to the existence of two isomeric states of the backbone chain [38]. A similar band in the FIR spectra of the other polymetacrylates, e.g. PBMA and PCHMA is displaced to 210 and 208 cm^{-1}, respectively [39]. This band similar to that of skeletal vibrations in PVA [40] at 232 cm^{-1} (Fig. 4, curve 1) is structureless, which agrees with the absence of rotational isomerism in these polymers.

In the FIR spectrum of PAN (Fig. 4, curve 4) the band at 259 cm^{-1} (the conformationally sensitive doublet band with maxima at 240 and 257 cm^{-1} according to our data) is assigned to skeletal chain vibrations and to the deformation of C–C–N and C–C–CN groups [9, 41].

The lowest frequencies of skeletal vibrations are observed for PVC and PVF. The spectrum of oriented PVC was reported in [5]; calculations permitted the assignment of absorption with the maxima of 160 and 182 cm^{-1} to skeletal vibrations of the backbone chain of this polymer. The strongest absorption in the spectra of isotropic PVC and PVF in this frequency range are the doublet at 190 cm^{-1} and the absorption band centered at 165 cm^{-1}, respectively. These frequencies of skeletal vibrations, characteristic of macromolecules with small side groups, are relatively close to the calculated value of the skeletal torsional vibration frequency at 140 cm^{-1} in PE [5], which indicates a low degree of perturbation of the main chain caused by fluorine and chlorine atoms.

3.3 Relationship Between Skeletal Torsional Vibrations and Intermolecular Interactions in Glassy Polymers

The preceding experimental data show that the position of the skeletal vibration bands in the region 150-300 cm^{-1} depends on the chemical structure of the carbon chain polymers. It was of interest to determine if there is a relationship between the spectral parameters of their bands and the specific molecular characteristics of the polymers, such as the constant dipole moment (μ_{eff}), the repeat unit mass (M), and the parameter of molecular interaction – the cohesion energy (E_{coh}).

As a result, an unequivocal relationship was found between the position of the maximum of the low-frequency skeletal vibration bands and E_{coh} (Fig. 5): $v_{max} \sim (E_{coh})^{1/2}$. This proportionality shows directly that the force constant of the low-frequency skeletal motion of the backbone chain depends on the level of molecular interactions in the polymers (the nonbonded interactions between the nearest units of both the chain and the neighboring chains).

A change in the energy of these interactions due to any reason should evidently affect the spectral parameters of the absorption considered. The changes observed in the FIR spectra of prestrained glassy polymers confirm this supposition. Actually, an inelastic strain of $\varepsilon = 15–50\%$ increasing the enthalpy [42] and stimulating the molecular motion, or the intensity of mechanical relaxation [43, 46], simultaneously leads to a broadening and slight increase in the intensity of skeletal torsional vibration bands [44]. This effect is evidently due to an increase in amplitudes (anharmonicity) of the vibrational motion of the chain segments owing to the decrease in cohesion forces after deformation. Meanwhile, the shifts of maxima of skeletal vibrational bands turned out to be small, no more than $3–4$ cm^{-1}.

A similar result was obtained in Ref. [45] where the effect of chemical cross-links on the molecular packing and properties of poly(dimethacrylates)

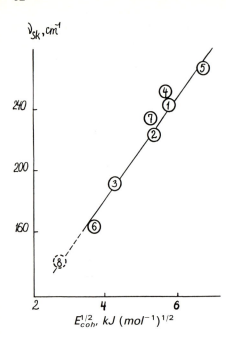

Fig. 5. Position of the maximum of the band assigned to low-frequency skeletal vibrations in the backbone chain as a function the polymer cohesion energy. (*1*) PS, (*2*) PMMA, (*3*) PVC, (*4*) PAN, (*5*) PSC, (*6*) PVF, (*7*) PVA, (*8*) PE-calculated in [5]. From Bershtein and Ryzhov, Ref. [31]

(PDiMA) was investigated, judging by the behaviour of skeletal vibration bands. It has been shown that these bands in the spectrum of such cross-linked PMMA are considerably wider than those in linear PMMA. Taking into account that the width of skeletal vibration bands is proportional to the amplitude and the anharmonicity coefficient of the vibrations, a conclusion was made that cross-links have led to an increase in the local free volume and a decrease in intermolecular interactions owing to a worse molecular packing in PDiMA. These changes inevitably enhance the local mobility of atomic groups in a polymer network. At the same time, the frequencies of skeletal vibrations in polymethacrylate chains of linear and cross-linked polymers differ only little. Therefore, although the loosening effect of the cross-links enhances considerably anharmonicity of torsional vibrations, it does not influence noticeably the relevant force constant of these vibrations.

Undoubtedly, the skeletal vibrations must control to a considerable extent the probability of relaxation transitions originating from a large-amplitude motion in a backbone chain, and requiring surmounting of the barriers to internal rotation.

Indeed, the FIR spectrum of prestrained PVC [46] shows the increase in intensity and its redistribution in a doublet at 190 cm^{-1} which is evidently due to a change in the isomeric composition of polymer after its deformation. The possibility of studying the conformational transformations and molecular mechanisms of relaxation transitions in polymers by means of FIR spectroscopy are discussed below.

In the above mentioned work of Chantry and co-workers [40] who studied the far infrared spectra of ethylene-vinyl acetate copolymers (CEV) the problem was discussed of assignment of the absorption band in this copolymer in the region of the calculated skeletal mode of PE. Figure 6 shows the most typical spectrum of copolymers compared to that of PVA and PE. Although no normal vibration calculations are available for PVA, it could be expected, (cf. Ref. [47]), that for a polymer molecule consisting of a main chain and side groups, the low-frequency vibrations and their associated infrared activity may be thought of as belonging to one of three broad types: (1) modes of vibration of the main chain which are only slightly perturbed but made active by the presence of attached side groups; (2) internal modes of side groups and, finally, if the intermolecular interactions in macromolecules are strong enough, (3) localized chain modes which may arise when the side groups perturb the basic chain backbone vibration modes.

In this case, the new modes include atomic movements over a region of the chain extending only a few repeat units either side of a defect such as conformations of the main chain. In these cases, an increased infrared absorption may be expected at points near to maxima in the density of states function for an unperturbed chain.

It is possible that the copolymer the spectrum of which we have shown here could be considered as a possible model for this behaviour. Since there are no changes in the peak position of the 232 cm^{-1} band in CEV as compared to PVA, apparently the internal mode of the acetate group also participated in forming this band.

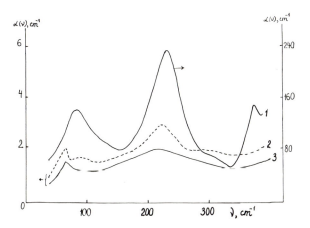

Fig. 6. Far infrared spectra of PVA (*1*); ethylene-vinyl acetate copolymer, 0, 2 per cent VA (*2*); and PE (*3*). From Chantry et al., Refs. [2, 40]

4 "Liquid–Lattice" Type Absorption in the Far Infrared Spectra of Polymers

4.1 A Simple Model of Hindered Rotation (Libration) and Librational Absorption

The theoretical predictions except the participation of internal modes in the forming of absorption bands in the range below $100 \, \text{cm}^{-1}$ [48], so for the skeletal torsional vibrations, e.g. around the C–O or C–C bonds or in the peptide chain [49, 182], it can be made beginning only from $v \approx 100–150 \, \text{cm}^{-1}$. At $v < 150 \, \text{cm}^{-1}$, the absorption bands can be observed either due to lattice vibrations, just as the $73 \, \text{cm}^{-1}$ band in PE, or to intermolecular bonds vibrations – the H-bond for example.

It is now well-established that the most general contribution in the FIR spectra of polymers in the range below $150 \, \text{cm}^{-1}$ originates from the broadband absorption which is sometimes the background of the sharp lattice modes. Chantry and coworkers were the first to observe this broad and weak underlying absorption in the FIR spectra of PE, PP and TPX (a polyolefine based upon poly(4-methyl pentene-1) [2, 50]. It was supposed to be analogous to liquid-lattice type absorption observed in the far infrared spectra of polar and nonpolar liquids [51–56].

Appearance of this absorption can be described by a model of damped rotational oscillations (librations) of polar molecules, i.e. as Poley-type librational absorption [52–54].

In a simple model proposed by Brot et al. [52, 57], the Poley absorption is considered in terms of librational motion of a polar molecule with a moment of inertia (I) within a potential well of depth (V) and semiangular aperture (ξ), due to its nearest neighbours. It is of the form:

$$U(\varphi) = V \sin^2(\pi\varphi/2\xi) \tag{8}$$

with an harmonic librational frequency

$$\omega_0 = \frac{\pi}{\xi}(V/2I)^{1/2} \tag{9}$$

For many liquids, the librational frequencies are close to the position of maximum broad-band absorption at $v \leq 100 \, \text{cm}^{-1}$ [53]. The total intensity of this absorption is determined by an expression [58]:

$$A = \int_0^\infty \alpha(v)\,dv = \frac{\pi}{3c^2}\sum \mu_z^2(1/I_x + 1/I_y), \tag{10}$$

where $\alpha(v)$ is the absorption coefficient, μ_z is the dipole moment along the molecular axis and I_x and I_y are the moments of inertia perpendicular to this axis. The sum is over all molecules. Thus, the band intensity is proportional to the square of the permanent dipole moment of the molecule and has an universal dependence on the moment of inertia. If I_x and I_y are of the same order, then the mean moment of inertia I, where I is taken as $(I_x + I_y)/2$, may be used as an approximation. The linear plot of $\alpha(v)_{max}$ vs. μ^2/I was demonstrated in Refs. [53, 59, 102].

The effect of the dipole moment magnitude of the librating molecule on the intensity of Poley-type absorption was discussed in Ref. [55] for nonpolar liquids where it was found to be significantly (by about an order of magnitude) lower than the observed intensities in the FIR spectra of polar liquids. In nonpolar liquids, the intensity of this absorption is usually related to the induced dipole moments.

An example (Fig. 7, curve 2) is the FIR spectrum of chlorobenzene at 290 K [56]. Here, the broad absorption band at $v_{max} = 45$ cm^{-1} was assigned to the quasi-lattice motion of chlorobenzene molecules arising from the fact that the FIR spectrum of crystallized chlorobenzene at 130 K exhibits several sharp absorption bands corresponding to lattice modes observed in the same frequency range (Fig. 7, curve 3).

According to Ref. [56], the experimentally-observed broad band might be composed of a number of overlapping absorption lines, the widths of which "would be determined by the lifetime of the pseudo–lattice state", and which would be expected to be of the order of the dielectric relaxation time. Microwave spectra and viscosity measurements in chlorobenzene yielded the value of the

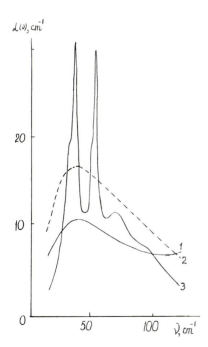

Fig. 7. Liquid-lattice band in far infrared spectra of PCS (*1*), chlorobenzene in liquid (*2*) and crystalline (at 130 K) states (*3*). From Evans et al., Ref. [74]

bandwidth of 10 cm^{-1}. This means that the "liquid–lattice" band consists of overlapping absorption lines of the width of 10 cm^{-1} each. The total intensity of this absorption is of the same order of magnitude as that of the lattice spectrum.

Liquid–lattice absorption bands are not unique for liquids but are found for many disordered solids e.g. glasses [54, 60].

4.2 Librational Absorption in the FIR Spectra of Glassy Polymers

Absorption of this type observed in the far-infrared spectra of semi-crystalline and amorphous polymers [2, 31, 61] gives us information essentially differing from that obtained from mid-infrared spectra because it is much more dependent on intermolecular dynamics.

It has been noted that the conditions for observation of this type of absorption are much more favourable in solid polymers than in liquids. The point is that the relaxation times of polar liquids are of the order of magnitude of 10^{-11} s; this means that the peak of Debye relaxation process will occur in the 1–10 cm^{-1} region and the broad absorption will extend through the region in which the Poley band is found. For polymers, however, the relaxation times are commonly much longer, typically ca 10^{-9} s, and the peak frequency of the Debye process is moved to a lower frequency; therefore, the probability of resolving it from the Poley absorption is much higher.

The ethylene-vinyl acetate copolymers (CEV) containing approximately 3, 6 and 10 mol % vinyl acetate are well suited for the study of the liquid–lattice type absorption in polymers [40, 61]. In the FIR spectra of these copolymers, the lattice mode of a polyethylene type of structure at 73 cm^{-1} was reduced in intensity and a broad underlying absorption appeared in this region. It might

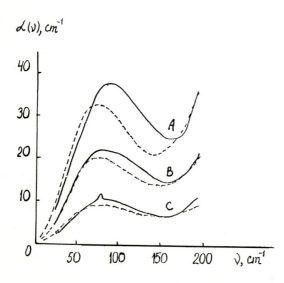

Fig. 8. Absorption spectra of ethylene-vinyl acetate copolymers at room (*dashed lines*) and liquid nitrogen (*solid lines*) temperatures. The curves are for samples of approximately 10% (*A*), 6% (*B*) and 3% VA (*C*). From Chantry et al., Ref. [61]

well be the Poley band arising from the amorphous regions in the sample, because its intensity increased substantially by increasing the concentration of polar VA units (Fig. 8).

It is interesting that the direct proportionality between the absorption coefficient and polar unit concentration applies even if the polar unit is chemically bound to its non-polar surroundings. This behaviour of the band exactly parallels that observed for polar liquids.

The assignment is also supported by the shift of frequency of maximum absorption to higher values as the temperature is lowered. Since the band is not reduced in intensity as the temperature falls, an assignment to the difference frequency absorption is ruled out and it is unlikely that any intramolecular mode of the acetate group can have the half-width as large as that observed. At low vinyl acetate concentrations (i.e. in the 0.2% M sample), the Poley band intensity is of the same order of magnitude as that of the intrinsic continuum absorption in PE. Furthermore, there are now clearly present crystalline regions in the copolymer as shown by the presence of the lattice mode of PE at 73 cm^{-1}. Typically, this band becomes larger and shifts to 78 cm^{-1} on cooling. At 3 mole % vinyl acetate, it is only possible to discern this band in samples cooled with liquid nitrogen (Fig. 8).

Another example of Poley-type absorption is a broad absorption band centered at 50 cm^{-1} in the far infrared spectra of PTFE [26]. In this region, the FIR spectrum of high-crystalline PTFE at low temperature (Fig. 3c) includes several sharp bands at 46, 55, 70 and 85 cm^{-1} close to those predicted for rotational and translational lattice modes [62]. The bands were interpreted by the authors of Ref. [63] in terms of the conformational disordering of chains. However, heating of the PTFE samples which substantially decreases its crystallinity does not result in an enhancement of the bands intensity due to an increase in conformational disordering but in their suppression. At the same time, the background absorption mentioned above or the Poley band due to the amorphous regions in the sample are observed.

In a review on the FIR spectra of polymers compiled by Chantry and Chamberlaine in 1972 [2], the possibility of the presence of Poley-type absorption in FIR spectra of amorphous polymers, namely, PS and PMMA at $v \leq 100$ cm^{-1} is only mentioned. Some more information may be extracted from the Raman spectra. For instance, in Ref. [36] the band at about 60 cm^{-1} in PS is assigned to phenyl group librations; a similar assignment is relevant to the absorption band with the peak position in the range of 40–65 cm^{-1} for substituted polystyrenes; the band also shows a decrease in intensity with increasing side group mass. The intensity of the Raman band at ca 65 cm^{-1} in the PMS spectrum does not decrease so fast above about 70 cm^{-1}. It is suggested that two Raman bands in this region may be caused by the influence of the backbone chain conformation on phenyl group motion.

Analogously, the librational band in the Raman spectrum of amorphous PMMA is located at 70 cm^{-1} [64] and at 80 cm^{-1} for PC [65]. In addition, the Raman spectra of PS, PMMA, PC and substituted PS [36, 64–66] exhibit an absorption at $v \leq 50 \, cm^{-1}$ assigned mostly to the translational oscillations of the chains.

Let us return to the FIR spectra of PMMA, PMA, PVA and PVC in the range of $v \leq 130 \, cm^{-1}$. The FIR spectra of isotactic, syndiotactic and atactic PMMA, showing broad doublet bands at 97 and $123 \, cm^{-1}$, 80 and $100 \, cm^{-1}$, 75, and $95 \, cm^{-1}$, respectively, have been reported in Ref. [38]. The bands were assigned to torsional vibrations of side groups, while their doublet structure was due to the existence of two isomeric states of the ester group. Similar bands at $83 \, cm^{-1}$ in the PMA spectrum and at $85 \, cm^{-1}$ in the PVA spectrum are structureless, which agrees with the absence of rotational isomerism in these polymers. An absence of absorption corresponding to high-symmetrical torsional vibration of methyl groups in this range has been observed.

The FIR spectrum of PVC has attracted comparatively little attention. Three low-frequency modes were observed by Krimm and coworkers [67, 70] at 67, 89 and $180 \, cm^{-1}$. The modes were compared with the results of the calculations, based on normal coordinates analyses [68, 69], which predicted bands at 54 and $117 \, cm^{-1}$. They involve vibrations that are predominantly torsional modes, and a third band at $127 \, cm^{-1}$ that corresponds to the C–Cl wagging modes. Somewhat later, Moore and Krimm [70] used a more realistic model, including an intermolecular force field to allow for weak interactions of the type C–H $\cdots$ Cl–C, and predicted bands at 29, 64 and $90 \, cm^{-1}$. The first of these is a C–Cl $\cdots$ H bending mode, the second is a hybrid torsion/bending mode of C–Cl $\cdots$ H and the third is a torsional mode of the backbone chain. More recently, the FIR spectrum of PVC has been measured at various temperatures in the range of 60 to 320 K [71]. In general, the results obtained confirm the experimental data of Krimm et al. However, the band at $65 \, cm^{-1}$ is attributed to a lattice mode and is not indicative of weak hydrogen bonds of the type C–H $\cdots$ Cl–C. Assignment of the $90 \, cm^{-1}$ band to a torsional mode appears wholly logical; this is supported by the dependence of spectral parameters of the band on the ordering degree and syndiotacticity of PVC.

The liquid–lattice band in the FIR spectra of polymers was studied systematically in Refs. [30, 31, 72, 73]. Various polymers, primarily the amorphous ones (PS, PMMA, PVC, PCS, PAN, PC, et al.) were examined. The peculiarity of these works was that the results of analyses of the spectra of the corresponding low-molecular-weight liquids [53, 74] were taken into account.

The comparison of the FIR spectra of amorphous polymers with the literature data on the FIR spectra of liquids shows that the absorption of liquids and glassy polymers over the frequency range below $150 \, cm^{-1}$ is of a common nature: the main contribution to the spectra of polymers in this range is evidently provided by torsional vibrations (librations) of polar molecules of liquid and monomer units (or polar side groups) in the macromolecule.

This relationship may be followed clearly, for example, in PCS, which has a chemical structure of monomer unit close to that of chlorobenzene. Figure 7 shows that a broad band at ca $47 \, cm^{-1}$ in PCS spectrum corresponds to the absorption band of liquid chlorobenzene with a maximum at approximately the

same frequency. It is assigned to the libration motion of chlorobenzene molecules [56]. The spectrum of crystalline chlorobenzene exhibits sharp bands at
37, 55 and 70 cm^{-1}.

The absorption with the maximum at ca 80 cm^{-1} in the PS spectrum is
mainly due to benzene ring librations [75]. A similar band at 75 cm^{-1} is
observed in the spectrum of liquid benzene [76] and the absorption with a
maximum at 60 cm^{-1} in the Raman spectrum of PS is assigned to this
mechanism [36]. Lattice vibration spectrum of crystalline benzene in the
frequency range under discussion exhibits bands at 65, 74, 92 and 116 cm^{-1}.

Specific are the FIR spectra of cross-linked PS [77] and poly-
(pentafluorostyrene) (PPFS) [78]. In cross-linked PS, the considerable number
of benzene rings are simultaneously chemically bound with two different chains
at a high concentration of the cross-linking agent (divinylbenzene in S-DVB
copolymer) and thus the rotation of the dipoles is more hindered which makes
the 80 cm^{-1} band shift by 10 cm^{-1} to higher frequencies. In contrast, in the FIR
spectrum of PPFS, the librational band is shifted to lower frequencies (Fig. 9)
owing to a greater moment of inertia of a fluorine-substituted benzene ring, in
conformity with the proportionality v_{max} to the value $(E/2I)^{1/2}$. Figure 9, curve 3
shows that a broad absorption band at 23 ± 4 cm^{-1} in PPFS corresponds to
the absorption band of liquid pentafluorobenzene which has a chemical structure close to that of a monomer unit of PPFS.

It follows from Ref. [71] that the absorption band with a maximum at
90 cm^{-1} in the FIR spectrum of PVC is due to the libration of atomic groups
containing chlorine atoms. In this case the authors of Ref. [71] consider the

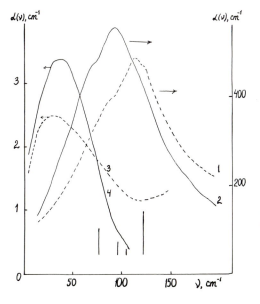

Fig. 9. Liquid-lattice band in far infrared spectra of PAN (*1*); liquid acrylonitrile (*2*), Ref. [81]; PPFS (*3*), Ref. [78] and pentafluorobenzene (*4*), Ref. [74]. The *vertical lines* indicate the position of bands in spectrum of crystalline acetonitrile at 38 K Ref. [81]

absorption near 65 cm^{-1} as being due to a lattice mode. A similar assignment may probably be made for PVF which has a structure close to that of PVC and has in this region an absorption band with a maximum at 88 cm^{-1} and a shoulder at 66 cm^{-1}.

The FIR spectrum of PAN (Fig. 4, curve 4 and Fig. 9, curve 1) in the region 25–150 cm^{-1} exhibits a strong band with a maximum at 127 cm^{-1} and a shoulder at 85 cm^{-1}. According to the assignment made in Ref. [41], the 127 cm^{-1} band in the PAN spectrum is due to torsional vibrations of the CN groups. However, the spectrum of liquid acetonitrile (with a chemical structure close to that of the PAN monomer unit) also exhibits a broad band with a maximum at ca 100 cm^{-1} which is assigned to the librational motion of the interacting molecules of acetonitrile [79]. The absorption in the same frequency range in the FIR spectra of simple nitriles [80] and acetonitrile at low temperature [81] is assigned to vibrations in their dimers of different configuration. Thus, the low-frequency absorption in PAN evidently characterizes the librational motion of monomer units and the intermolecular vibrations of their dipole–dipole complexes.

As it has been noted above, in the FIR spectrum of PMMA a broad absorption band at 95 cm^{-1} is assigned [38] to torsional motion of side ester groups. As seen in Fig. 21, curve 1, the band has a complicated contour evidently

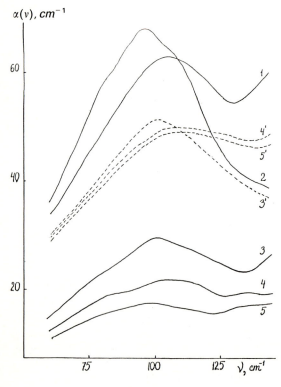

Fig. 10. Liquid–lattice band in far infrared spectra of PCHMA (*1*), PMMA (*2*), PBMA (*3*), POMA (*4*) and PDMA (*5*). In 3′, 4′ and 5′ the intensities of bands were calculated without a contribution from the polymethylene side groups, Ref. [39]

due to the existence of two isomeric states of the ester group. The out-of-plane skeletal bending mode $v_{12} = 73$ cm^{-1} [27] was probably also responsible for the existence of this band.

It is interesting that the FIR spectrum at frequencies below 130 cm^{-1} remains almost unchanged by passing from PMMA to other polymethacrylates, namely, PBMA, POMA, PDMA, PCHMA, i.e. by increasing the polymethylene side group (Fig. 10) [39], although the peak position of the librational band might be expected to shift to lower frequencies because of a great moment of inertia of these groups. This allowed us to assume that nonpolar polymethylene groups joined to the backbone chains via the oxygen "hinges" move independently and do not contribute much to the absorption considered. The Poley-type absorption in the FIR spectra of polymers of this series is mainly due to libration realized within the $-CH_2-\underset{\underset{-C=O}{\overset{|}{C}}}{\overset{CH_3}{}}-$ group. Nonpolar polymethylene side chains cause no more than a small decrease of the libration amplitude and the resultant shift of the band from 95 cm^{-1} to 105 cm^{-1} or to 110 cm^{-1} in the FIR spectra POMA and PDMA, respectively.

Finally, the changes in parameters of the libration band by varying the polymerization degree are also relatively small. In the measurements performed with MMA oligomers [72], the broad absorption band at 95 cm^{-1} was

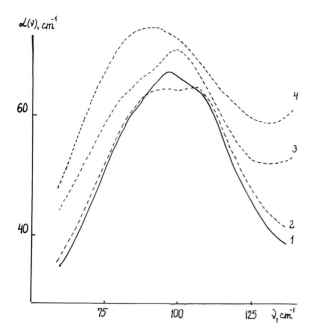

Fig. 11. Liquid–lattice band in far infrared spectra of PMMA (*1*) and oligomethacrylates with *n* = 50 (*2*), *n* = 7 (*3*), and n = 2 (*4*), Ref. [72]

observed in oligomers with $n = 2, 7, 9$ and 50 where n is the number of monomer units in the backbone chain. The spectrum of the oligomer with $n = 9$ was already practically identical with that of the polymer (Fig. 11). The same was observed in Ref. [82] for PMS and its oligomers, (Fig. 12a).

The shape of the Poley absorption band in PMS, which is mainly due to librational motion of the benzene rings, contains two bands at $75\ cm^{-1}$ and

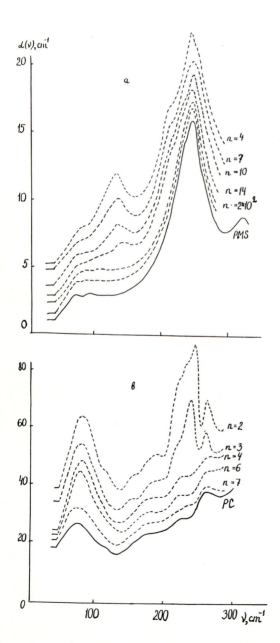

Fig. 12a. Far infrared spectra of PMC and its oligomers, with $n = 2 \times 10^2$, 14, 10, 7 and 4, Ref. [82]; **(b)** far infrared spectra of PC and its oligomers, with $n = 7, 6, 4, 3, 2$, Ref. [82]. Spectra of Oligomers are shifted along the ordinate

95 cm^{-1}. In a speculative way, it was suggested that two bands in this region might be caused by the influence of the backbone chain conformation on phenyl group motion. For example, as shown in Ref. [83] for this side group rotation in PS about the bond connecting it with the backbone chain, the equilibrium position of the phenyl ring plane for the most skeletal configurations bisects the backbone chain valence angle. However, for the most probable backbone chain conformations of PMS, the equilibrium position of the phenyl group was shown to be determined by the skeletal conformations involved. In the configurations tttt and ttg$^+$g$^+$, the plane of the phenyl ring again bisects the valence angle, but in the other two conformations the phenyl ring lies in the plane of this angle. These two conditions may, therefore, be expected to give rise to two different distributions of phenyl group librational frequencies.

The Poley-type absorption band in the FIR spectra of polymers without large side groups, may be also observed, for instance, in PC [82] and poly(dimethylsiloxane) (PDMS) [84]. In PC and oligocarbonates, the libration of the benzene rings by the angles of the order $\pm 20°$ gives rise to a broad absorption band centered at 75 cm^{-1} (Fig. 12b). Assignment of the band at 80 cm^{-1} in the Raman spectrum of PC is the same [163]. With increasing degree of polymerization, the peak position of the 75 cm^{-1} band in the spectrum of PC does not change by frequency, only a decrease in intensity, especially in comparison with its values in spectra of the dimer and trimer, can be detected.

The intensity of an analogous band due to librational motion of monomer units in PDMS [84] with $n = 2, 50$ and 1000 increases with rising n. Apparently, the constraints imposed by the neighbouring monomer units in the chains of this very flexible-chain polymer to the motion of the monomer units is not significant. That this is indeed so can also be concluded from the position of the maxima of spectra. The shift of the maxima toward lower frequencies with increasing molecular weight as in case of PC is very small, indicating that the motion observed does not differ very much irrespective of whether the liquid consists of monomeric or polymeric molecules.

This is an additional indication of the local character of the observed mode. Despite this, however, the large width of the librational bands in the spectra of PC and PDMS reflects the significant perturbation of the motion by the interaction with the environment.

4.3 Relationships Between Parameters of Librational Absorption and Molecular Characteristics of Polymers

In Refs. [30, 31], the characteristics of low-frequency absorption bands assigned to the Poley-type absorption due to the libration of polar groups were compared with the molecular parameters of polymers. We proceeded from the following concepts:

1. The frequency of torsional vibrations ν_{libr} is proportional to $(U/2I)^{1/2}$ where U is the height of the potential barrier of the librator motion and I is its

moment of inertia [57]. This relationship is indeed observed for a large number of simple liquids [53] and molecular crystals; in the latter the frequency of rotational oscillations, v_R, according [85] should be correlatable with the lattice energy and molecular structure through the proportionality $v_R = \dfrac{1}{2\pi}(2\Delta H_s/I)^{1/2}$ where ΔH_s is the characteristic quantity for the lattice energy and I is the corresponding principal moment of inertia. Thus, the librational frequency is governed by the cohesion forces which can be measured directly in molecular crystals as the heat of sublimation ΔH_s.

2. By analogy with liquids [53], it could be assumed that for polymers a proportionality may exist between the low-frequency absorption coefficient $\alpha(v)$ and μ_{eff}^2/I, where μ_{eff} is the effective dipole moment of the monomer unit (a librator).

3. A relationship should be expected between U and the cohesion energy, E_{coh}, characterizing the molecular interactions in the polymer.

The values of μ_{eff} for PVC, PCS, PMMA, PBMA and for PVA were taken from Refs. [86, 87]. Because of intermolecular interactions between neighbouring dipoles in the macromolecule, these values are lower than those of dipole moments of the corresponding low-molecular liquids μ_0: the correlation factor $g = \mu_{eff}^2/\mu_0^2 \approx 0.7 \pm 0.1$ [87]. This value was used in the evaluation of μ_{eff} for PVF, PAN, and PS. For calculating $I = \dfrac{2}{5}MR_{eq}^2$ where M is the molecular weight of the librator, the equivalent radius R_{eq} of the librational unit was determined on the basis of its van der Waals volume [88] by assuming it to be in the form of a sphere.

It can be seen from Fig. 13a that for the carbon chain polymers in question, as in the case of liquids [53], a proportional dependence between $\alpha(v)$ and μ_{eff}^2/I

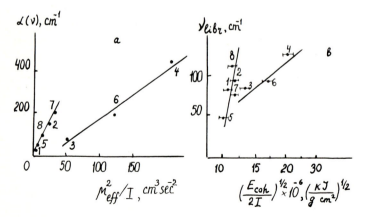

Fig. 13a. Intensity and (**b**) frequency of the liquid–lattice band at a maximum versus molecular parameters of polymers: *1*, PS; *2*, PMMA; *3*, PVC; *4*, PAN; *5*, PCS; *6*, PVF; *7*, PVA; *8*, PBMA, Ref. [31]

is observed which supports this assignment. The dependence between v_{libr} and $(E_{coh}/2I)^{1/2}$ (Fig. 13b) reflecting the relationship of potential barriers to libration to the level of molecular interactions in the polymer was also tentatively determined.

For PVC, PVF, and PAN, the experimental points have different dependences $\alpha(v)$ vs. μ_{eff}^2/I and v_{libr} vs. $(E_{coh}/2I)^{1/2}$. This deviation seems to be due to large μ_{eff} and to some specificities of these polymers. As known, PVC, PAN and PVF chains have an increased rigidity, namely, their Kuhn segment consists of 10–12 monomer units whereas for the other polymers here considered it is composed of only only 6–8 units. This leads probably to a certain coupling between the librations of the neighbours. Apparently, the librator parameters in these polymers must be somewhat different from those calculated for a separate monomeric unit.

The data discussed above show that the FIR spectra of glassy polymers at frequencies below 130 cm^{-1} exhibit, just as liquids do, an absorption band due to hindered rotation (libration) of polar units, i.e. a Poley-type absorption. Spectral characteristics of this absorption are determined by the molecular parameters of polymers, such as the effective dipole moment of a monomeric unit, its geometry, and cohesion energy. Thus, this region of the far infrared spectrum may be a direct source of information about the dynamics of monomer units dependent on its chemical structure and non-chemical interactions with the environment.

5 For Infrared Spectra and Molecular Interpretation of Relaxation Processes in Glassy Polymers

5.1 Submillimeter and Millimeter Spectra and Dielectric Relaxation

In the far infrared spectra of polymer as well as in the mid-infrared spectra, certain frequencies at which the resonance absorption of radiation by a system of oscillating atoms or molecules occurs are analyzed. However, in condensed matter, polymers in particular, at certain temperatures and frequencies the processes are possible which also cause dissipation of energy of the electromagnetic field through configurational and other rearrangements. Various methods of dielectric or mechanical spectroscopy allow these processes to be detected in the form of the relaxation type absorption: an α peak of losses immediately above T_g (the glass transition temperature) and the β, γ and δ peaks at temperatures below T_g [89, 90].

It is important to emphasize the physical difference between relaxational and resonance motions because confusion frequently arises due to the use of similar words to describe physically very dissimilar phenomena.

Consider a short segment of polymer molecule which for steric reasons has adopted a particular configuration; as a result of this, the overall dipole moment of the segment is aligned in a fixed direction. The segment can absorb infrared radiation by the excitation of any of its normal modes of vibration which are optically active. The result is resonance-type absorption bands whose frequencies of maximum absorption correspond to molecular vibration frequencies. The frequencies at which these absorptions are greatest are, to a first-order approximation, independent of temperature since they are determined only by the shape of the molecular potential energy surface and not by the distribution of molecules amongst the possible energy levels.

Now, however, consider a flip of the polymer segment from its initial position into another which is essentially irreversible. The orientation of the dipole will alter and, if it experiences frictional drag during the motion, energy will be absorbed from the electromagnetic field. The result is a relaxational type of absorption. The potential barriers opposing the segment reorientation will be temperature sensitive and, as a result, the relaxation modes will vary with temperature.

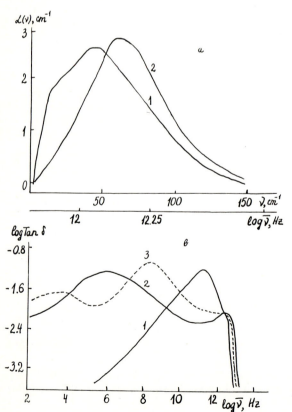

Fig. 14a. Absorption coefficient $\alpha(\nu)$ vs. ν (cm^{-1}) and log $\bar{\nu}$ (Hz) (**b**) Dielectric loss spectrum of fluorobenzene at 293 K (*1*), 110 K (*2*) and 143 K (*3*). From Reid and Evans, Ref. [60]

In relaxation measurements, it is desirable to cover the frequency region as broad as possible. The modern dielectric spectroscopy techniques operate in the range from 10^{-2} to 10^{12} Hz. Therefore, the region of overlap of radiotechniques with optical methods is the submillimeter region. Just over this region in liquids, the Debye type relaxational and the infrared resonant processes will merge and their resolution is a matter of conjecture. For polymers, however, relaxation times are commonly much longer; the peak frequency of the Debye process is moved to a lower frequency; therefore the probability of resolving it from resonant absorption is much increased.

Figure 14a shows the power absorption coefficient $\alpha(v)$ (neper cm^{-1} or cm^{-1}) vs. v (cm^{-1}) and log $\bar{v}$ (Hz) (where $\bar{v} = cv$ and c is the velocity of light) for fluorobenzene in decalin (10 mole % concentration) at 110 and 293 K. Figure 14b shows the dielectric loss spectrum for the same substance in the range $\bar{v}$ = 10^2–$10^{12.5}$ Hz [60].

The dielectric loss spectrum at 293 K consists of the main microwave peak at 10^{11} Hz (3.3 cm^{-1}) with a barely resolved shoulder at $\bar{v}$ ca 1.5×10^{12} Hz corresponding to the maximum of the power absorption coefficient at v ca 50 cm^{-1} in the far infrared spectrum. If the measurement temperature is lowered, the microwave loss peak, ascribed by the authors to β relaxation process, moves out considerably to lower frequencies and becomes resolved from the far infrared peak which is shifted with decreasing temperature in the opposite direction.

At 143 K, the loss spectrum shows one more peak in the range 10^3–10^4 Hz which corresponds to α-relaxation. It is due to the slowest cooperative motion. The β-process is the result of the influence of nearest neighbour cage fluctuations on libration, creating diffusion of the encaged molecule from one energy well into another. It has been described by Johari et al. [91] and Williams [92] as a remnant of the liquid-like diffusional reorientation observed at ambient temperatures; the loss process first explained by Debye.

The high-frequency part of the loss in the far infrared region is characterized by the authors [60] for the first time as the γ-process due to short time torsional oscillations of the solute dipoles. This occurs universally and is the high-frequency adjunct of the α and β-processes. This part of loss in the far infrared region can be studied in detail because of the relation between the dielectric loss $\varepsilon''(\omega)$ and the optical power absorption coefficient:

$$\varepsilon''(\omega) = \alpha(\omega) \cdot n(\omega) \cdot c/\omega \qquad (11)$$

where $\omega = 2\pi vc$ and $n(\omega)$ is the refractive index at frequency $\bar{v} = \omega/2\pi$.

The β- and γ-relaxations obey Arrhenius equation:

$$\bar{v}_{max}(T) = \bar{v}_0 \exp\left(-\frac{E}{kT}\right)$$ where $\bar{v}_0 = 10^{13}$ Hz and E is the activation energy

(barrier height) for the process. Estimations of values of E_γ, E_β and E_α have shown [60, 93] that E_γ for librational motion of a molecule is of the order of magnitude of several kJ mole^{-1}, the β-process (reorientation) has an activation energy of the order of magnitude of few tens of kJ mole^{-1} and E_α, i.e. the barriers for cooperative (inclusive of the nearest neighbours) diffusional reorientation of the cage molecules, are associated with high activation energies of about 100 kJ mole^{-1}.

One more example of the study of the dielectric dispersion spectrum over an extended frequency range (including infrared) also indicates that n-alkanes (C_5–C_{14}) show a broad dispersion in the microwave frequency range followed by a second dispersion in the far infrared [94]. The frequency dependence of tan δ for n-alkanes (Fig. 15) shows clearly two separate peaks:

1) The high-frequency absorption appears to correspond to resonant Poley-type absorption (γ-process following the notation [60] – the counterpart of δ-relaxation in polymers) and it is due to short time and small-angle torsional oscillations of a dipoles within a one well. The shape and position of the Poley absorption seems to be independent of the number of carbon atoms in the chain.

2) The broad low-frequency absorption (in the range $10^9 - 5 \times 10^{11}$ Hz) appears to correspond to relaxational Debye absorption (β-process) and is due to a much slower diffusion of the dipoles through large angles from one energy well to another. Tan δ maximum of this absorption shifts towards lower frequencies again with an increase in the length of the chain. The Debye relaxation time calculated by using equation $\tau_d = 1/2\pi c v_{max}$ increases as an exponential function of the number of carbon atoms in a molecule.

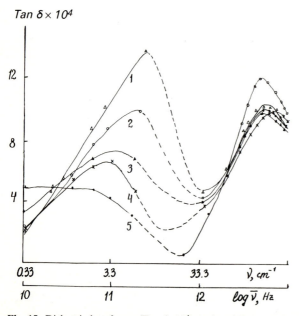

Fig. 15. Dielectric loss factor (Tan $\delta \times 10^4$) vs. log $\bar{v}$ (Hz) or v (cm^{-1}) for n-alkanes: n-pentane (*1*), n-hexane (*2*), n-heptane (*3*), n-nonane (*4*) and n-tetradecane (*5*). From *Vij, Ref.* {94}

The broad β peak (or γ, following the notation [61]) of the Debye relaxation process for PE occurs in the range 10^9 Hz and is due to diffusional reorientation of small segments of the polymer chain. It corresponds to a 170 K peak in the temperature dependence of mechanical losses spectrum at 1 kHz [95]. Its losses can be seen in Fig. 2e which shows the dielectric data for PE extend over a wide frequency range including the far infrared region with the well-known 73 cm^{-1} band of the v_5^a lattice mode and the absorption band centered at 200 cm^{-1} due to skeletal (torsional and bending) vibrations in the backbone chain.

The Poley-type absorption in the region of 100 cm^{-1} is due to the librational motion of the monomer units of the backbone chain and characterizes the δ-process in this nonpolar polymer (see below). The origin of dielectric loss in PE at microwaves is dipolar impurities, end groups, chain folds and branch points.

Dielectric measurements have been made by Buckingham and Belling [96] on ethylene/propylene copolymers. The data show an increase in the loss in range of 10^8–10^9 Hz due to some more polarity of PP. Unlike low dielectric losses in PE, the copolymer spectrum exhibits a α-relaxation peak at 10^4 Hz at room temperature. The effect of oxidation on the microwave behaviour of PE can be seen in the data of Ref. [95]. A loss peak at 10^8 Hz (room temperature) increased with time of oxidation.

In Refs. [2, 78], the microwave spectra of dielectric losses in PE, PP, and TPX near the low-frequency edge of δ-relaxation are compared (Fig. 16b). Here again, these data are presented as spectroscopic absorption coefficients against v (cm^{-1}), and thus show a rising coefficient rather than the familiar maximum in

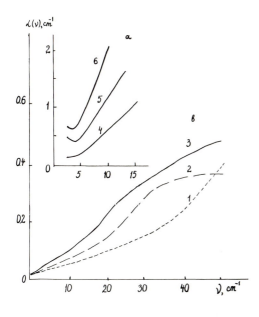

Fig. 16a, b. Millimeter (**a**) and submillimeter (**b**) absorption spectra of PE (*1*), PP (*2*), TPX (*3*) (after Chantry and Chemberlain, Ref. [2]) and ethylene-vinyl acetate copolymers (CEV) with 3% (*4*), 6% (*5*) and 10% (*6*) VA (from Chantry et al., Ref. [61])

ε''. However, the higher absorption of the more substituted chains is quite evident. Since $\tan \delta$ at the peak of the β-process can increase by some orders of magnitude, for example, on oxidation, we might expect an effect, at least near the low-frequency end of the FIR spectra of these nonpolar polymers.

In more polar ethylene-vinyl acetate copolymer (CEV), the Debye relaxation and Poley-type absorption can be observed separately. As we discussed above, a broad absorption band observed for this copolymer in the 70–80 cm^{-1} region (Figs. 6 and 8) arises from a "liquid–lattice" or Poley-type mechanism. According to Refs. [60, 93], the Poley absorption is observed in the dielectric loss spectra of the glassy solutions as the γ dispersion (corresponds to δ-process in polymers). For polymers, the β loss peak is expected at 10^9 Hz, i.e. at microwaves. Figure 16a shows the spectra of CEV observed in the region below 15 cm^{-1}. Dependences for nearly all spectra observed in the submillimetre region would pass through the origin if smoothly extrapolated back to lower frequency. This is clearly not the case for the spectra shown in Fig. 16a. These spectra show a change of slope below 5 cm^{-1} and this can only mean that there is another absorption band at a still lower frequency. Apparently, this lower-frequency absorption arises as the result of the relaxation process. If the absorption below 3 cm^{-1} is indeed due to the high-frequency tail of the β-process, then a clear separation between that and the Poley band is observed.

In Figure 17 (curve 1), the dielectric loss spectrum for PS at room temperature as taken from Bur [18] is presented. There are no pronounced relaxation loss peaks due to α- and β-processes in this polymer which is considered to be "nonpolar", although in fact it possesses a small dipole moment due to the asymmetry at the phenyl side group. The loss tangent is seen to be constant and relatively small over a very broad frequency range from subaudio to 10^{11} Hz. A loss peak occurs at $v \cong 2 \times 10^{12}$ Hz, a very high frequency for Debye relaxation dispersion. It appears to be the δ-peak which has been measured by McCammon et al. at 46 K (1 kHz) with an activation energy of 12 kJ mole^{-1}

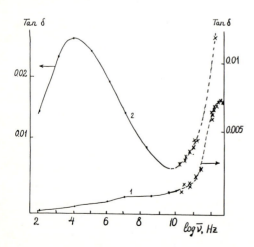

Fig. 17. Loss tangent vs log $\bar{v}$ plot for PS (*1*) and PVC (*2*) at room temperature. From Bur, Ref. [18]

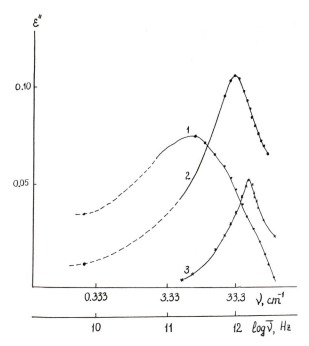

Fig. 18. Dielectric loss ε'' vs. ν (cm^{-1}) or log $\bar{\nu}$ (Hz) for PPFS (*1*), PTFS (*2*) and PS (*3*). From North, Ref. [78]

[97]. In the FIR spectrum of PS, this peak corresponds to the Poley absorption band which is mainly due to libration of the phenyl side groups [31, 75, 98].

The dielectric losses at high frequencies are found to increase in intensity with increasing polarity of librating atomic groups in substituted polystyrenes (Fig. 18). In Ref. [78], the observations have been made on PPFS in which the aromatic nucleus contains the five fluorine atoms. As expected for heavier groups, the frequency of maximum loss drops to 3 cm^{-1} ($\sim 10^{11}$ Hz) and the maximum loss amplitude rises compared to PS.

In poly(trifluorostyrene) (PTFS) in which the backbone hydrogen atoms are substituted and we have again the phenyl nucleus, the frequency of maximum losses occurs at about 30 cm^{-1} (compared with 60 cm^{-1} in PS) and the loss amplitude is almost twice that in PS. This shows that the δ-process in polymers, at least of the given series, is primarily due to libration of the side group dipole, but also some properties of the backbone are important.

The loss tangent vs. log $\bar{\nu}$ plotted for PVC at room temperature within a wide frequency range together with the infrared absorption data in the range 3×10^{10}–4.5×10^{11} Hz is shown in Fig. 17, curve 2 [18]. These data show that for PVC at room temperature the β-dispersion due to the local segmental motion in the backbone chain occurs at ca 10^4 Hz. It corresponds to the loss peak which has been observed in mechanical measurements at $T_\beta = 230$–250 K and $\bar{v} = 1\,\mathrm{Hz}$ [99]. The dielectric loss at very high frequency apparently corresponds to the low-frequency tail of the $90\,\mathrm{cm}^{-1}$ absorption band which is due to the libration in the main chain, i.e. to the δ-process.

5.2 Librations and Low-Temperature δ-Relaxation

As we have seen above, the modes of molecular and dipole motion which contribute to losses associated with α- and β-relaxations in polymers manifest themselves at audio and microwave frequencies. Nevertheless, far infrared spectroscopy may also be an important source of information about the molecular mechanisms of these relaxational processes.

We first focus our attention on the losses (or the absorption) in the frequency range where the relaxational type of spectrum and the resonant type tend to coexist, because from there the information about molecular mobility 'preparing' α- and β-relaxations may be extracted. We mean the small-angle torsional oscillations (librations) leading to Poley-type absorption at $10–130\,\mathrm{cm}^{-1}$. Relationships between the spectral parameters of this type of absorption and molecular characteristics of polymers were discussed above.

As shown in a series of papers [30, 31, 72, 73], the analysis of the Poley-type absorption in the FIR spectra of linear polymers, oligomers and low-molecular liquids allows us to estimate the potential barriers to libration of monomer units. Moreover, the underlying nature of these barriers can be identified as mostly intermolecular, i.e. cohesional. Finally, there exists a correlation between the heights of these barriers and the activation energies of δ-relaxation in polymers. On this basis, the molecular interpretation of δ-relaxation can be addressed.

The libration barriers were estimated from the far infrared spectra using the Brot–Darmon model (see Sect. 4.1). Taking in Eq. (8) $U(\varphi) = Q_{libr}$, one gets

$$v_{libr} = \omega_0/2\pi c = \frac{1}{c\pi\varphi}(Q_{libr}/2I)^{1/2} \qquad (12)$$

at small librational angles. It should be noted that the Brot–Darmon model for polar liquids provides good agreement with the experiment and the potential barriers of libration calculated using this model are close to those obtained in other ways [57, 102].

An analysis was performed [72] proceeding from the far infrared spectra of flexible-chain linear polymers of various structure: their macromolecules either included massive or small side atomic groups or no side groups at all but strongly polar groups in the main chain. Calculations were performed both for a side group and a wholly monomer unit as a librator. The van der Waals volume of a librator was calculated by assuming it to be in the form of a sphere, then the moment of inertia $I = \frac{2}{5}MR_{eq}^2$ was evaluated where M is the molecular weight and R_{eq} is the equivalent radius of the repeat unit. Based on the theory [100,

101], the libration angle for polymers was taken to be equal to 15°. The libration angles for molecules of low-molecular liquids are tabulated in Refs. [59, 102].

In Table 2, the values of Q_{libr} calculated from the far infrared spectra are compared to cohesion energies E_{coh} taken from Ref. [88, 103]. Besides, the activation energies of low-temperature δ-transition in polymers [104–108] are also listed. If the libration of a monomer unit as a whole is supposed, a satisfactory agreement is observed: $Q_{libr} \approx Q_\delta$. Consequently, the low-temperature δ-relaxation in polymers results from the small-angle torsional vibrations of a monomer units. Of course, in the case of a massive side group it is rather difficult to distinguish between the libration of a side group and a monomer unit as a whole within the scope of this approach (see data for PS and PMS).

A remarkable relationship between Q_{libr} and E_{coh} was obtained, namely, $Q_{libr} \approx E_{coh}/3$, which points to the essential contribution to libration barriers from intermolecular interactions. The contribution into Q_{libr} from the barrier to internal rotation Q_{C-C} is substantially less. This is apparently what accounts for an approximate coincidence between the peak position of librational absorption in the far infrared spectra of polymers and low-molecular-weight liquids close to the structure to the respective monomer units, e.g. of PS and C_6H_6 or PCS and C_6H_5Cl (Table 2).

The relationship revealed between Q_{libr}, $E_{coh}/3$ and Q_δ is illustrated by Fig. 19 which confirms the assignment of a low-temperature δ-relaxation to the small-angle torsional vibrations of some molecular unit close in size to a repeat unit of macromolecule. Table 2 gives both experimental and calculated values of the δ-transition temperature T_δ using the Arrhenius equation $\bar{v} \simeq 10^{13} \exp\{-Q/RT\}$ at $\bar{v} = 1$ Hz and $Q_{libr} = E_{coh}/3$. These values are of the order of 20–70 K, i.e. in the temperature range typical for excess (super-Debye) specific heat and δ loss peak observed in dynamic mechanical measurements.

The authors of Refs. [72, 73] demonstrated the relation between δ-relaxation and librations in macromolecules noting the following: First, the analogous relation can also be established for low-molecular-weight liquids. As discussed in Refs. [60, 93], the high-frequency part of the losses in the far infrared region, which are due to Poley-type absorption, may be studied in detail because of the relation between the dielectric loss and the optical power absorption coefficient. Secondly, for polymers with complicated structure or long repeating unit, the assignment of the δ-process to the monomer unit libration may need some corrections. For instance, for poly(alkyl methacrylates) with long side alkyl radicals, the relationship between Q_{libr} and $E_{coh}/3$ still holds provided that the calculations are normalized for the section of the monomer unit up to the oxygen "hinge", i.e. without account of an alkyl radical. Just this section is apparently a librator responsible for the Poley-type absorption in this case. As for the motion of the alkyl radical itself, it determines a specific type of γ-relaxation (see below).

It should be mentioned that the δ-relaxation in polymers is sometimes attributed to rotation of a methyl side group. Indeed, in polymers containing

Table 2. Parameters of the librational motion and low-temperature δ-relaxation in polymers and liquids [72]

No	Material	Librational motion v_{libr} (cm^{-1})	Librational motion φ (rad)	I, side group	I, monomer unit or molecule	Q_{libr}, side group	Q_{libr} monomer unit or molecule	$\dfrac{E_{coh}}{3}$	Q_δ	T_δ, K
				×10^{40} g cm^2	×10^{40} g cm^2	kJ mole^{-1}	kJ mole^{-1}			
1	CH$_3$CCl$_3$	27	0.61	—	715	—	9	9.5	—	35*
2	C$_6$H$_6$	75	0.44	—	340	—	14	11	—	44*
3	CHCl$_3$	35	0.52	—	525	—	7	8	—	30*
4	C$_6$H$_5$Cl	44	0.52	—	570	—	13	14	—	55*
5	(CH$_3$)$_3$CCl$_3$	23	0.61	—	520	—	5	9	—	33*
6	C$_6$H$_5$N	45	0.49	—	390	—	10	12	—	47*
7	C$_6$H$_5$Br	36	0.52	—	950	—	13	14.5	—	57*
8	CH$_2$Cl$_2$	78	0.84	—	50	—	9	9	—	33*
9	PE	100 [2]	0.35 [101]	—	76	—	2.7	2.6	2.2	40 (1 Hz) [107], 10*
10	PP	50, 100 [2]	0.26 [101]	—	155	—	1.2, 4.8	4.5	5	10–15 (1 Hz) [108], 18*
11	POE	105 [178]	0.26 [101]	—	140	—	4.4	3.2	—	13*
12	PVC	86 [31]	0.26 [101]	52	210	2.5	5	6	5	18 (1 Hz) [105], 23*
13	PS	80 [31]	0.26 [101]	345	590	7	11	12	9	45 (1 kHz) [104], 48*
14	PCS	45 [31]	0.26 [101]	600	950	7	12	14	12	50 (1 kHz) [108], 51*
15	PMMA	95 [31]	0.26 [101]	288	528	8	14	11	10	45 (300 Hz) [104], 44*
16	PMA	83 [38]	0.26 [101]	190	400	4	8	8	—	32*
17	PVA	86 [61]	0.26 [101]	190	400	4	9	10	—	40*
18	PAN	127 [31]	0.26 [101]	55	203	3	10	11	—	44*
19	PVF	88 [31]	0.26 [101]	—	138	—	3	5	—	20*
20	PTFE	52 [25]	0.26 [101]	—	372	—	3.2	2.5	—	10*
21	PPFS	24 [78]	0.26 [101]	1210	1600	7	9	10–16	—	62*
22	PDMS	45 [84]	0.26 [101]	—	500	—	3.5	5	5.5 [190]	23*
23	PMS	80 [82]	0.26 [101]	345	815	10	16	15	10	35 (1 kHz) [106], 60*

Notes: (1) The values of v_{libr} and φ for liquids have been taken from Refs. [53, 59, 102 and 74], and E_{coh} from Refs. [88, 103]. (2) The values of Q_δ have been estimated using Arrhenius equation $\bar{v} = 10^{13}\exp(-Q/RT)$ at $\bar{v} = 1$ Hz and T_δ from published experimental data. The temperature T_δ with sign (*) was predicted from the conformity of Q_δ and $E_{coh}/3$, obtained from Ref. [72].

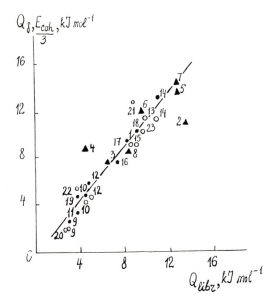

Fig. 19. Relationship between the values of Q_δ ($\bigcirc$), $E_{coh}/3$ for the polymers ($\bullet$), $E_{coh}/3$ for low-molecular liquids ($\blacktriangle$) and Q_{libr}. See numeration in Table 2, Ref. [72]

this group the peaks of mechanical losses at low temperatures are sometimes detected, e.g. at 10 K and 1 Hz in PVA [109] or at 10–20 K and 300 Hz in PMMA [104]. As seen from the Table 2, these temperatures lie below the region of relaxation due to librations of a repeating unit of polymer molecule.

5.3 Skeletal Torsional Vibrations and the Concept of β-relaxation

The study of the β-relaxation process in polymers occurring at temperature of $T_\beta < T_g$ the nearest to glass transition temperature T_g is a subject of continuous and great scientific and technological interest. Until recently, there has been no unambiguous explanation of the origin of the β-relaxation. This could be explained either by the absence of a universal mechanism of β-transition for polymers of various molecular structure or by a lack of understanding of this phenomenon owing to the lack of sufficient and direct experimental evidence.

Lately, new results have been published obtained using FIR spectroscopy, DSC, and other methods which allow a formulation of a common concept for β-transition. According to Refs. [110–112], the β-process is associated with rotational (reorientational) motion of backbone chain segments close in size to the correlation length, or statistical (Kuhn) segment. The barriers opposing these movements are mostly the intermolecular ones, though a one-barrier $t \leftrightarrow g$ transition also participates in this event.

It has been shown that β-relaxation is of particular importance, because the relevant quasi-independent or cooperative motion of segments in fact predetermines the segmental mobility in polymers at any temperature $T \geq T_\beta$

including α-(glass) transition, rubberlike elasticity and flow regions and thus all the physical properties and chemical processes controlled by segmental dynamics [111, 112, 123]. Recently, the experimental facts concerning β-relaxation have been summarized by Bershtein and Egorov [114].

Below, we report some results concerning the molecular interpretation of β-relaxation obtained by means of FIR spectroscopy.

In Ref. [77], the low-frequency skeletal vibrations were studied by modifying the linear polymers with chemical cross-links. The measurements were carried out for two series of cross-linked polymers, namely, copolymers of styrene and divinylbenzene (S-DVB) and those of methyl methacrylate and dimethacrylate of ethylene glycol (MMA-DMEG). In these copolymers, the concentration of the second (cross-linking) comonomer (giving rigid cross-links in the case of DVB and relatively flexible cross-links in that of DMEG containing oxygen "hinges") was varied from 1 mole % to 80–100 mole %.

The difference between the skeletal vibration bands at 245 and 225 cm^{-1} in the spectra of PS and PMMA, respectively, on one hand, and those of copolymers investigated, on the other hand, consists of a decrease in intensity of these bands in the spectra of copolymers (Figs. 20 and 21). This is particularly pronounced at a concentration of the second comonomer exceeding 5–10 mole % and caused evidently by both the decrease in the amplitude of these vibrations and in the number of vibrating units (absorption centers).

In addition, with an increasing degree cross-linking the absorption at 120–150 cm^{-1} grows, especially at $C_{DVB} \geq 25$ mole % and $C_{DMEG} = 33$ and 100 mole %. Analysis has shown that this absorption cannot be attributed to

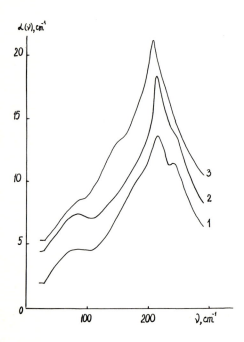

Fig. 20. Far infrared spectra of PS and S-DVB copolymers: Curves *1*, PS; 2, 3 mole % and 3, 25 mole % DVB. Spectra *2* and *3* are shifted along the ordinate, Ref. [31]

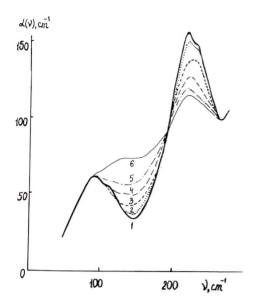

Fig. 21. Far infrared spectra of PMMA, Curve *1* and copolymers of MMA with DMEG at various concentrations of the latter: *2*, 4 mole %; *3*, 5 mole %; *4*, 11 mole %; *5*, 33 mole % and *6*, 100 mole %, (Curves 2–6), Ref. [31]

oscillations of the transverse bridges formed by cross-linking or to defects of the network formed.

As noted in Sect. 2, normal vibration analysis predicts absorption bands due to the low-frequency skeletal modes of a chain of the carbon–carbon backbone in the range 100–200 cm^{-1}. Apparently, the suppression of the segmental motion leads to a decrease of the absorption by which the short range order in the backbone chain is reflected and only the modes allowed by the selection rules are kept.

Figure 22 shows the normalized changes in the intensity of the skeletal vibration band A_{sk} plotted (as a semilogarithmic plot) against the average cross-link spacing expressed in the number of monomer units (N). The values of $N \approx 0.5$ and 0.1 correspond to the most tightly cross-linked networks with concentrations of $C_{DVB} = 82$ mole % and $C_{DMEG} = 100$ mole %. The case of $A_{sk} = 1$ corresponds to the difference between the band intensities for linear polymer and that for copolymer with the most tightly cross-linked network. The $A_{sk} < 1$ characterizes the decrease in the intensity of bands at 245 and 225 cm^{-1} at various cross-linking degrees.

It can be seen that the dependence $A_{sk}(N)$ exhibits a transitional range extending from $N \leq 15$–20 (the same band intensity as that designated A_{sk} for linear polymers) to $N \geq 5$ (the largest decrease in intensity in copolymer with the most tightly cross-linked network). This range is relatively wide, evidently, as a result of the inevitable dispersion of cross-link spacings. The center of this transition range is at $N \simeq 8$–10 units. This value is close to that of the correlation length characterizing the equilibrium chain rigidity, i.e. the Kuhn segment size which for PS and PMMA is equal to $S = 6$–8 monomer units [115].

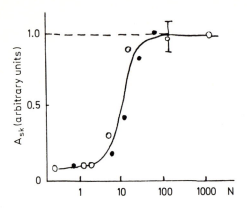

Fig. 22. Intensity (A_{sk}) of the 245 and 225 cm^{-1} bands as a function of the mean spacing between chemical cross-links expressed in the number of monomeric units (N). *Open circles*, copolymers of S-DVB; *filled circles*, copolymers of MMA-DMEG, Ref. [77]

Thus, the observed decrease in the intensity of bands of low-frequency skeletal vibrations indicates that the vibrational motion considered is gradually suppressed as the cross-link spacing approaches the length of the correlation chain part. This result agrees well with the molecular dynamics analysis of a polymer chain with a rigid cross-link [116]. Presumably, these chain vibrations are related to equilibrium chain rigidity, and the chain part involved in this motion is commensurate with the Kuhn segment.

The relationship between the parameters of the infrared absorption band and the length of the polymer chain was discussed by Gribov and co-workers [117] and Zerbi et al. [118]. It has been demonstrated that the most profound variation of the spectral parameters may be expected for the oligomers containing only a few repeating units; the following growth of the backbone chain length results in insignificant changes of the absorption spectrum. The IR spectra were used for estimation of the segment of conformational regularity [11] and the size of the vibrational segment in oligomers of PMMA [12] and PS [119]. The evaluation of the vibrational segment from the mid-infrared spectra related to intramolecular oscillations alone is, however, indirect.

We present below our original estimates of the value of a vibrational segment of the backbone chain based on the FIR spectra of polymer and oligomer samples with various degree of polymerization (n). Experiments were performed for oligomers of PMC with $n = 4, 7, 10, 14$ and 2×10^2; MMA oligomers with $n = 2, 7, 9$ and 50 and oligocarbonates with $n = 2, 3, 4, 6, 7$ and 1.3×10^3.

As seen in Fig. 23, the most radical changes in the MMA oligomers spectra at $v > 130$ cm^{-1} occur when passing to low molecular weight oligomers with $n = 2$ and 7. In the range of 130–280 cm^{-1} their spectra become close to polymer spectrum beginning from $n = 9$. The particularly pronounced changes occur with the band at 245 cm^{-1} assigned to the torsional vibrations in the backbone chain which not only decreases in intensity but has a more complicated shape as well. In addition, a new absorption in the range 130–160 cm^{-1}

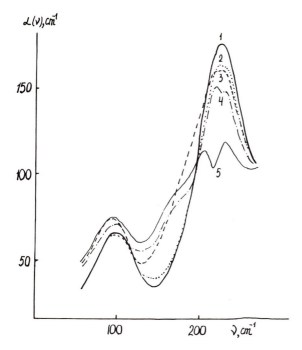

Fig. 23. Far infrared spectra of PMMA (*1*) and oligomethacrylates with $n = 50$ (*2*), $n = 9$ (*3*), $n = 7$ (*4*) and $n = 2$ (*5*) at the temperature of liquid nitrogen, Ref. [135]

appears in approximately the same spectral range as with cross-linking of linear polymers. Note that the length of the Kuhn segment for PMMA is $S = 6$.

For oligocarbonates (Fig. 12b), only the FIR spectra of the dimer and trimer are essentially different. Beginning from the tetramer, the spectra are practically identical to that of PC in the range of $25–300$ cm^{-1}. The FIR spectra of PMS oligomers almost coincide with the polymer spectrum beginning from $n \approx 10$, (Fig. 12a).

Figure 24 shows the semilogarithmic plot $\alpha_{sk}(v)$ against n, where $\alpha_{sk}(v)$ is the skeletal torsional vibration bands intensity in the FIR spectra of PC and PMMA and n is the average chain length of oligomers expressed in the number of monomer units. The dependences $\alpha_{sk}(n)$ exhibit a region of a sharp decrease in intensity. The shaded bands in Fig. 24 correspond to the correlation chain part determined by the equilibrium chain rigidity, i.e. the Kuhn segment size which for PMS and PC is equal to 8 and 2 monomer units, respectively [86]. This result indicates that the absorption bands of skeletal vibrations in these cases become unaltered in their spectral parameters after the length of the backbone chain exceeds the size of the statistical segment.

Important information about the nature of β-relaxation in flexible-chain polymers was offered in Ref. [120] when estimating the potential barriers Q_{sk} and motional units from the FIR spectra. The reasoning of the authors was as follows:

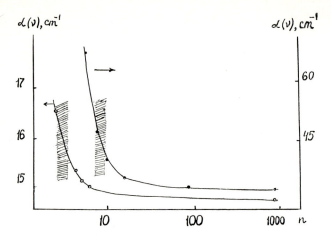

Fig. 24. Dependence of the intensity of low-frequency skeletal band in the far infrared spectra of PMC (●) and PC (○) on the degree of polymerization (n). The *shaded areas* correspond to the Kuhn segment value, Ref. [82]

In the simplest hydrocarbon molecules, e.g. substituted ethane, the potential hindering rotation through the angle ϕ may be represented as [100]:

$$Q(\phi) = \frac{Q_0}{2}(1 - \mathrm{Cos}\,3\phi) + Q(r_{ik}) \tag{13}$$

where Q_0 is the torsional (orientational) energy barrier hindering rotation around a covalent bond in the absence of nonbonded interactions and $Q(r_{ik})$ is a steric barrier which hinders rotation by nonbonded interactions between substituents separated by a distance r_{ik} and attached to the respective atoms joined by the chemical bond.

In the simplest case of ethane when the intermolecular potential may be neglected due to a small van der Waals radius of the hydrogen atom, the total barrier of hindered rotation around the C–C bond is mainly controlled by the value of

$$Q(\phi) = \frac{Q_0}{2}(1 - \cos 3\phi) \tag{14}$$

where $Q_0 = Q_{C-C} \approx 15$ kJ mole^{-1} [101, 109].

In this case assuming a simple harmonic motion, the frequency of the first torsional transition has the form $v = \frac{1}{2}\pi c (f/I)^{1/2}$, where $f = \partial^2 Q(\phi)/\partial\phi^2$ is the force constant for torsional motion, and I is the reduced moment of inertia of the molecule. The frequencies of torsional vibrations for the ethane-like molecules calculated from a trivial potential (14) lie in the range of 120–280 cm^{-1}. This is in agreement with values observed spectroscopically [101].

For carbon chain polymers, the calculations proceeding from $Q_0 = Q_{C-C}$ predict absorption bands in the range of 100–200 cm^{-1} due to torsional vibrations in a chain of carbon–carbon backbone [121], in particular at $v_{calc} = 140$ cm^{-1} for PE [5]. However, the absorption due to skeletal torsional vibrations exhibits a broad band centered at 200 cm^{-1} in the PE spectrum, and at frequencies $v_{sk} = 170$–260 cm^{-1} in the other linear carbon chain polymers (Table 3).

This discrepancy between the experiment and the above calculations may be naturally explained by the fact that the latter do not take into account the steric potential, i.e. the term $Q(r_{ik})$ in the expression (13) for $Q(\phi)$. This term is a result of indispensable involvement (because of correlation in the motion) of the groups adjacent to the rotation axis into the event of torsional motion, especially for rotation at large angles when the torsional vibrations of the neighbouring monomeric units cannot be considered as independent ones. Thus, the real potential barrier for torsional vibrations in the macromolecules is rather controlled by nonbonded (intermolecular) interactions between the nearest units of the chain involved in the motion rather than by the value of Q_0, i.e. $Q_{sk} = Q_0 + \sum Q(r_{ik})$, where the sum extends over all units of the correlation chain part. One more piece of evidence for the importance of nonbonded molecular interactions is the proportionality between the skeletal torsional vibration frequency and cohesion energy E_{coh} observed in the experiment (Fig. 5).

Since the torsional frequency v_{sk} is known from the FIR spectra, the value of Q_{sk} can be estimated from the comparison of v_{sk} with calculated frequency v_{calc}

Table 3. Comparison of parameters of low-frequency skeletal vibrations and parameters of β-relaxation [120]

No	Polymers	v_{sk} (cm^{-1})	Q_{libr}	Q_{sk}	Q_β	T_β, K at 10^{-2}–1 Hz	N (monomer units)	S^a
			kJ mole^{-1}					
1	PS	245 [31]	11	86	100 ± 10	300–330	7	8
2	PMS	245 [82]	12	80	110 ± 10	330–340	7	8
3	PCS	250–260 [31]	14	100	90 ± 10	320	6	8
4	PAN	250 [31]	10	90	110 ± 10	330–340	7	9
5	PVA	232 [61]	9	67	58 ± 6	200–240	6	7
6	PVC	195 [31]	5	44	62 ± 6	230–250	6	12
7	PE	208 [9]	2.7	32	36 ± 8	140–170	7	8
8	PP	169, 251 [21]	1.2, 4.8	28, 62	45 ± 5	170–200	3, 9	9
9	PMA	200 [38]	8	49	42 ± 8	150	4	6
10	PMMA	225 [31]	11	77	85 ± 12	280–290	5	6
11*	PBMA	220 [73]	11	70	90 ± 10	280–290	5	6–7
12*	POMA	216 [73]	11	63	70–75	250–260	4–5	7–8
13*	PDMA	218 [73]	11	66	70–75	240–250	5	7–8
14	POE^b	169 [119]	4.4	20	25–30	120–140	4	4
15	PDMS^b	190 [188]	3.5	48	38 ± 4	135–140	8	5

* In this case Q_{sk} and Q_{libr} have been estimated without the contribution of the alkyl radical of side group;
^{a} Number of monomer units in the statistical segment;
^{b} Calculation was made using $Q_0 = Q_{C-O} \cong 4$ kJ mole^{-1} and $v_{calc} = 105$ cm^{-1} [178]

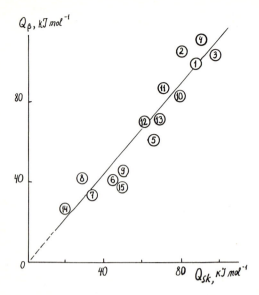

Fig. 25. Relationship between the activation energy of β-transition and the potential barriers of low-frequency skeletal vibrations. Points *1–15* as in Table 3, Ref. [120]

from the ratio: $(v_{sk}/v_{calc})^2 = Q_{sk}/Q_0$. Taking $Q(r_{ik}) = E_{coh}/3$, i.e. equal to the barrier of torsional vibrations of a monomer unit Q_{libr} (the possibility of estimation of Q_{libr} from the FIR spectra and an approximate equality Q_{libr} and $E_{coh}/3$ were discussed above) and assuming $\sum Q(r_{ik}) = NQ_{libr}$, we get $Q_{sk} = Q_0 + NQ_{libr}$. From this result, the number of units affected by skeletal motion, i.e. the correlation chain part (the torsional segment length) $N = (Q_{sk} - Q_0)/Q_{libr}$ can be calculated.

The experimental values of v_{sk} for 15 flexible-chain polymers varying in structure are given in Table 3 along with the activation energies Q_β and temperatures T_β of β-transition for the same polymers obtained from DSC and conventional relaxation data [112, 113]. Besides, the potential barriers of skeletal vibrations, Q_{sk}, the values of torsional segments, (N), and the values of Kuhn segments (S) are also listed in Table 3. It is seen that almost in all cases the barriers of Q_{sk} appear to be close to the activation energies Q_β (Fig. 25) with an accuracy of 20%; so are the torsional segment and statistical segment sizes. Thus, the results of Ref. [120] help to elucidate the molecular mechanism responsible for the β-transition common for the flexible-chain polymers of various structure.

5.4 Onset of Conformational Mobility in Chains and Characteristic Temperatures in Glassy Polymers

The model of β-relaxation in flexible-chain polymers [111, 112, 114] under discussion implies, in addition to rotation through different angles of the adjacent monomeric units, also the participation of a one-barrier *trans$\Leftrightarrow$gauche* transition in the motional event. The possibility of conformational transition at $T < T_g$ has been treated in Ref. [122] by combining the theoretical analysis and IR spectroscopy data in the mid-infrared region on the example of dynamics of intermolecular hydrogen bonds in glassy polymers. The result is in agreement with the fact of the onset of physical ageing related to molecular rearrangements starting from the β-relaxation region [113, 114].

The relations between the low-frequency torsional vibrations in chains and reorientational motion of a chain part close in size to the statistical segment characterizing the thermodynamic chain rigidity allowed us to expect that the conformational state of chains also influences the shape of low-frequency skeletal vibration bands. The conformational sensitivity of these bands was noted in Refs. [40, 124].

In Ref. [33], the FIR spectra of PVC, PMMA and PAN have been obtained in the temperature range from 90 K to T_g and from intensity changes in doublets of skeletal vibration bands, the temperatures of the onset of the conformational mobility unfreezing in these glassy polymers.

Figure 26 shows fragments of the FIR spectra of PVC and PMMA at various temperatures. Band contours consist of two subbands with maxima at 185 and 195 cm^{-1} for PVC and 220 and 230 cm^{-1} for PMMA at room temperature. These bands and the doublet with maxima at 240 and 257 cm^{-1}, corresponding to the skeletal vibration band in the spectrum of PAN, may be caused by the influence of the backbone chain conformation on torsional vibrations, i.e. by the presence of two rotational isomers. Actually, according to Ref. [101], the energy difference between isomeric (*gauche-trans*) states is equal to

$$\Delta E = E_g - E_t = \frac{RT_1 T_2}{T_1 - T_2} \ln \left[\frac{\alpha(v_2, T_2) \cdot \alpha(v_1, T_1)}{\alpha(v_1, T_2) \cdot \alpha(v_2, T_1)} \right] \tag{15}$$

where $\alpha(v_1, T_{1,2})$ and $\alpha(v_2, T_{1,2})$ are the intensities of the subbands measured at two temperatures $(T_1 > T_2)$. The estimates yielded values of $\Delta E = 5$–10 kJ mole^{-1}, which were close to what was expected for the polymers under investigation [125].

The parameter of $\alpha(v_1)/\alpha(v_2)$ is presented as a function of temperature in Fig. 27. This plot displays a very similar behaviour for all polymers investigated. The parameter remains constant up to a certain temperature and then starts rising due to an increase in the intensity of the lower-frequency component of the skeletal band. It starts to rise beginning from T^* which is equal to 220 K for PVC, 250 K for PMMA and 330 K for PAN, these values are much lower than T_g which is equal to 350, 380 and 390 K at 1 Hz, respectively. The temperatures T^*, however, were found to be rather close to β-transition temperatures equal to 230–250, 280–290 and 330–340 K at 10^{-2}–1 Hz, respectively [112, 114].

Some other temperatures also appeared to be close to T^*. First, the temperature at which translational motion of the chain segments begins displayed in the low-frequency Raman spectra [66]. Secondly, the temperature at which the torsional oscillations amplitude in chains increases sharply to give rise to variations of conformation of macromolecules under mechanical loading [126].

The approximate coincidence found in Ref. [33] between the temperatures T^*, T_β and the characteristic temperatures for glassy polymers, namely, T_0- the constant in the Fulcher–Fogel–Tamman and Williams–Landel–Ferry equations

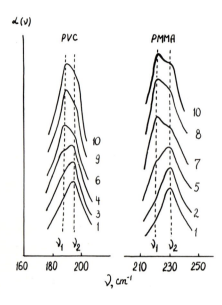

Fig. 26. Changes in the band shape of the skeletal torsional vibration in the far infrared spectra of PVC and PMMA in dependence on the temperature: (1) 85 K, (2) 145 K, (3) 205 K, (4) 225 K, (5) 250 K, (6) 270 K, (7) 273 K, (8) 285 K, (9) 298 K, (10) 315 K, Ref. [33]

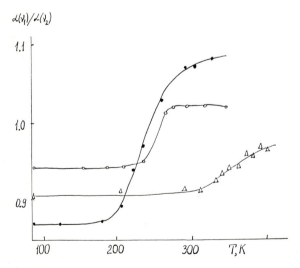

Fig. 27. Temperature dependence of the absorbance ratio in doublet of skeletal band (Fig. 26) for PMMA (○), PAN(△) and PVC (●), Ref. [33]

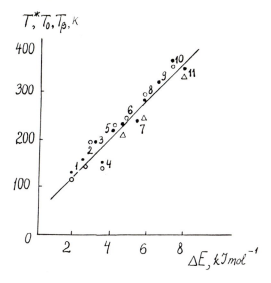

Fig. 28. Correlation of the temperatures T^* ($\triangle$), T_0 ($\bigcirc$) and T_β ($\bullet$) with the energy difference $\Delta E = E_g - E_t$ for different linear polymers, Ref. [33]

and temperature $\tilde{T}_2$ in the Gibbs–DiMarzio thermodynamic glass transition theory [127, 128] should be noted. Just as it has been supposed for $\tilde{T}_2$ and T_0, the temperature T^* corresponds to the stopping of conformational rearrangements in chains on cooling and, as it was supposed for $\tilde{T}_2$ [127], it depends on the energy difference between rotational isomeric states ΔE, (Fig. 28).

Skeletal torsional vibration bands in the FIR spectra of PVC, PMMA and PAN arise from the oscillatory motions at the bottom of the potential energy well corresponding, at low temperatures, to a more stable (*trans*) isomeric state which gives rise to the single band at ν_2, (Fig. 26). With increasing temperature, the amplitude of these oscillations increases as well and the subsequent excitation of the torsional states occurs. As a result, at T^* the barrier to internal rotation can be overcome and the potential well corresponding to the higher-energy (*gauche*) conformer is occupied; it can be illustrated by scheme:

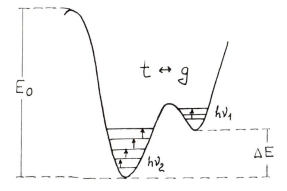

When the degree of occupancy of the potential well corresponding to the *gauche* conformer is high enough, an additional lower-frequency component v_1 of the doublet is observed in the skeletal vibration band under consideration.

Therefore, T^* is a quasi-thermodynamic temperature at which the correlated torsional oscillations in chain are large enough by the amplitude for $t \Leftrightarrow g$ transitions to occur and thus also for the local unfreezing of segmental motion, i.e. for the β-relaxation manifestation. The β-relaxation kinetics, known as being determined by the total potential barrier height of rotational-isomeric transition and seemingly by the proportionality of T_β to ΔE (Fig. 28), looks surprising.

Nevertheless, there is no sharp boundary between the thermodynamics and kinetics of rotational-isomeric transitions in macromolecules. Thermodynamic characteristics that govern the ability of macromolecules to change their conformation are related to the total potential barrier height and thus to kinetic properties of macromolecules. Following Volkenshtein [101], the average cosine of the angle of hindered rotation, which determines the size of a statistical segment as a measure of thermodynamic flexibility of the chain, is equal to

$$\overline{\text{Cos }\phi} = [1 - \exp(-\Delta E/kT)]/[1 + 2\exp(-\Delta E/kT)] \tag{16}$$

On the other hand, according to the Bresler and Frenkel approach [129]

$$\overline{\text{Cos }\phi} = \text{Cot}(E_0/2kT) - 2kT/E_0 \tag{17}$$

where E_0 is the total barrier height of the rotational-isomeric transition. The comparison of these expressions reveals the relationship between the value of a statistical segment, total potential barrier height, and ΔE.

The thermodynamic chain rigidity along with the potential barriers of inter- and intramolecular interactions control the temperature of the β-transition. For a large number of flexible-chain linear polymers, the regular relationships for $Q_\beta \cong (0.3 \pm 0.05) E_{coh} S + Q_0$ and, according to the Arrhenius equation, for $T_\beta = Q_\beta$ (kJ mole^{-1})/(0.25–0.019 lg $\bar{v}$) were demonstrated using DSC [112, 114].

Consequently, a close interdependence between the equilibrium and relaxation properties of macromolecules related to rotational-isomeric transitions is revealed.

The results obtained support the mechanism supposed for the β-relaxation event, including the displacement during a *trans-gauche* transition of adjacent monomer units due to the accumulation of the torsional vibrations in the chain part, commensurate with the length of the statistical segment; the latter appears to be a kinetic motional unit active in β-transition.

5.5 Gamma-Relaxation

In flexible-chain polymers, the relaxation loss peak is frequently observed at temperatures intermediate with respect to those between β- and low-temperature δ-transitions. It is known as a γ-relaxation, its physical origin still remains in question. Its molecular assignment is well established in two specific cases only. First, it is a so-called "chair–chair" type conformational transition in the cyclohexyl group of poly(cyclohexyl methacrylate) and other cyclohexyl-containing polymers exhibited at 190 ± 10 K and 1 Hz [130]. Secondly, it is a motion of long alkyl side groups independent of the main chain in poly(alkyl methacrylates) and other comb-like polymers [109].

In the more general cases, γ-transition is attributed to oscillations of the end or side groups [109, 131–133] or to the torsional vibrations of short chain segments without overcoming the barriers to internal rotation [134]. The former of the proposed explanations seems to be doubtful since the contribution from the end groups in high-molecular-weight polymers must be negligible. Besides, the potential barriers of torsional vibrations of side groups are much lower than the activation energies of γ-transition.

In Ref. [135], the origin of γ-relaxation in polymers was studied using FIR spectroscopy and dynamic mechanical measurements. The authors proceeded from the suggestion that, apart from librational motion of polar groups or monomer units and correlated torsional vibrations in a statistical segment which give rise to conformational mobility of the main chain, some other localized torsional oscillations in the main chain are possible without overcoming the barriers to internal rotation. The relative role of these oscillations must increase when the segmental motion is suppressed or becomes impossible, e.g. due to cross-linking of chains with the average distance between the cross-links $N < S$ or, in the case of oligomers, with the average length of chains $n < S$. It was supposed that just these uncorrelated oscillations give rise to γ-relaxation in most flexible-chain polymers.

To verify this supposition, the FIR spectra of MMA and α-methyl styrene oligomers and also of cross-linked PS and PMMA have been considered. In addition, the FIR spectra of cross-linked PMMA (MMA-DMEG copolymers) were compared with the mechanical loss spectra measured by torsional pendulum technique at 1 Hz in the range 100–450 K.

Now, we recall Figs. 20, 21 and 12, 23 and examine absorption in the range of $120–160$ cm^{-1} that is intermediate between the ranges of librational absorption and absorption due to low-frequency skeletal torsional vibrations in the correlation chain part (statistical segment). It can be seen that the absorption at these frequencies is more pronounced in spectra of oligomers and cross-linked polymers. For example, in the PMS series, the absorption in the range $120–150$ cm^{-1} appears and rises in intensity by passing from polymers to oligomers (Fig. 12a). A priori, the absorption increase could be assigned either to skeletal vibrations of the backbone chain or to vibrational modes of the end groups. In fact, the last suggestion should be rejected since there is no correlation between the concentration of these groups and the intensity of the intermediate absorption and, in addition, the absorption of PS in this frequency range increases as cross-linking develops (Fig. 20).

Analogous changes were observed in this frequency range by cross-linking PMMA chains (Fig. 21) or in the series of PMMA oligomers (Fig. 23). The FIR spectra of PMMA and of the 50-mer of MMA were practically identical, but even for the nonamer this intermediate absorption starts to grow; the spectra of the heptamer and dimer exhibit a sharp increase in absorption at 140–160 cm^{-1}. A new absorption band at $v = 150 \pm 15$ cm^{-1} increases also in cross-linked PMMA with an increasing degree of cross-linking. Meanwhile, the analysis of the spectra of the cross-linked systems has shown that the effect cannot be related to oscillations in the cross-links themselves or defects of the network.

Let us assume that the absorption at the intermediate frequencies v_{loc} both of cross-linked polymers and oligomers is due to the torsional vibrations in the short (with $N < S$) backbone chain parts. The potential barrier of this motion can be evaluated from the frequency v_{sk} of the skeletal torsional vibrations in the corresponding polymers and the frequency v_{loc} of the intermediate absorption according to the relation [73, 135]:

$$Q_{loc} = Q_{sk}(v_{loc}/v_{sk})^2 \qquad\qquad (18)$$

where $Q_{sk} \approx Q_\beta \approx Q_0 + Q_{libr}S$ involved the librational barriers of S monomeric units of the correlation chain part and the barrier to internal rotation Q_0. Formally, $Q_{loc} = Q_0 + Q_{libr}N$ and $N = (Q_{loc} - Q_0)/Q_{libr}$. The values of Q_{loc} and N estimated in this way are listed in Table 4. Comparison of these values with activation energies of γ-transition shows that $Q_{loc} \approx Q_\gamma$ in the cases considered and only two neighbouring monomer units take part in the torsional motion in the backbone chain considered here. Consequently, when segmental motion is suppressed, the key role goes over to local (uncorrelated) skeletal torsional vibrations that govern γ-relaxation.

This conclusion is also supported by identical changes in the FIR spectra of oligomethacrylates and cross-linked PMMA as well as in the internal friction spectra of the latter measured by a torsion pendulum. As the cross-linking of PMMA proceeds, a growth of absorption at 120–150 cm^{-1} and a decrease of

Table 4. Comparison of parameters of low-frequency skeletal vibrations and parameters of γ-relaxation in glassy polymers [135]

Polymers	Q_{libr} * (kJ mole^{-1})	v_{sk}^*	v_{loc}	Q_{loc}	Q_γ	T_γ, K (~1 Hz)	N
		(cm^{-1})		(kJ mole^{-1})			
Cross-linked PMMA	14	225	140–160	28–42	43 [140]	170	2
Oligomers of PMMA	14	225–230	160–190	42–56	43 [140]	–	2–3
Oligomers of PMS	13	245	120–150	32 ± 7	35 [109]	140	2
Cross-linked PS	11	245	140	28 ± 5	33 [189]	132	2

* From Ref. [72, 120]

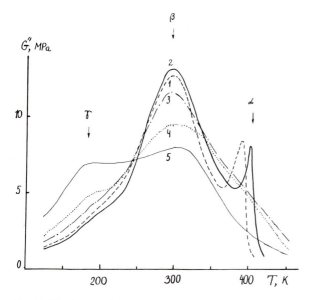

Fig. 29. The loss modulus as a function of temperature for PMMA (*1*) and the MMA-DMEG copolymers with 5 mole % (*2*), 10 mole % (*3*), 50 mole % (*4*) and 100 mole % (*5*) DMEG. Torsional pendulum, 1 Hz, Ref. [135]

absorption at 225–230 cm^{-1} (Fig. 21) are observed. The same effect is observed when PMMA chains get shorter: in the FIR spectra of oligomers with $n = 2$ and 7, the absorption at 120–150 cm^{-1} appears as a new band (Fig. 23).

In the spectra of mechanical losses (Fig. 29) the peak of β-relaxation ($T_\beta \approx 300$ K) decreases in intensity as the cross-linking of PMMA chains proceeds, and simultaneously the peak due to a α-(glass) transition at $T_g \approx 400$ K, related to intermolecularly cooperative motion of the same segments as in β-transition, disappears [114]. At the same time, the loss peak at $T \cong 170$ K associated with the γ-process appears and rises.

The increase of absorption mentioned above in the intermediate region of the FIR spectrum with increasing chemical cross-linking density of PS is accompanied by the increase of the mechanical losses at $T \approx 140$–160 K, or in γ-relaxation [136]; the same was observed by radiation cross-linking of PS [137].

In Fig. 30, the dependences of the intensities of the skeletal absorption bands in the FIR spectra and of loss intensities in the dynamic mechanical measurements plotted against the average distance between cross-links (N) or the length of the oligomer molecules (polymerization degree n) for PMMA samples are presented. It is seen that in all cases the dependences are qualitatively similar. The decrease in intensity of the β peak and the 225–230 cm^{-1} absorption

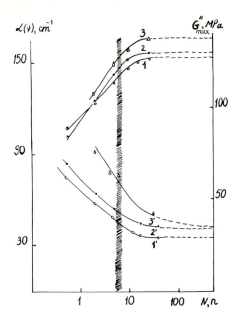

Fig. 30. Intensity $\alpha(v)$ of the skeletal bands at 225 cm^{-1} (*1*), 140 ± 20 cm^{-1} (*1'*) in far infrared spectra of MMA-DMEG copolymers and at 225 cm^{-1} (*3*), 175 ± 10 cm^{-1} (*3'*) in far infrared spectra of oligomethacrylates and the loss modulus G''_{max} (in the ranges of β- (*2*) and γ- (*2'*) -transitions) as a function of mean spacing between chemical cross-links expressed in the number of monomer units (*N*) or average length of oligomer molecule (*n*), Refs. [114, 135]

accompanied by an increase in intensity of the γ peak and the 140–190 cm^{-1} absorption is observed exactly when the values of N or n become less than the length of the statistical segment (for PMMA $S \approx 6$, see shaded band).

Finally, one more piece of evidence in favour of the proposed mechanism of γ-relaxation is the estimate of the activation volume V_a of this process in PMMA. For β-transition, $V_a \approx 0.9$ nm^3 which is approximately equal to the volume of the statistical segment [111, 138], whereas at the γ-relaxation temperature $V_a \approx (0.4\text{–}0.5)$ nm^3 is equal to that of 2–3 units [139].

It is noteworthy that sufficiently strong intermolecular H-bonds similarly to cross-links can promote γ-relaxation. Therefore, the molecular interpretation of the γ-transition in polyamides and polyurethanes detected at 120–160 K and 1 Hz is associated with the torsional motion in the short polymethylene chains between H-bridges [109] and does not contradict the above model. Moreover, when H-bridges were formed in polymethacrylates by sorbed water the peak of γ-relaxation also increases [140].

Thus, the data presented above give evidence of the common mechanism of γ-relaxation in the flexible-chain polymers as the local torsional vibrations in the backbone chain.

All the results presented in this chapter show that FIR spectroscopy allows us to clarify molecular mechanisms of relaxational δ, β and γ-transitions in glassy polymers and to establish the relations of these processes with molecular characteristics of polymers such as the structure of a monomer unit, the cohesion energy, and thermodynamic chain rigidity.

6 Far Infrared Spectra as a Source of Direct Information About Interchain Interactions in Polymers

6.1 The Hydrogen Bond

Hydrogen bonding is usually considered the strongest secondary force in the solid state of polymers. The specificity and magnitude of this type of interaction may strongly influence both chain conformation and packing. Although the energies of hydrogen bonds are weak in comparison to covalent bonds, this type of molecular interaction is large enough to produce appreciable frequency and intensity changes in the vibrational spectra. In fact, the disturbances are so significant that IR spectroscopy is a very valuable source of information on the presence of hydrogen bonds.

Hydrogen bonding involves the interaction between a proton-donating group R_1-X-H and a proton acceptor $Y-R_2$ and may be described schematically by $R_1-X-H \cdots Y -R_2$. The formation of a H-bond is generally favoured by highly electronegative X and Y atoms with relatively small atomic radii (for example, O, N, F) and requires Y atoms with a free electron pair. As a consequence of hydrogen bonding forces, the $v(XH)$- and $v(YR_2)$-stretching frequencies will be lowered, whereas the deformation frequencies associated with the motion of the H and Y atoms perpendicular to their X–H and $Y-R_2$ bonds, respectively, will be increased. The energy of hydrogen bond is directly reflected in the $v(XH \cdots Y)$-stretching and $\delta(XH \cdots Y)$-deformation vibrations. These vibrations are of low frequency located in the far infrared region.

A direct analysis of the low-frequency vibrations of the hydrogen bonds in synthetic polymers is rare. Up till now, a considerable amount of information on FIR spectra of liquids with H-bonds (carboxylic acids, alcohols, phenols, etc.) has been accumulated [14, 141], which significantly facilitates the identification of absorption bands of H-bond vibration in the far infrared spectra. In low-molecular weight systems, e.g. alcohols and phenols, the H-bond stretching vibrations are manifested usually at 110–180 cm^{-1} while in carboxylic acids at 190–250 cm^{-1}; the frequencies of deformation and torsional vibrations of H-bonds are 100–150 cm^{-1} and 40–60 cm^{-1}, respectively [142]. We present below some examples of FIR spectra of polymers with H-bonds.

In Ref. [143], the FIR spectra of a series of polyamides (Nylon 2, 4, 6, 8, 10 and 12) were studied which differed both in a number of CH_2 groups between the amide groups and in the way the chains were packed. In principle, the polyamides occur in two conformations forming extremely different crystalline modifications named α and γ. In the α-conformation, the molecules form a simple all-*trans* chain. In order to form a maximum number of H-bridges in the bulk, the all-*trans* form is only possible if the neighbouring chains are antiparallel. In the case of γ-conformation, the plane of the CONH group is perpendicular to that of the CH_2 groups and in this form the neighbouring chains are parallel.

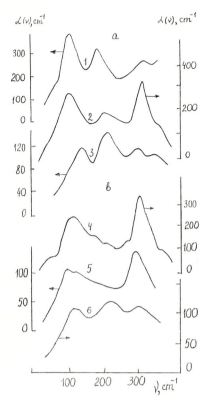

Fig. 31. Absorption spectra of Nylons in α-(a) and γ-(b) conformations (3)-Ny2, (1)-Ny4, (2, 4)-Ny6, (5)-Ny10, (6)-Ny12. From Frank and Fiedler, Ref. [143]

It was found that Nylons 2, 4 and 6 are normally in α-conformation, while Nylon 8, 10, and 12 presumably occur in γ-form.

The FIR spectra of Nylons are shown in Fig. 31. A complete assignment of the low-frequency spectra of polyamides needs group-theoretical considerations as well as potential function calculations. This work has been done only for Nylon 6 by Jakes and Krimm [144].

The assignment of the observed bands made in Ref. [143] is based on a simple model. From the bonded and nonbonded N–H group frequencies reported in Ref. [145], the authors could calculate the force constants belonging to the N–H bond ($k_1 = 704$ N/m) and for the H-bridge coupling to the C–O group in the neighbouring chain ($k_2 = 56$ N/m). It was also assumed that the amide group is rigid and can move as a whole under the influence of the H-bridge force (constant k_2). A model describing the translational and torsional motion (the force constant k_t) of this group is shown in the following scheme. Finally, assuming that the translational motion of the amide group is nearly free in direction perpendicular to the chain backbone they obtained the frequency from

$$\nu_{O...H} = 130.3 \ (k_2/2m)^{1/2} \tag{19}$$

with $m = m_C + m_O + m_N + m_H = 43$ AMU equal to $\nu_{O...H} = 105$ cm^{-1}.

[Scheme: hydrogen-bonded amide groups showing CH_2, $C=O$, $H{-}N$ units linked by force constants k_1, k_2, k_t.]

Therefore, the peak observed in the range of 100–110 cm^{-1} in all spectra of Nylons has been assigned to a translational vibration of the amide group. The shift of this band to 116 cm^{-1} in Nylon 2 (polyglycine) can be understood because only one CH_2 group is centered between two amide groups. The FIR spectrum of Nylon 6 deuterated at the amide group shows no band shift [146]: this observation is in agreement with the above model and the assignment.

Another strong band in the FIR spectra of polyamides observed at 630 cm^{-1} in Nylon 2; at 374 cm^{-1} in Nylon 4, and at 291 cm^{-1} in Nylon 6 (see Fig. 31) is assigned to a librational motion of the rigid amide group with the N–C bond as an axis of rotation. The evidence for the librational origin of this band is a correlation of its characteristics with the mass, the moment of inertia and dipole moment of the group involved in the motion shown in the above scheme.

Similar studies for polyurethanes are described in Ref. [147]. The absorption bands in the FIR spectrum of poly(methacrylic acid) (PMAA) were assigned from the comparison of its spectrum with that of methacrylic acid (MAA) and other carboxylic acids [148]. Two groups of absorption bands are observed in the MAA spectrum at $v \leq 400$ cm^{-1}, (Fig. 32, curves 1, 2). The first group, which is located in the range 300–400 cm^{-1}, was assigned to out-of-plane and in-plane skeletal bending modes, similar to the assignment in the FIR spectrum of acrylic acid [149]. Similar modes apparently contribute to the 350 and 400 cm^{-1} absorption bands in the FIR spectrum of PMAA (Fig. 32, curve 3).

The other group of bands at 100–200 cm^{-1} should be assigned to oscillations of the H-bond. It is known that H-bonds can form planar cyclic MAA dimers and three of the six hydrogen bond vibrational modes of such dimers (torsional, asymmetric deformation and asymmetric stretching) must be active in IR absorption. From the calculations and the experiment [150], the stretching mode of H-bond may be assigned to the band observed at 136 cm^{-1} and the deformation and torsional modes to the bands at 97 cm^{-1} and about 50 cm^{-1}, respectively (the latter is not shown in Fig. 32). The weak band observed at 113 cm^{-1} is also tentatively assigned to the torsional mode.

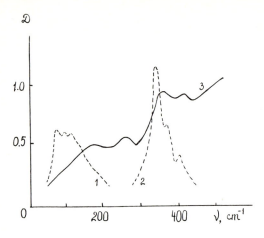

Fig. 32. Far infrared spectra of MAA (Curves *1* and *2*; samples of thickness 200 μm and 90 μm, respectively) and PMAA (Curve *3*; sample of thickness 200 μm). From Belopolskaya, Ref. [148]

While in the spectrum of MAA the absorption due to stretching and deformation modes of H-bond has been observed in the range $100-150$ cm^{-1}, in the spectrum of PMAA, an analogous absorption was shifted to higher frequencies. It looks like a broad structureless band with a maximum at 185 cm^{-1}. It should be noted that a similar broad-band absorption is typical also for liquids molecules of which form chains of H-bridges [151]. The intensity of this band was found to decrease considerably on heating up to 500 K which agrees with the results of the study of the temperature dependence of v_{O-H} absorption band in the mid-infrared spectra of PMAA [152]. This fact gives convincing evidence for breaking of a certain part of H-bridges under these conditions.

Another example of the manifestation of the H-bond vibrations in the far infrared region is the FIR spectrum of poly(vinyl alcohol) (PVAl) [5]. The absorption band at 360 cm^{-1} in this spectrum was assigned to the overtone mode $2v_{OH...O}$.

The above assignments in the FIR spectra of Nylons, PMAA and PVAl were used for interpretation of the FIR spectra of copolymers of styrene with methacrylic acid (S-MAA) [44, 153]. In these spectra, in addition to the absorption bands assigned to PS modes there are the three H-bond absorption bands located in the ranges of $100-120$ cm^{-1}, $150-200$ cm^{-1} and about 360 cm^{-1}. The band around 110 cm^{-1} is assigned to the vibration of the hydrogen bond between OH and C=O groups. The absorption band at ca 180 cm^{-1} apparently gives evidence of the H-bonds of polymeric type in the copolymers, by analogy with the absorption band at 185 cm^{-1} in the FIR spectrum of PMAA; the band about 360 cm^{-1} is the overtone mode. The strong and sharp band at 273 cm^{-1} is apparently like that observed in the spectra of Nylon 6 and Nylon 10 at ca 290 cm^{-1}; it is due to the librational motion of the H-bonded rigid COOH groups. In the spectrum of PMAA, the respective band is found at 267 cm^{-1} (Fig. 32, curve 3).

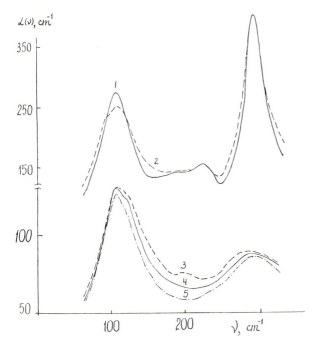

Fig. 33. Far infrared spectra of Ny6 (*1, 2*) and Ny12 (*3–5*) in initial state (*1, 4*) and after prestraining by 20–30% (*2, 3*); (*5*)-the spectrum of Ny12 under tensile stress $\delta = 27$ MN/m^2, Ref. [110]

Proceeding from the above assignments, we shall consider the changes in FIR spectra of the hydrogen bonds containing polymers under loading or after inelastic prestraining. A decrease in the intensity and broadening of the absorption band at about 105 cm^{-1} assigned to the vibration of the hydrogen bond between NH and C=O groups (Fig. 33, curves 1, 2) gives evidence of breaking of some H-bonds in Nylon 6 prestrained by 30%. At the same time, the 295 cm^{-1} band is broadened due to a somewhat stimulated mobility in macromolecules of the prestrained polymer. The same changes were observed in the FIR spectrum of S-MAA copolymer after its inelastic prestraining [44].

The spectral changes in the frequency range of 150–280 cm^{-1}, which characterizes a torsional vibrations in the CH$_2$-sequences connected with the CONH group, also point to an elevated mobility of chains in Nylon 12 prestrained, while the breaking of H-bonds (110 cm^{-1} band) in this polymer may be observed only for loaded samples (Fig. 33, curve 5). Also the FIR spectra of loaded Nylon 12 show a narrowing of the translational and torsional vibration bands which is connected with a decrease in the amplitudes (and anharmonicity) of these motions because of an increase in the chain rigidity under tension on account of so-called "mechanical vitrification" [126].

6.2 Ionic Interactions

The ion-containing copolymers (ionomers) have been the subject of extensive research which has established that the mechanical and rheological properties of these materials depend on the state of aggregation of ions [154, 155]. Most studies have been concerned with ionomers having ionic comonomer concentrations up to 10–15 mole %. In such copolymers, the low dielectric constant of the major component leads to the formation of ion pairs which are thought to aggregate further into multiplets of a few ion pairs or into somewhat larger formations called clusters [156].

In Refs. [157–161], the possibilities of FIR spectroscopy to study coulombic (electrostatic) forces acting in ionomers and to detect the presence of multiplets and clusters in ionomers have been shown. Using far infrared investigations, it was shown how in ion-containing copolymers of styrene [157–159] and ethylene [160] with salts of acrylic and methacrylic acids as well as in poly(styrene sulfonic acid) (PSSA) ionomers [161] and ion-containing inorganic glasses [162] the magnitude of the cation–anion site interactions depends on the degree of clustering, the extent of hydration, and the nature of the cationic and anionic polymers.

The FIR spectra of copolymers of styrene and methacrylic acid (PSMAA ionomers) with a chemical composition represented by

$$-(CH_2-CH)_n-(-CH_2-\underset{\underset{COO^{(-)}\ M^{(+)}}{|}}{\overset{\overset{CH_3}{|}}{C}}-)_m-$$
$$\underset{C_6H_5}{|}$$

where M^+ is a metal cation (Na^+, Cs^+, or Ba^{2+}), and the value of $m/(m + n)$ represents the mole fraction of carboxylate groups and of the un-ionized form of these copolymers, are shown in Fig. 34. The dominant feature of these spectra below 300 cm^{-1} is the presence of a broad, well-defined band, which is absent in the spectrum of the un-ionized copolymer. This band is centered at about 250 $\pm$ 5 cm^{-1} for the Na^+ ionomer, at about 185 $\pm$ 5 cm^{-1} for the Ba^{2+} ionomer and at about 115 $\pm$ 5 cm^{-1} for the Cs^+ ionomer. Since these bands are shifted markedly with cation mass, they must be due to cation–anion site vibrations, i.e. they should be assigned to cation motion in the anionic field of the copolymer.

Figure 34b shows the spectra of ionomers containing increasing concentration of sodium methacrylate. When the concentration of sodium methacrylate increases, a new band at ca 170 cm^{-1} appears on the side of the main Na^+-motion band. The band at ca 220 cm^{-1} may be assigned to a weak internal mode of the polymer. The band at about 250 cm^{-1} is the primary cation motion band; it is present at low ionic concentrations and may be assigned to the vibration of an aggregate involving a few ions (low order multiplet). The band at ca 170 cm^{-1} may be assigned to the vibrations of the aggregates involving many cations and anionic sites together. This may correspond to the formation of higher aggregates or clusters. Because in such ionic aggregates like the

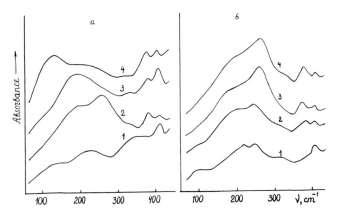

Fig. 34a. Far infrared spectra of S-MAA copolymer (*1*) and ionomers containing Na^+ (*2*), Ba^{2+} (*3*) and Cs^+ (*4*); (**b**) Far infrared spectra of a series PSMAA-Na^+ ionomers with various content of carboxylate groups: (*1*) 0.6; (*2*) 3.8; (*3*) 6.2; (*4*) 8.6 mole %. From Rouse et al., Ref. [158]

cation–anion site the attraction becomes increasingly screened, the vibrational frequency becomes lower than that of the simple cation–anion sites or multiplets. This is consistent with the observation that in solution the ion motion frequencies for simple ion pairs are higher than those for higher aggregates [164]. The large half-width of the cation motion band indicates that there may be several environments for the cation, which differ somewhat in the nature of forces [160].

The spectra of all investigated PSMAA ionomers and of un-ionized co-polymers have a band at ca 405 cm^{-1} with intensity practically independent of the sodium methacrylate content of the copolymer. By analogy with PS, it may be assigned to an out-of-plane bend of the benzene ring or to a C–C–C bend along the polymer chain. A new band which appears only in spectra of ionomers at ca 380–390 cm^{-1} may be assigned to bending motions of the carboxylate groups and connected with the effect of metal ions on the group vibrations.

In the case of styrene-based ionomers containing sodium acrylate, the band assigned to the formation of higher aggregates was found at ca 155 cm^{-1} [159]. The different positions of these bands in the spectra of ionomers containing sodium acrylate and sodium methacrylate may be due to attractions of a somewhat different type creating different structures of methacrylic and acrylic units.

Another example of the use of FIR spectroscopy for investigation of the cation–anion site interactions and state of aggregation in ionomers was described by Mattera and Risen [161]. They obtained the FIR spectra of poly(styrene sulfonic acid) (PSSA) ionomers containing alkali and alkaline earth ions. Strong broad bands the frequencies of which are cation dependent have been observed and assigned to cation motion (Fig. 35, Table 5). Figure 36 shows plots of the cation motion frequency v_0 vs. $(M_{C^{n+}})^{-1/2}$, where $M_{C^{n+}}$ is the mass of the mono- or bivalent cation; the plots are nearly linear. These results show that the vibration considered is due primarily to the cation-anion site vibration.

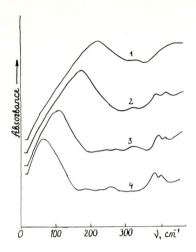

Fig. 35. Far infrared spectra of PSSA ionomers with 6.9 mole % of Na (*1*), K (*2*), Rb (*3*), and Cs sulfonate (*4*). From Mattera and Risen, Ref. [161]

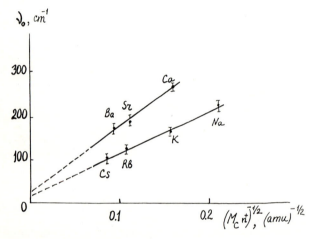

Fig. 36. Plots of the cation motion frequency v_0 vs. $(M_{C^{n+}})^{-1/2}$ for different dehydrated PSSA (6.9 mole % sulfonate) ionomer films, Ref. [161]

In other words, it is a motion of the cation under the influence of the coulombic field of the sulfonate group; moreover, the values of the force constants are of the same order of magnitude for ions of the given group; for the alkaline earth ions the values are larger than for the alkaline ions.

The effects of dehydration and annealing on FIR spectra of PSSA ionomers were also discussed. It was shown that the cation motion frequency decreases and the absorption bands become more pronounced and better defined upon dehydration. This is interpreted as being due to an increased interaction between the sulfonate groups and the cation as solvent is removed. Also the vibrational bands become better defined after the samples of PSSA ionomers

Table 5. Observed cation motion bands (cm^{-1}) in PSSA ionomers [161]

Cation	Average number of unsulfonated styrene units between sulfonate groups			
	30	15	8	6
Na^+	220	210	–	210
K^+	170	162	170	155
Rb^+	120	120	120	116
Cs^+	100	100	100	95
Ca^{2+}	–	270	270	–
Sr^{2+}	180	177	175	170
Ba^{2+}	170	160	148	145

have been annealed above T_g. These results are interpreted in terms of decreased intermolecular interactions at elevated temperature and are consistent with the formation of multiplets from ionic clusters at higher temperatures [165].

It was of interest to compare the spectra of PSSA ionomers as a function of sulfonate composition in light of the observation of a cluster mode present in their FIR spectra. As shown in Table 5, the value of v_0 falls somewhat as the content of sulfonate increases. This is consistent with increasing numbers of cations in large ionic clusters at higher sulfonate concentrations, because the vibrations of the cations in such domains are expected to have both a larger effective mass and lower effective force constant since increased screening lowers the cation-sulfonate site attraction.

6.3 Effect of Crystallinity, Hot Bands, Identification of Biological Macromolecules

The crystalline bands arise from intermolecular forces in the crystal lattice where the polymer molecules pack together in a regular three-dimensional arrangement. These bands are relatively rare in the mid-infrared spectra but PE and PS do exhibit such bands at 730 and 985 cm^{-1}, respectively [21].

However, the determination of the volume fraction of the crystalline phase in a polymer using the mid-infrared spectra faces difficulties. The latter are controlled mostly by molecular structure and conformation of macromolecules, and the usual estimation of the crystallinity degree from the absorbance ratio of the bands corresponding to *trans* and *gauche* conformers is not perfectly correct, because the disordered regions of the polymer also contain the *trans* conformers.

FIR spectroscopy allows us a neater approach to the investigation of crystallinity of polymers because most of the low-frequency vibrations are intermolecular ones and are determined to a large extent by the packing of chains in the lattice.

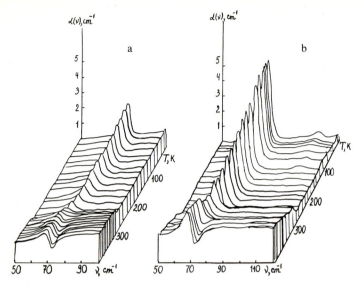

Fig. 37a, b. Change in far infrared spectra with temperature: (a) for low-density PE and (b) for linear high-density PE. From Frank and Leute, Ref. [4]

The effects of varying both crystallinity and temperature can be seen in Fig. 37. The absorption spectra were plotted as a function of temperature for two typical kinds of PE. The results for a low-density PE containing many branching points in the backbone chain are shown in Fig. 37a. In this example, approximately 40 branch points per 1000 carbon atoms occur which results in the presence of side chains of different lengths. The branches inhibit crystallization, so that this type of PE has a relatively low volume fraction of crystallinity ($\sim 40\%$). Figure 37b shows the results for a linear PE of high density. The number of branching points is only about 3 per 1000 carbon atoms; therefore, the degree of crystallinity is comparatively high (about 80%). The different degree of crystallinity of these samples is reflected in different peak heights of the v_5^a mode. The 108 cm^{-1} band (v_5^b mode) is weak and can be seen only at low temperatures. With decreasing temperature the peak maximum shifts to a higher wavenumber; it also has been classified as a crystallinity band [166, 167].

In the FIR spectra of PP, with an increasing degree of crystallinity narrowing of some absorption bands takes place [20] (Fig. 38). The higher intermolecular interactions within the crystalline regions of the polymer caused some bands to split. On decreasing the degree of crystallinity the intensity of the relevant bands decreased, whereas the background absorption related to disordered regions increased.

In the FIR spectra of high-crystalline PTFE at low temperatures, four narrow absorption bands are observed (Fig. 3c). The three of them are due to translational vibrations of the lattice and the fourth band, according to Ref. [62], is due to torsional lattice vibrations.

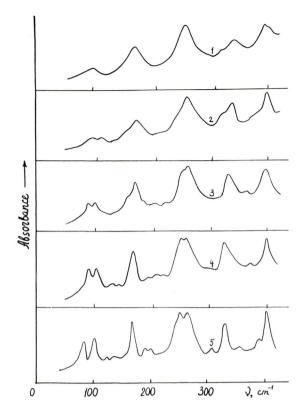

Fig. 38. Far infrared spectra of isotactic PP with different degrees of crystallinity: (*1*) 52%; (*2*) 58%; (*3*) 66%; (*4*) 70%; (*5*) 75%. From Goldstein et al., Ref. [20]

In Refs. [168, 169], it was shown that, properly used, FIR spectroscopy can be applied, for example, to follow the microstructural transformation of poly(vinylidene fluoride) (PVF$_2$) in a static electric field at various temperatures. At 65 °C, the transformation from α to β phase can be easily accomplished by a low field at a high rate and characterized by the bands of crystallinity at 70 cm^{-1} (the librational mode of lattice) and 176 cm^{-1} (the torsional mode CH–CF–CH–CF).

Other results of the far infrared studies of crystallinity of polymers are presented in the papers [22, 170–172].

Here, we almost do not touch the determinations of the conformations of the polymer chains in the solid state by means of FIR spectroscopy [173–175] since they do not exhibit any specificity compared to the mid-infrared spectra. In the FIR spectra of polymers, the changes in conformational composition induced by variations of temperature, pressure, or the uniaxial drawing are also exhibited as redistribution of intensities in doublet of conformation-sensitive absorption bands.

It is important to emphasize one more feature of the FIR spectra: The bands referred to as 'hot transitions' bands, or simply hot bands which are due to transitions between the higher vibrational levels. One should distinguish between the hot bands and overtone bands. The latter are due to transitions from the ground level to a higher one. Overtone bands are especially useful in studies of anharmonicity of oscillations; they lie in a range of frequencies above the conventional infrared region, namely, in the so-called overtone region (the near-infrared region of the spectrum) [176]. In contrast, the hot bands can be detected only in the far infrared, since it is exactly there that the transitions between the highest vibrational levels are most probable.

Let n_1 and n_2 be the numbers of molecules in the ground (l_0) and the first vibrational level (l_1), respectively. Then

$$n_2/n_1 = \exp(-hcv/kT) = \exp(-1.43v/T) \tag{20}$$

where v is the frequency of the fundamental oscillation and T is the temperature in Kelvins. Estimation of the population of the first vibrational level at room temperature shows that $n_2/n_1 \cong 0.4$ at $v = 150$ cm^{-1} which is sufficient for the appearance of the absorption due to $l_1 \to l_2$ transitions. This mode must have a frequency which is lower than the frequency mode of transition $l_0 \to l_1$, because of anharmonicity of low-frequency vibrations.

Figure 39 shows the FIR spectra of poly(dioxolane) $(CH_2CH_2OCH_2O)_n$ and poly(tetramethylene oxide) $(CH_2CH_2CH_2CH_2O)_n$. Intensive bands at ca 145 cm^{-1} are due to torsional vibrations in the backbone chains of these polymers containing oxygen "hinges" [177]. For both polymers, the bands exhibit shoulders on their low-frequency side assigned to $l_1 \to l_2$ type transitions.

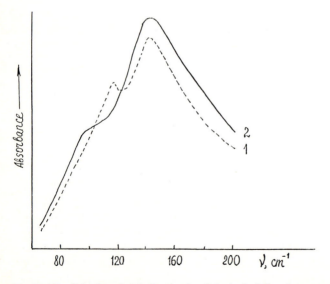

Fig. 39. The skeletal torsional vibration bands in the far infrared spectra of 0.5 mm thick films of the poly(dioxolane) (Curve *1*) and poly(tetramethylene oxide) (Curve *2*) at room temperature. From Roshchupkin et al., Ref. [177]

The difference between the frequencies of the fundamental and "hot" modes is $20 \, \text{cm}^{-1}$ for poly(dioxolane) and $40 \, \text{cm}^{-1}$ for poly(tetramethylene oxide). It means that anharmonicity of torsional vibrations is much larger for the latter polymer which is consistent with the nuclear magnetic resonance measurements of potential barriers of molecular motion in these polymers [178]. In order to assign unambiguously the lower-frequency bands considered to be hot transitions modes, temperature measurements are required.

It should be noted that a considerable population of the higher levels of low-frequency oscillations even at room temperature makes a correction for induced emission into the FIR spectra necessary. According to Refs. [179, 180], this correction must result in profound changes in the shape of absorption band and its shift to zero frequency. According to Ref. [181], a correction of the FIR spectra for induced emission seems impossible at present. Moreover, taking into account many indirect data the effect must be subtle. In our review, we strictly follow the original results without any corrections.

In conclusion, we present the FIR spectra of the most complicated and important polymers–biological polymers–the spectra of which are very rich, especially at low temperatures when sharpening of the bands decreases the overlap between the bands and the background absorption. It somewhat favours separation of the bands and their subsequent identification.

Most work concerning biomacromolecules has been confined to the wavenumber region above about $300 \, \text{cm}^{-1}$ because such spectra are easier to interpret and also because of increasing experimental difficulties at lower frequencies. In general, the low-frequency region is more important for conformational studies of biomacromolecules, because the vibrational modes occurring there are more affected by a change of the further surroundings than are the localized high-frequency group vibrations. This statement is based on the fact that low-frequency modes are more likely associated with larger reduced masses of the corresponding normal modes.

In Ref. [182] the FIR spectra of amino acids and polypeptides were studied. Figure 40a shows the spectra of poly-L-alanine (PLA). Absorption bands for this polymer can be identified at room temperature as follows: the $248 \, \text{cm}^{-1}$ band was assigned to skeletal deformation vibrations, the $205 \, \text{cm}^{-1}$ band to skeletal torsional vibrations, and the $103 \, \text{cm}^{-1}$ band to intermolecular hydrogen bond (between the CO- and NH-groups) vibrations. The lowest-frequency absorption bands remained unidentified because of the probable contribution from such factors as the presence of various conformations, namely the α-helix and β-form of flexible and rigid segments, heavy metals, etc. In the FIR spectrum of poly-α-glycine (Fig. 40b), the $203 \, \text{cm}^{-1}$ band is assigned to torsional skeletal vibrations and the $120 \, \text{cm}^{-1}$ band to stretching of H-bond between the $C=O$ and OH-groups. The $117 \, \text{cm}^{-1}$ band the intensity of which decreases with lowering temperature is perhaps the hot band.

Concerning the other features of the FIR spectra of polypeptides we would like to draw attention to the remarkably steep onset of the absorption at about $50 \, \text{cm}^{-1}$ and to the increase which is not quadratic with frequency as expected

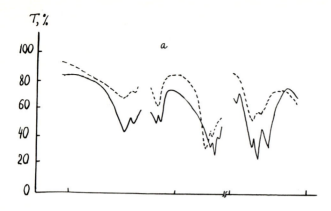

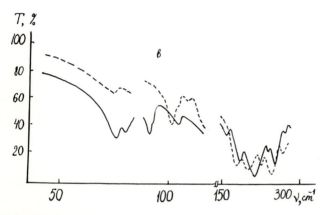

Fig. 40a, b. Transmission spectra of poly-L-alanine (**a**) and poly-α-glycine (**b**) films at room temperature (*dashed lines*) and at 113 K (*solid lines*). The thickness of the samples was about 80 μm. From Feairheller and Miller, Ref. [182]

from a purely disorder-induced one-phonone absorption due to acoustic modes. This indicates the existence of a large number of individual, more localized, vibrational modes causing an in-homogeneously broadened absorption in addition to the background phonon absorption.

In Ref. [183] the dispersion curves for right α-helix of PLA were calculated. Unfortunately, these data are rather difficult to apply for identification at low frequencies because acoustic and optical oscillations are not completely separated in this case.

The FIR spectra of biopolymers are also discussed in Refs. [184–187]. As examples, we show the FIR spectrum of polysaccharide-cellulose (Fig. 41b) and monosaccharide-ascorbic acid (Fig. 41a, Husain et al., Ref. [185]). Some absorption bands in the spectrum of cellulose can be interpreted in terms of the monosaccharide characteristic modes which may be calculated by the normal mode analysis.

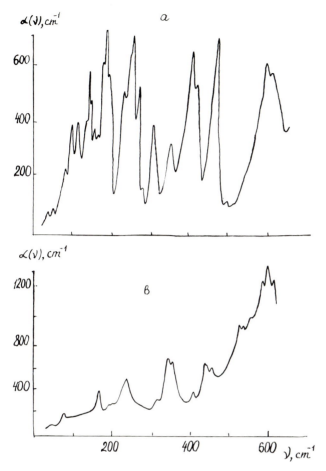

Fig. 41a, b. Absorption spectra of the ascorbic acid (**a**) and cellulose (**b**) at 100 K. From Husain et al., Ref. [185]

The other spectra of polysaccharides, such as starch, pectin, and agar show a general resemblance to that of liquid H_2O, having a broad maximum in the region of 200 cm^{-1} and then increasing rapidly with increasing wavenumber [185].

7 Conclusions

The FIR studies constitute only a small part of molecular spectroscopy but they demonstrate their unique possibilities for investigating polymers.

IR spectroscopy is traditionally applied for investigation of the high-frequency group of vibrations situated in a spectral region where only the main valences are deformed; these vibrations are purely intramolecular and, therefore, not influenced by changes in the polymer system. In contrast, most of the low-frequency far-infrared vibrations are intermolecular, which means that they consist of collective excitations and are determined by the packing of chains in the lattice.

The problems of the lattice and quasi-lattice vibrations, sensitivity to the conformational state of a chain and anharmonicity of vibrations, intrinsic vibrations of hydrogen bonds, the motion of heavy ions in ionomers and coulombic interactions have been widely discussed in this review; they all look attractive for future investigations.

For polymers, the far infrared region is of crucial interest because there the relaxational type of spectrum and the resonant type coexist. Special attention was given to the way how information about molecular motions giving rise to β, γ and δ-transitions in glassy polymers can be obtained from FIR spectra in addition to the other experimental data, e.g. by differential scanning calorimetry [114]. The experimental evidence of the mechanism of δ-relaxation, as a hindered rotation (libration) of monomeric units, as well as of β-relaxation as a reorientational motion of the chain part (close to the magnitude of the statistical segment) surmounting primarily the intermolecular potential barriers are essential.

The main characteristics of these transitions were found to be related to main molecular parameters of polymers such as the structure of the monomeric unit, thermodynamic chain rigidity, cohesion energy, and the potential barrier to internal rotation.

The limited extent of this paper did not allow us to consider many of the interesting papers on the FIR spectra of polymers. These papers include studies on PE [191–194], PP [195–196], PET [197], PTFE [198], polyethers [199, 200], polyamides [201, 202], PMMA [203, 204], PVF_2 [205, 206], TPX [207], polyacetylene [208, 209], poly(N-vinylcarbazole) [210], copolymers [211], polypeptides and proteins [212], and liquid-crystalline polymers [213, 214].

The discussions in this review stress the usefulness of the FIR spectra for the study of the chemical and physical structure and for the elucidation of the molecular dynamics in polymeric materials. We are sure that many more applications will appear as the instrumental techniques are improved and our theoretical understanding increases.

8 References

1. Amrhein EM, Muller FH (1969) Kolloid ZZ Polym 234: 1078
2. Chantry GW, Chemberlain JE (1972) In: Jenkins AD (ed) Polymer science. A materials science handbook. North-Holland, Amsterdam, p 1373
3. Hummel DO (1966) Infrared spectra of polymers in the medium and long wavelength regions. Interscience, New York

4. Frank WFX, Leute U (1983) In: Button KL (ed) Infrared and millimeter waves, vol 8, p 51. Academic, New York
5. Krimm S (1960) Adv Polym Sci. 2: 51
6. Tasumi M, Krimm S (1967) J Chem Phys 46: 755
7. Gribov LA (1977) Teoriya infrakrasnykh spektrov polimerov. "Nauka", Moskow, p 163
8. Zbinden R (1964) Infrared spectroscopy of high polymers. Academic, New York
9. Krimm S (1963) In: Davies M (ed) Infrared spectroscopy and molecular structure. Elsevier, Amsterdam
10. Zerbi G (1975) In: Califano S (ed) Lattice dynamics and intermolecular forces. Academic, New York, p 384
11. Kumpanenko IV, Tshuhanov NV (1981) Uspehi khimii 50: 1627
12. Ozerkovskii BV, Roshchupkin VP (1980) Dokl Akad Nauk SSSR 254: 157
13. Frank WFX, Rabus G (1974) Colloid Polym Sci, 252: 1003
14. Möller KD, Rothschild WG (1971) Far infrared spectroscopy. Wiley, New York, p 495
15. Frank WFX (1979) In: Reed RP, Hartwig G (eds) Nonmetallic materials and composites at low temperatures. Plenum, New York, p 51
16. Leute U, Grossman HP (1981) Polymer 22: 1333
17. Fleming JM, Chantry GW, Turner PA, Nicol EA, Willis HA, Cudby MEA (1972) Chem Phys Lett 72: 84
18. Bur AJ (1985) Polymer 26: 963
19. Amrhein EM (1972) Ann NY Acad Sci 196: 179
20. Goldstein M, Seeley ME, Willis HA, Zichy VJI (1973) Polymer 14: 530
21. Dechant J (1972) Ultrarotspektroskopische untersuchungen an Polymeren. Akademie, Berlin
22. Tadokoro H (1979) Structure of crystalline polymers. Wiley, New York
23. Amrhein EM, Frischkorn H (1973) Kolloid Z Z Polym 251: 369
24. Hannon MJ, Boerio FJ, Koenig JL (1969) J Chem Phys 50: 2829
25. Chantry GW, Fleming JW, Nicol EA, Willis MA, Cudby MEA, Boerio FJ (1974) Polymer 15: 69
26. Chantry GW, Fleming JW, Nicol EA, Willis HA, Cudby MEA, Boerio FJ (1977) Polymer 18: 37
27. Manley TR, Martin CG (1971) Polymer 12: 524
28. Krimm S, Folt VL, Shipmen JJ, Berens A (1963) J Polym Sci, Polym Phys Ed 1: 2621
29. Jasse B, Monnerie L (1975) J Phys D 8: 863
30. Bershtein VA, Ryzhov VA (1982) Fizika Tv Tela 24: 162
31. Bershtein VA, Ryzhov VA (1984) J Macromol Sci – Phys B 23: 271
32. Ryzhov VA, Tonkov MV (1973) In: Denisov GS (ed) Molecular spectroscopy, 2 ser. (in Russian). University Press, Leningrad
33. Ryzhov VA, Bershtein VA (1987) Vysokomol Soedin A 29: 1852
34. Neppel A, Butter IS (1984) Spectrochim Acta A 40: 1095
35. Wright CJ (1976) In: Ivin KJ (ed) Structural studies of macromolecules by spectroscopic methods. Wiley, London
36. Spell SJ, Shepherd IW, Wright CJ (1977) Polymer 18: 905
37. Hoffman V, Frank W, Zeil W (1970) Kolloid ZZ Polym 241: 1044
38. Belopolskaya TV (1968) Dokl Akad Nauk SSSR 180: 1588
39. Berhstein VA, Egorov BM, Ryzhov VA (1986) Vysokomol Soedin B 28: 268
40. Chantry GW, Fleming JW, Cook RJ, Moss DG, Nicol EA, Willis HA, Cudby MEA (1973) Infrared Phys 13: 157
41. Liang CJ, Krimm S (1958) J Polym Sci 31: 513
42. Godovskii YuC (1976) Teplofizicheskie metody issledovaniya polimerov. Ed "Khimiya", Moscow
43. Bershtein VA, Emelyanov YuA, Stepanov VA (1984) Vysokomol. Soedin. A 26: 2272
44. Bershtein VA, Ryzhov VA (1982) Vysokomol. Soedin B 24: 495
45. Ozerkovskii BV, Roshchupkin VP (1979) Dokl Akad Nauk SSSR 248: 659
46. Bershtein VA, Egorov VM, Ryzhov VA, Sinani AB, Stepanov VA (1981) Fizika Tv Tela 23: 1611
47. Zerbi G, Piseri L, Cabassi F (1971) Molec Phys 22: 241
48. Sverdlov GW, Kovner MA, Krainov EP (1970) Kolebatelnye spektry mnogoatomnykh molecul. Ed. Nauka, Moscow
49. Shimanouchi T, Koyama J, Itoh K (1974) Progr Polym Sci Japan 7: 273
50. Chantry GW, Evans MW, Chamberlin J, Gebbie HA (1969) Infrared Phys 9: 85

51. Davis M, Pardoe GWF, Chamberlin JE, Gebbie HA (1970) Trans Farad Soc II 66: 273
52. Brot C, Darmon I (1971) Molec Phys 21: 785
53. Jain SR, Walker S (1971) J Chem Phys 75: 2942
54. Chantry G (1983) In: Button KL (ed) Infrared and millimeter waves, vol 8, chap 1. Academic, New York
55. Chantry GW (1967) Nature 214: 163
56. Chantry GW, Gebbie HA (1965) Nature 208: 378
57. Larkin I (1973) Trans Faraday Soc II 69: 1278
58. Hill N, Vangan W, Price A, Davies M (1969) Dielectric properties and molecular behaviour. Van Nostrand, Princeton, New York
59. Higasi K, Minami R, Takahashi H, Ohio A (1973) Trans Faraday Soc II 69: 1579
60. Reid CJ, Evans MW (1982) J Chem Phys 76: 2576
61. Chantry GW, Fleming JW, Nicol EA (1972) Infrared Phys. 12: 101
62. Johnson KW, Rabolt JF (1973) J Chem Phys 58: 4536
63. Piseri L, Powell BM, Dalling G (1973) J Chem Phys 58: 158
64. Viras F, King TA (1984) Polymer 25: 899
65. Viras F, King TA (1990) J Non-Crystalline Solids 119: 65
66. Wendorff J (1977) In: Proc 4th Internat Conf Phys Non-Cryst Solids Aedermansdorf, p 94
67. Warrier AVR, Krimm S (1970) Macromolecules 3: 709
68. Tasumi M, Shimanouchi T (1971) Polymer J 2: 62
69. Moore WH, Krimm S (1973) J Chem Phys 59: 5195
70. Moore WH, Krimm S (1975) Makromol Chem Supplm 1: 491
71. Goldstein M, Stephenson D, Maddamas WF (1983) Polymer 24: 823
72. Ryzhov VA, Bershtein VA (1989) Vysokomol Soedin A 31: 451
73. Bershtein VA, Ryzhov VA (1985) Dokl Akad Nauk SSSR 284: 890
74. Evans MW, Evans GJ, Coffey WT, Gricolini P (1982) Molecular dynamics and theory of broad band spectroscopy. Wiley-Interscience, New York
75. McLelan CK, Walker S (1977) Canad J Chem 55: 2411
76. Pardoe GWF (1970) Trans Faraday Soc II 66: 2699
77. Bershtein VA, Ryzhov VA, Ganicheva CH, Ginzburg LI (1983) Vysokomol Soedin A 25: 1385
78. North AM (1975) J Polym Sci, Polym Symp 50: 345
79. Bossis G (1982) Physica A 110: 408
80. Jakobsen RJ, Brasch JW (1964) J Amer Chem Soc 86: 3371
81. Knözinger E, Jacox M (1978) Ber Bunsenges Phys Chem 88: 57
82. Ryzhov VA, Bershtein VA (1988) In: Tüdós F (ed) 8th Europ Symp Polym Spectr (ESOPS 8), 23–26 Aug 1988. Budapest, Hungary
83. Tonelli AE (1971) Macromolecules 6: 682
84. Arning HJ, Dorfmüller Th (1984) Infrared Phys 24: 221
85. Bondi A (1968) Physical properties of molecular crystals, liquids and glasses. Wiley, New York, p 29
86. Entsiklopediya polimerov, vol 1 (1974) ed Sov. entsiklopediya, Moscow, p 726
87. Tager AA (1978) Fiziko-khimiya polimerov Ed. Khimiya, Moscow
88. Askadskii AA, Kolmakova LK, Tager AA, Slonimskii GL, Korshak VV (1977) Vysokomol Soedin A 19: 1004
89. McCrum NG, Read BE, Williams G (1967) Anelastic and dielectric effects in polymeric solids. John Wiley, New York
90. Heijboer J (1977) Intern J Polym Mater 6: 11
91. Johari G and Goldstein M (1973) J Chem Phys 58: 1766
92. Williams G (1977) Dielectric and related molecular processes, vol 2 (Spec Period Report, The Chemical Society, London)
93. Colin J, Reid CJ, Evans MW (1979) Trans Faraday Soc II 75: 1218
94. Vij JK (1983) Nuovo Cimento D 2: 751
95. Ashcraft CS, Boyd RH (1976) J Polym Sci Polym Phys Ed 14: 2153
96. Buckingham KA, Belling JW (1981) Proc IEE 128: 215
97. McCammon RD, Saba RG, Work RN (1969) J Polym Sci, Polym Phys Ed 7: 1721
98. North AM, Parker TG (1972) Trans Faraday Soc II 68: 1
99. Amrhein EM (1967) Kolloid ZZ Polym 38: 216
100. Birshtein TM, Ptitsyn OB (1964) Konformatsia makromolecul. Ed. Nauka, Moscow
101. Volkenstein MV (1963) Conformational statistics of polymer chains. Interscience, New York
102. Kalmykov JuP, Gaiduk VI (1981) J Fiz Khimii 55: 305

103. Van Krevelen DW (1972) Properties of polymers: Correlation with chemical structure. Elsevier, Amsterdam
104. Miller S, Tomazawa M, McCrone RK (1972) Molec Cryst 7: 54
105. Hayler L, Goldstein M (1977) J Chem Phys 66: 4336
106. Irvine JD, Work RN (1973) J Polym Sci, Polym Phys Ed 11: 179
107. Hedvig P (1987) Dielectric properties of polymers. Wiley, New York
108. Sauer O (1971) J Polym Sci Polym Symp 2: 69
109. Cowie JWG (1980) J Macromol Sci – Phys B 18: 568
110. Bershtein VA, Ryzhov VA (1982) In: Lindberg J (ed) 6th Europ Symp Polym Spectr (ESOPS 6), 11–13 Aug 1982 Hämeenlinna, Aulanko
111. Bershtein VA, Egorov BM, Emelyanov YuA, Stepanov BA (1983) Polymer Bull 9: 98
112. Bershtein VA, Egorov BM (1985) Vysokomol Soedin A 27: 2440
113. Struik L (1976) Ann NY Acad Sci 279: 78
114. Bershtein VA, Egorov VM (1990) Differential scanning calorimetry in the physics and chemistry of polymers. "Chemistry" Publ House, Leningrad [Translation: Differential Scanning Calorimetry of Polymers. Ellis Horwood, Chichester, 1993]
115. Tsvetkov VN, Eskin VE, Frenkel SYa (1964) Struktura makromolecul v rastvorah. Ed. Nauka, Moscow
116. Neelov NM, Darinskii AA, Gotlib YuYa, Balabaev NR (1980) Vysokomol Soedin A 22: 1761
117. Gribov LA, Zubkova OB, Shabadash AN (1973) Z Prikl Spektr 18: 1028
118. Zerbi G (1977) In: Barnes AJ, Orville-Thomas WJ (eds) Vibrational spectroscopy-modern trends. Elsevier, New York
119. Smirnov BR, Plotnikov VD, Ozerkovskii BV, Roshchupkin VP, Enikolopian NS (1981) Vysokomol Soedin A 23: 2588
120. Ryzhov VA, Bershtein VA (1989) Vysokomol Soedin A 31: 458
121. Gotlib Yu Ya, Darinskii AA, Svetlov Yu E (1986) Fizischeskaya kinetika makromolecul. Ed. Khimiya, Leningrad
122. Bershtein VA, Pertsev NA (1984) Acta Polym 35: 575
123. Bershtein VA, Egorov VM, Emelyanov YuA (1985) Vysokomol Soedin A 27: 2451
124. Belopolskaya TV (1969) Opt Spectr 26: 476
125. O'Reilly J (1977) J Appl Phys 48: 4043
126. Egorov EA, Zhizhenkov VV (1982) J Polym Sci, Polym Phys Ed 20: 1089
127. Gibbs J, DiMarzio E (1957) J Chem Phys 28: 373
128. Adams G, Gibbs J (1965) J Chem Phys 43: 139
129. Bresler SE, Ierysalimskii TV (1965) Fizika i khimiya makromolecul. Ed. Nauka, Moscow-Leningrad, p 81
130. Heijboer J (1965) In: Heijboer J (ed) Physics of non-crystalline solids. New Holland, Amsterdam, p 231
131. Boyer R (1976) Polymer 17: 996
132. Bartenev GM (1979) Struktura i relaksatsionnyia svoistva elastomerov. Izd Khimiya, Moscow
133. Kolařík J (1982) Adv Polym Sci 46: 119
134. Hayakawa K, Wada J (1974) J Polym Sci, Polym Phys Ed 12: 2119
135. Ryzhov VA, Bershtein VA, Sinani AB (1990) Vysokomol Soedin A 32: 90
136. Aras L, Baysal BM (1964) J Polym Sci, Polym Phys Ed 22: 1453
137. Baccaradda M, Butti E, Frosini V (1966) J Appl Polym Sci 10: 399
138. Bershtein VA, Egorov VM, Podolsky AF, Stepanov VA (1985) J Polym Sci, Polym Lett Ed 23: 371
139. Bershtein VA, Emelyanov YuA, Stepanov VA (1981) Mehan Komposit Mater 1: 9
140. Desando MA, Rashem M, Siddiqui NA, Walker S (1984) Trans Faraday Soc II 80: 747
141. Finch A, Gates P, Radcliffe K, Dickson F, Bentley F (1970) Chemical application of far infrared spectroscopy. Academic, London
142. Jakobsen R, Brasch JW (1965) Spectrochim Acta 20: 1753
143. Frank WFX, Fiedler H (1979) Infrared Phys 19: 481
144. Jakeš J, Krimm S (1971) Spectrochim Acta 27: 35
145. Bessler E, Bier G (1969) Makromol Chem 20: 122
146. Tadokoro H, Kobayashi M, Yoshidome H, Tai K, Makino D (1968) J Chem Phys 49: 3359
147. Shen DY, Pollack SK, Hsu SL (1989) Macromolecules 22: 2564
148. Belopolskaya TV (1976) Vestnik LGU 22: 44
149. Feairheller WR, Kotton JE (1967) Spectrochim. Acta 23: 2225
150. Stanevich AE (1964) Opt Spectr 16: 243

151. Jakobsen RJ, Mikawa J, Brasch JW (1967) Spectrochim. Acta 23: 2199
152. Belopolskaya TV, Trapeznikova ON (1966) Opt Spectr 20: 246
153. Ryzhov VA, Bershtein VA (1992) In: Skorohodov S (ed) 10th Europ Symp Polym Spectr (ESOPS 10), 29 Sept–2 Oct. St Petersburg, Russia
154. Holliday L (1975) Ionic polymers. Applied Science, London, p 35
155. Eisenberg A, King M (1977) Ion-containing polymers: Physical properties and structure. Academic, New York, p 141
156. Eisenberg A (1970) Macromolecules 3: 147
157. Neppel A, Bulter IS, Eisenberg A (1979) Can J Chem 57: 2518
158. Rouse GB, Risen WM, Tsatsas AT, Eisenberg A (1979) J Polym Sci, Polym Phys Ed 17: 81
159. Suchocka-Galas K (1987) Europ Polym J 12: 951
160. Tsatsas AT, Reed JW, Risen WM (1971) J Chem Phys 55: 3260
161. Mattera VD, Risen WM (1984) J Polym Sci, Polym Phys Ed 22: 67
162. Exarhos GJ, Miller PJ, Risen WM (1974) J Chem Phys 60: 4145
163. Viras F, King TA (1984) Polymer 25: 1411
164. Tsatsas AT, Risen WM (1970) J Am Chem Soc 92: 1789
165. Neppel A, Bulter IS, Eisenberg A (1979) Macromolecules 12: 948
166. Grossman HP Frank WFX (1977) Polymer 18: 341
167. Frank WFX, Schmidt H, Wulf W (1977) J Polym Sci, Polym Symp 61: 317
168. Latour M, Abo Dorra H, Galigne JL (1984) J Polym Sci, Polym Phys Ed 22: 345
169. Lu FJ, Waldman DA, Hsu SL (1984) J Polym Sci, Polym Phys Ed 22: 827
170. Bank AS, Krimm S (1968) J Appl Phys 39: 4951
171. Frank WFX (1976) Proc 2nd Int Conf and Winter School on Submillimeter Waves, San Pierto Rico, p 111
172. Willis HA, Cudby MEA (1976) In: Ivin KJ (ed) Structural studies of macromolecules by spectroscopic methods. Wiley-Interscience, London
173. Willis HA, Cudby MEA (1975) Polymer 16: 74
174. Manley TR, Williams DA, (1969) Polymer 10: 339
175. Bruno F (1980) Ann Rev Phys Chem 31: 265
176. Siesler HW, Holland-Moritz K (1980) Infrared and Raman spectroscopy of polymers. Marcel Dekker, New York, p 320
177. Roshchupkin VP, Andreev IS, Roshchupkina TA, Chplakhÿn GM, Starobunskii AB (1976) Uch Zap Univ of Erevan 2: 46
178. Roshchupkin VP, Lubovskii AB, Kochervinskii VV, Roshchupkina TA (1969) Vysokomol Soedin A 11: 2505
179. Bosomworth DR, Gush HP (1965) Canad J Phys 43: 751
180. Tonkov MV (1970) In: Bulanin MO (ed) Spectroscopy of interactions of molecules. University of Leningrad Press, Leningrad
181. Libov VS, Perova GS (1976) Opt-mehan prom 5: 52
182. Feairheller WR, Willler JI (1971) Appl Spectr 25: 175
183. Itoh K, Shimanouchi T (1970) Biopolymers 9: 383
184. Shen S, Santo L, Genzel L (1981) Canad J Spectr 23: 126
185. Husain SR, Hasted JB, Rosen D, Nicol E, Birsh JR (1984) Infrared Phys 24: 209
186. Belmont ML, Ambarek AN, Sik V (1984) Infrared Phys. 24: 215
187. Itoh K, Hinomoto H, Shimanouchi T (1974) Biopolymers 13: 307
188. Kovalev PF, Shevchenko IV, Voronkov MG, Kozlov NV (1973) Dokl Akad Nauk SSSR 212: 101
189. Jano O, Wada J (1971) J Polym Sci, Polym Phys Ed 9: 669
190. Lazarev AN (1968) Kolebatelnie spektry i struktura silikatov Ed. Nauka, Leningrad
191. Leute U, Frank WFX (1980) Infrared Phys 20: 327
192. Frank WFX, Schmidt H, Heise B, Chantry GW, Nicol EA, Willis HA (1981) Polymer 22: 17
193. Schlotter NE, Rabolt JE (1984) Macromolecules 17: 1581
194. Birch JR (1990) Infrared Phys 30: 195
195. Chantry GW, Fleming JW, Pardoe GWF, Reddish W, Willis HA (1971) Infrared Phys 11: 109
196. Beckett DR, Chalmers JM, Mackenzie MW, Willis HA, Edwards HGM, Lees JS, Lord DA (1985) Europ Polym J 21: 849
197. Frank W, Knaupp D (1975) Ber Bunsenges Phys Chem 79: 1041
198. Willis HA, Reddish W, Buckingham KA, Llewellyn-Jones DT, Knight KJ, Gebbie HA (1981) Polymer 22: 20

199. Chantry GW, Fleming JW, Smith PM, Cudby M, Willis HA (1971) Chem Phys Lett 10: 473
200. Kobayashi M, Morishita H, Shimomura M (1989) Macromolecules 22: 3726
201. Matsubara I, Magill JH (1973) J Polym Sci, Polym Phys Ed 11: 1172
202. Nie CS, Kremer F, Poglitsch A, Bechtold (1985) J Polym Sci, Polym Phys Ed 23: 1247
203. Tadokoro H, Chatani J, Kusanagi N, Yokayama M (1970) Macromolecules 3: 441
204. Manley TR (1976) In: Jones DW (ed) Introduction to the spectroscopy of biological polymers. Academic, New York, p 119
205. Yang DC, Tomas EL (1984) J Mater Sci 3: 928
206. Kochervinskii VV, Glukhov VA, Romandin VF, Sokolov VG, Lokshin BV (1988) Vysokomol Soedin A 30: 1916
207. Birch JR, Nicol EA (1984) Infrared Phys 24: 573
208. Genzel L, Kremer F, Poglitsch A (1984) Phys Rev B29: 4595
209. Bechtold G, Gensel L, Roth S (1985) Solid St Commun 53: 1
210. Fujno M, Mikawa H, Yokoyama M (1984) J Non-Cryst Solids 64: 163
211. Latour H, Moreira RL (1987) J Polym Phys, Polym Phys Ed 25: 1913
212. Hayest W, Praff FL, Wong KS, Raneto K, Yoshino K (1985) J Phys Solid St Phys 19: L555
213. Blumstein A (1985) Polymer liquid crystals. Academic, New York
214. Evans GJ, McCicki AS, Evans MW (1986) J Mol Liquids 32: 149

Editor: K. Dušek
Received 26 October 1992

Selective Spectroscopy of Chromophore Doped Polymers and Glasses

I. S. Osad'ko
Department of Physics, V.I. Lenin State Pedagogical University,
119435, Moscow, Russia

Selective spectroscopy methods such as photochemical and photophysical spectral hole-burning, fluorescence line narrowing and photon echoes are discussed. These methods enable one to remove inhomogeneous broadening which hides the fine structure of the homogeneous impurity optical bands in polymers and glasses. The influence of the various types of the electron-phonon interaction on a homogeneous optical band is considered. Theoretical treatment methods for realistic homogeneous bands in solids are discussed. Tunneling degrees of freedom of polymers and glasses which are responsible for the spectral diffusion and for the anomalous low-temperature hole broadening in polymers and glasses are considered.

Advances in Polymer Science, Vol. 114
© Springer-Verlag Berlin Heidelberg 1994

1 Introduction

Optical spectroscopy has been employed for a long time as an effective tool to investigate molecular dynamics and relaxations in molecular impurity crystals. The spectroscopic methods work effectively in crystals because impurity bands have a well-resolved structure. Optical bands of such type can be treated with the help of the comprehensively developed theory for homogeneous optical band shape. The theory enables one to get information concerning both the electron-phonon interaction and the molecular dynamics in crystals.

Unfortunately, conventional methods of optical spectroscopy work poorly when applied to polymers and glasses. This occurs because optical bands of chromophore doped polymers and glasses have, as a rule, strong inhomogeneous broadening. This broadening hides the fine structure of a homogeneous optical band and, therefore, the application of the theory for homogeneous bands becomes ineffective.

The situation changed ten years ago since when the methods of selective laser spectroscopy have been applied extensively to polymers and glasses. These methods enable one to remove inhomogeneous broadening by using a selective excitation permitting one to select a homogeneous molecular ensemble the molecules of which have a single electronic frequency. The optical band of this homogeneous ensemble exhibits a well-resolved structure and it can be treated using the theory developed to analyze homogeneous optical bands in crystals.

Selective spectroscopy applied to low-temperature polymers and glasses has shown that a special kind of low-frequency excitations exists in these solids. These excitations manifest themselves when the molecular structure of a polymer or glass is changed. These changes of the polymer structure have tunneling characteristics at low temperatures. These tunneling degrees of freedom exist in addition to the conventional vibrational degrees of freedom and they have very low excitation energies. Therefore, low-temperature physical properties of polymers and glasses differ considerably from those of crystals.

The tunneling degrees of freedom were first discovered in glasses twenty years ago. For the first decade they were investigated by means of nonoptical methods. However, the first spectroscopic experiments made ten years ago revealed that the tunneling degrees of freedom exist in both glasses and polymers. Since then, the tunneling systems of polymers have been investigated solely by means of spectroscopic methods. All the novel experimental data and theories concerning low-temperature optical dynamics of polymers and glasses will be discussed in this review.

By now, there exist a number of monographs [1–3] and reviews [4–8] which discuss the application of selective spectroscopy to polymers and glasses. However, they mainly address the specialists in the field of solid state spectroscopy. This review addresses a wider circle of readers and, therefore, it is written at the tutorial level. Methods of selective spectroscopy are discussed in detail and, in analyzing experimental data, only the simplest variants of the theories are used.

2 Principles of Selective Spectroscopy

First of all, we need to discuss main ideas concerning homogeneous and inhomogeneous band broadening in solids and to explain the terminology which will be used in this review.

2.1 Homogeneous and Inhomogeneous Broadening of Optical Bands

It is known that the dipolar moment $d(t)$ of a molecule oscillating with time is a source of electro-magnetic radiation. If this oscillation is of a harmonic type, so that $d(t) = d \cdot \exp(-i\omega_0 t)$, the dipolar moment absorbs and radiates monochromatic light with a frequency ω_0. This means that the optical line is described by a Lorentzian with a vanishing half-width i.e. by a δ-function. By using the integral representation for δ-function, we can write for the optical line shape function the following expression:

$$I(\omega) = \delta(\omega - \omega_0) = \frac{1}{2\pi} \cdot \int_{-\infty}^{\infty} dt \cdot e^{i(\omega - \omega_0)t} . \tag{1}$$

In condensed matter, the dipolar moment of every molecule is influenced by the neighbour molecules. This leads to complicated characteristics of dipolar oscillations and, therefore, the resonant frequency ω_0 becomes a function of time: $\omega_0 \rightarrow \omega_0(t)$. In this case, we can write:

$$d(t) = d \cdot \exp\left(-i \cdot \int_0^t d\tau \cdot \omega_0(\tau) \right)$$

and Eq. (1) should be rewritten as follows:

$$I(\omega) \propto \int_{-\infty}^{\infty} dt \cdot \exp(i\omega t) \cdot \exp\left(-i \int_0^t d\tau \cdot \omega_0(\tau) \right). \tag{2}$$

This line has a nonvanishing halfwidth. The line shape depends on the character of the temporal behaviour of the resonant frequency, ω_0. There are two theoretical approaches to the line broadening problem.

Stochastic approach. Within the scope of this approach, the resonant frequency $\omega_0(\tau)$ is a classical function of time [9, 10]. Random changes of the frequency $\omega_0(\tau)$ can be fast or slow compared with the time scale of the experiment (Fig. 1). Fast changes lead to line broadening which is of a *homogeneous* type. The slow evolution of the resonant frequency will not manifest itself on the timescale of the experiment. However, if we make the timescale longer this slow evolution will be appreciable (Fig. 1). Every molecule of the ensemble has a temporal evolution like that depicted in Fig. 1. Returning to the former time scale, we can

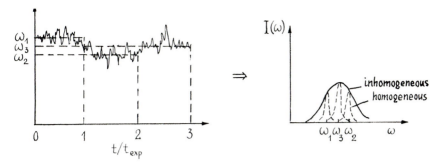

Fig. 1. Stochastic picture of inhomogeneous and homogeneous line broadening. t_{exp} is the time scale of an experiment

consider that every molecule of the ensemble has its own temporal history. Therefore, the first molecule has the average frequency ω_1, the second molecule the frequency ω_2, etc. On the time scale of the experiment, the j-th molecule has its own average frequency ω_j. A distribution over ω_j looks like *inhomogeneous* broadening (Fig. 1).

Within the scope of the stochastic approach, there is no strict boundary between homogeneous and inhomogeneous broadening. The boundary depends on the timescale of the experiment. This approach is convenient for considering the line broadening problem in vapour or liquid phases. However, stochastic theory cannot explain the optical band shape in solids. Therefore, we shall only use a dynamical approach to the line broadening problem.

Dynamical approach. Within the scope of this approach, an operator $\hat{\Lambda}$ plays the role of the classical resonant frequency ω_0. This operator is the difference between adiabatic Hamiltonians of an impurity chromophore in the excited and ground electronic states [8, 11, 12]. The operator $\hat{\Lambda}$ is that of the electron-phonon interaction. We shall discuss this operator in detail in Sect. 3.1. Quantum mechanical calculations yield the following expression for the absorption (g) and fluorescence (e) band shape functions [8, 12]:

$$I^{g,e}(\omega) \propto \int_{-\infty}^{\infty} dt \cdot e^{i(\omega - \omega_0)t + \varphi^{g,e}(t)} . \tag{3}$$

Here, ω_0 is the resonant electronic frequency when the interaction with phonons is absent. The cumulant functions $\varphi^{g,e}(t)$ describe the influence of the electron-phonon interaction on the resonant frequency. In contrast to the classical Eq. (2), the quantum mechanical dynamical theory yields different expressions $I^{g,e}$ for the absorption and the emission band. A value of the resonant frequency ω_0 is determined solely by electrostatic interaction between the chromophore and its neighbours. All the dynamical interactions only influence the cumulant functions $\varphi^{g,e}(t)$ which cause a homogeneous broadening.

Local fields in different places in polymers and glasses are quite different. Therefore, chromophores embedded in the polymer or glass have various resonant frequencies ω_0. This distribution over ω_0 describes an *inhomogeneous* broadening of the molecular ensemble. This broadening does not depend on the time scale of the experiment. The dynamical theory predicts different expressions for the homogeneous absorption and fluorescence band. Therefore, it describes optical bands in solids better than the stochastic theory does.

2.2 Frequency and Time Domain Selection

The main task of selective spectroscopy is to measure a homogeneous optical band because the shape of this band includes important information concerning the molecular dynamics and the electron-phonon interaction. By now, there are three main experimental methods for selective spectroscopy: fluorescence line narrowing (FLN), spectral hole-burning (HB) and various types of photon echoes (PE).

Fluorescence line narrowing. An inhomogeneously broadened optical band has no structure because it is a sum of many homogeneous optical lines as shown in Fig. 2a. If we irradiate this sample by monochromatic light with a frequency ω_b the molecules, whose resonant frequencies ω_0 coincide with ω_b, are excited more effectively than others. It is obvious that the majority of the excited molecules have $\omega_0 \simeq \omega_b$. Therefore, the fluorescence line will almost coincide with the homogeneous line (Fig. 2b). The first applications of the FLN method to inorganic and organic systems have been carried out in [13, 14].

Spectral hole burning. If radiationless processes dominate in a chromophore it does not radiate light and therefore the FLN method cannot be employed. In this case, we can use the spectral hole burning method. Let us turn to Fig. 2 again. After molecules are excited with the frequency $\omega_0 \simeq \omega_b$, they cannot

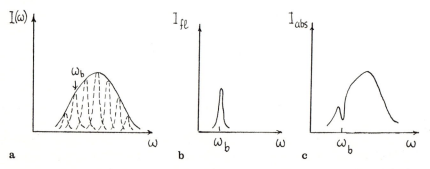

Fig. 2a–c. a Inhomogeneously broadened absorption band. **b** Fluorescence line narrowing. **c** Spectral hole burning ω_b is a frequency of burning light

absorb light any more. Therefore, the optical density has a narrow hole at the frequency $\omega_0 \simeq \omega_b$. Its shape is close to the homogeneous line shape (Fig. 2c). This hole lasts until molecules come back to the ground electronic state. If the excited molecule go to the ground electronic state directly, the hole lifetime coincides with the fluorescence lifetime T_1. If the molecular path from the excited state to the ground state goes through a triplet level, the hole lifetime coincides with the triplet lifetime. The spectral holes of such type are called "transient holes". They have been known in laser spectroscopy for a long time [15]. However, if the excited molecule undergoes a photochemical conversion and its resonant frequency changes after this conversion, the hole lifetime coincides with that for the photoproduct. These longliving holes are called "persistent holes". The first applications of the persistent hole-burning methods as a tool in selective spectroscopy have been carried out in [16, 17]. Today, photochemical hole-burning is the most popular method of selective spectroscopy.

Photon echo. FLN and HB methods realize a frequency selection of chromophores by preparing a homogeneous ensemble of chromophores. Therefore, both radiated light and burnt hole have a spectral width close to that in the homogeneous spectra. In contrast to FLN and HB methods, which are special kinds of spectral methods, in PE experiments we measure a temporal dependence of the dipolar moment of a molecular ensemble. Therefore, the duration of exciting pulses is of great importance in PE experiments. The shorter the exciting pulses are the better the resolution in PE experiments. In contrast to FLN and HB, the excitation in PE experiments can be of nonmonochromatic type.

There is a variety of PE experiments: two-pulse PE (2PE), three-pulse PE (3PE), accumulated PE (APE), incoherent PE (IPE), etc. In every type of PE, after several pulses applied to the sample we measure a spontaneously emerging coherent echo pulse. Let us consider for example 2PE [15, 18].

The picture of 2PE looks as follows. After we apply to the sample a series of two short laser pulses, a spontaneous coherent pulse emerges (Fig. 3). The distance between echo and second pulse equals that between first and second pulse. Why then does the echo signal emerge?

In order to answer this question, we should take into account that, after the first laser pulse is applied, the following induced dipolar moment appears in the sample:

$$D(\tau) = \sum_j d_j \cdot e^{i\omega_j \tau} \tag{4}$$

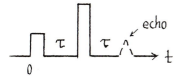

Fig. 3. Train of pulses yielding echo signal

where

$$d_j \cdot e^{i\omega_j\tau} = \int dx \cdot \varphi_{2j}(x) \cdot e^{iE_{2j}\tau} \cdot \hat{d} \cdot \varphi_{1j}(x) \cdot e^{-iE_{1j}\tau}. \tag{5}$$

Here φ_{1j} and φ_{2j} are the electronic functions of j-th atom which describe the initial (ground) and final (excited) electronic states with the energies E_{1j} and E_{2j}. At the instant when the second laser pulse is applied to the j-th atom, it finds itself in the excited (initial) electronic state. Now, the ground electronic state is the final one. Therefore, by considering the dipolar moment after the second pulse has been applied we must make the following substitution in Eq. (5): $E_{2j} \rightleftharpoons E_{1j}$ and $\omega_j\tau \rightarrow -\omega_j\tau'$. The dipolar moment of the sample after two laser pulses have already passed is described by the following expression:

$$D(\tau',\tau) = \sum_j d_j \cdot e^{i\omega_j(\tau-\tau')} \simeq d \cdot \sum_j e^{i\omega_j(\tau-\tau')}. \tag{6}$$

It is obvious that $D(\tau',\tau)$ has a maximum at $\tau' = \tau$. This maximum corresponds to the maximum of radiated light and is called the echo signal.

Up to this moment we ignored we electron-photon interaction. Taking it into account we should substitute $\exp(-i\omega\tau)$ in Eq. (6) by $\exp(-i \cdot \int_0^\tau dx \cdot \omega(x))$. Then, Eq. (6) takes the following form [6, 19]:

$$D(\tau',\tau) \propto \left\langle \exp\left(i \cdot \int_0^\tau dx \cdot \omega(x) - i \cdot \int_\tau^{\tau+\tau'} dx \cdot \omega(x) \right) \right\rangle \tag{7}$$

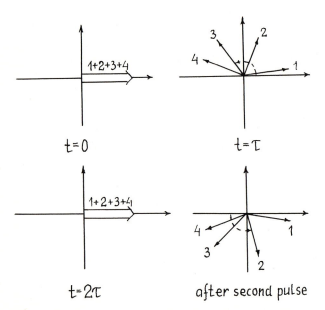

Fig. 4. The temporal evolution of a dipolar ensemble in the phase plane ωt during two-pulse echo experiment. *Arrows* denote molecular dipoles

where $\langle \cdots \rangle$ means the sum over chromophores. Figure 4 shows the phase plane of molecules. We see dephasing in a molecular ensemble after the first pulse and phasing after the second. The amplitude of the 2PE signal is given by

$$
E(2\tau) \propto \left\langle \left| \exp\left(-i \cdot \int_0^\tau dx \cdot \omega(x) \right) \right| \right\rangle^2 .
\tag{8}
$$

Equation (8) includes the temporal function whose Fourier transform describes a homogeneous optical band – see Eq. (2).

3 Manifestation of Solvent Dynamics in Homogeneous Optical Bands

We have already seen that selective spectroscopy methods enable one to find homogeneous optical bands. Therefore, we consider now what type of information can be obtained from a homogeneous optical band.

3.1 Electron-Phonon Interaction

It is known that optical bands of chromophores in condensed phase have a visible frequency shift as compared with the bands of the same chromophores in vapours [20]. This means that an interaction with other molecules influences the electronic subsystem of a chromophore. Therefore, the electronic wave functions and the distance between electronic levels of the chromophore depend on its environment. Hence the energy of the electronic excitation of the chromophore is a function of the coordinates $\vec{R}$ of the neighbouring molecules:

$$
E^f(\vec{R}) = \varepsilon^f + U^f(\vec{R}) .
\tag{9}
$$

Here, ε^f is the energy of the f-th electronic excitation of a molecule in the vapour phase; $U^f(\vec{R})$ is the additional energy due to the interaction with other molecules.

Equation (9) can be derived within the framework of the Born–Oppenheimer approach [21]. This approach is universal and it can be applied to an arbitrary system of interacting nuclei and electrons. The total Hamiltonian of such a system averaged over fast electronic motion is as follows:

$$
\langle f | H(r\vec{R}) | f \rangle = H^f(\vec{R}) = T(\vec{R}) + \varepsilon^f + U^f(\vec{R}) .
\tag{10}
$$

Here, $T(\vec{R})$ is an operator of the nuclear kinetic energy. The Hamiltonian $H^f(\vec{R})$ is called the adiabatic Hamiltonian. It determines the nuclear dynamics. The wave functions which describe these dynamics can be found from the

Schrödinger equation:

$$H^f(\vec{R}) \cdot \Psi_v^f(\vec{R}) = \hbar \cdot \Omega_v^f \cdot \Psi_v^f(\vec{R}) \,. \tag{11}$$

If the function $U^f(\vec{R})$ can be approximated by a quadratic form of the variable $\vec{R}$, Eq. (11) is one for a multidimensional harmonic oscillator. The function $U^f(\vec{R})$ plays the role of potential energy. It is called Franck–Condon potential surface. This surface depends on the electronic state and it changes if the electronic state of a chromophore changes. The difference between two potential surfaces is the Franck–Condon electron-phonon interaction. In harmonic approximations, the electron-phonon interaction is as follows:

$$H^e(R) - H^g(R) = V(R) = \varepsilon^e - \varepsilon^g + (\vec{R} + \vec{a}) \cdot \frac{\hat{U}^e}{2} \cdot (\vec{R} + \vec{a}) - \vec{R} \cdot \frac{\hat{U}^g}{2} \cdot \vec{R}$$

$$= \varepsilon + \vec{a} \cdot \frac{\hat{U}^e}{2} \cdot \vec{a} + (\vec{a} \cdot \hat{U}^e)\vec{R} + \vec{R} \cdot \frac{\hat{U}^e - \hat{U}^g}{2} \cdot \vec{R} \,. \tag{12}$$

The vector $\vec{a}$ describes a change of the equilibrium position of harmonic oscillators. The matrix $U^e - U^g = W$ describes a change of the force-matrix upon electronic excitation of a chromophore. Figure 5 shows the change in adiabatic potentials described by Eq. (12).

The first and second terms in Eq. (12) do not depend on the dynamical variable R. Here ε determines the energy of a pure electronic excitation and it depends on an electrostatic interaction of a chromophore with other molecules. A distribution over values of ε leads to the appearance of an inhomogeneous broadening. The linear and quadratic equations in variable $\vec{R}$ terms describe the linear and quadratic electron-phonon interaction which transforms δ-like optical lines into optical bands. Let us consider first the effect of the linear interaction.

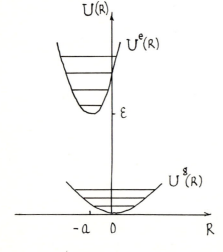

Fig. 5. Changes of the adiabatic potential upon electronic excitation of a chromophore (electron-phonon interaction)

3.2 Zero-Phonon Lines (ZPL) and Phonon Side-Bands (PSB)

We start our consideration with a case of the interaction with a single phonon mode q_0. In this case all the a_q except $a_{q_0} = a$ equal zero. The cumulant function φ^g in Eq. (3) is given by [8, 12]:

$$\varphi^g(t) = f(e^{-ivt} - 1) \tag{13}$$

where v is a phonon frequency. Here $f = (a/2a_0)^2$ where $a_0 = (\hbar/\mu v)^{1/2}$ is the amplitude of zero-point vibrations. Substituting $\varphi^g(t)$ from Eq. (13) in Eq. (3) and expanding the function $\exp(f \cdot \exp(-ivt))$ into the series of the function $f \cdot \exp(-ivt)$ we find after carrying out integration over t the following formula for the absorption band:

$$I^g(\omega) = e^{-f} \cdot \delta(\omega - \omega_0) + \Phi^g(\omega - \omega_0) \tag{14}$$

where

$$\Phi^g(\omega - \omega_0) = e^{-f} \cdot \sum_{n=1}^{\infty} \frac{f^n}{n!} \cdot \delta(\omega - \omega_0 - nv) . \tag{15}$$

The first term in Eq. (14) describes a photo-transition without creation of phonons. Therefore, this optical line is called a zero-phonon line (ZPL). The n-th term in Eq. (15) describes a photo-transition which is accompanied by the creation of n phonons. The sum over n describes the so-called phonon side band (PSB).

A fluorescence band can be considered with the help of Eq. (3) as well. Taking into account that

$$\varphi^e(t) = \varphi^{*g}(t) \tag{16}$$

we obtain with the help of Eq. (3) the following expression for the fluorescence band:

$$I^e(\omega) = e^{-f} \cdot \delta(\omega - \omega_0) + \Phi^e(\omega - \omega_0) \tag{17}$$

where $\Phi^e(\omega - \omega_0) = \Phi^g(\omega_0 - \omega)$. It is obvious that ZPLs in the fluorescence and absorption bands are in resonance but PSBs are not in resonance. The intensity distributions described by Φ^e and Φ^g are similar. There is a mirror symmetry between the functions Φ^e and Φ^g regarding ZPL.

The integrated intensity of ZPL and PSB is described by simple formulae:

$$I_{ZPL}^{g,e} = e^{-f}, \quad \Phi_{PSB}^{g,e} = 1 - e^{-f} . \tag{18}$$

The intensity distribution in absorption and emission bands depends on the only parameter f. It equals the ratio Φ_1/I_{ZPL} where Φ_1 is the integral intensity of the one-phonon transition. The parameter f characterizes a strength of the linear electron-phonon interaction. At strong coupling we have $f \gg 1$. In this case, the intensity of ZPL is too weak. Therefore, the optical band consists of PSB only. The ratio

$$\alpha = \frac{I_{ZPL}}{I_{ZPL} + I_{PSB}} = e^{-f} \tag{19}$$

is called the Debye–Waller factor. This factor can be measured if the homogeneous band is known. The linear electron-phonon interaction results in the optical band which is determined by a single parameter f.

This simple structure of the theoretical formula for an optical band is not changed if we take into account the interaction with many phonon modes. In the multimode case, we have many parameters $f_q = (a_q/2a_{0q})^2$ where $a_{0q} = (\hbar/\mu v_q)^{1/2}$. The one-phonon function plays the role of the parameter f

$$f(v) = \sum_{q=1}^{N} f_q \cdot \delta(v - v_q) \tag{20}$$

where N is the total number of phonon modes. If we substitute f_q by $1/N$, we find another one-phonon function

$$\rho_{ph}(v) = \frac{1}{N} \cdot \sum_{q=1}^{N} \delta(v - v_q) \tag{21}$$

which describes the phonon density of states of a solvent. The function $\rho_{ph}(v)$ is the important characteristic of a solid. It manifests itself in a variety of the physical experiments (for instance see Sect. 5.1). The function $f(v)$ describes the phonon density of states weighted by coupling parameters f_q. Therefore, this function includes information both on solvent dynamics and electronic properties of a guest chromophore. The function $f(v)$, in contrast with the function $\rho_{hp}(v)$, is characteristic of a pair: guest chromophore + solvent.

If sample temperature $T \neq 0$, the phototransition is accompanied by processes of creation and annihilation of phonons. In this realistic case, the cumulant function $\varphi^g(t)$ is [8, 12]:

$$\varphi^g(t) = f(t, T) - f(0, T) \tag{22}$$

where

$$f(t, T) = \sum_{q=1}^{N} f_q \cdot [(n_q + 1) \cdot e^{-iv_q t} + n_q \cdot e^{iv_q t}] \,. \tag{23}$$

Here $n_q = n(v_q) = [\exp(hv_q/kT) - 1]^{-1}$. For a single mode and $T = 0$, Eq. (22) is transformed into Eq. (13). By carrying out the integration in Eq. (3) over t, we find Eq. (14) again. However, now f is given by:

$$f(T) = \int_{-\infty}^{\infty} dv \cdot f(v, T) \tag{24}$$

where

$$f(v, T) = (n(v) + 1) \cdot f(v) + n(-v) \cdot f(-v) . \tag{25}$$

The one-phonon function $f(v)$ is described by Eq. (20). It is obvious that we may consider $f(T = 0) = \sum_q f_q$ as a coupling constant which characterizes the strength of the linear electron-phonon interaction in the multimode case. The n-th term in Eq. (15) is given in multimode case by

$$\Phi_n^g(\omega - \omega_0) = e^{-f} \cdot \frac{1}{n!} \cdot \int_{-\infty}^{\infty} dv_1 \ldots \int_{-\infty}^{\infty} dv_n \cdot f(v_1, T) \ldots f(v_n, T)$$

$$\cdot \delta(\omega - \omega_0 - v_1 \ldots - v_n) . \tag{26}$$

The function Φ_n^g is non-vanishing over the n-phonon frequency interval. Hence, the function $\Phi^g(\omega - \omega_0)$ describes a broad PSB.

The one-phonon function $f(v, T)$ has Stokes and anti-Stokes terms which are proportional to $n(v) + 1$ and $n(-v)$, respectively. The Pekar–Huang factor f is the integrated value of the one-phonon function and it depends on temperature. The probability of one-phonon transitions is described by

$$\Phi_1^g(\omega - \omega_0) = e^{-f(T)} \cdot f(\omega - \omega_0, T) . \tag{27}$$

The probability of n-phonon transitions is expressed solely via one-phonon function. This means that the one-phonon function $f(v, T)$ determines all the details of the whole optical band consisting of ZPL and PSB.

The Debye–Waller factor $\alpha = \exp(-f(T))$ depends on temperature because $f(T)$ is given by

$$f(T) = \int_0^{\infty} dv \cdot f(v) \cdot \text{cth} \frac{hv}{2kT} . \tag{28}$$

The Pekar–Huang factor $f(T)$ is increased and the intensity of ZPL is decreased with increasing temperature. Figure 6 shows the changes in the homogeneous optical band occurring when temperature increases. At $T > 40$ K, ZPL disappears and the homogeneous optical band consists solely of PSB. In this case, the band shape function can be derived from Eq. (3) if we expand Eqs. (22) and (23) as a power series of time t. We find

$$\varphi^{g,e}(t) = \mp iA \cdot t - \frac{B(t)}{2} \cdot t^2 + \ldots \tag{29}$$

where

$$A = \int_0^{\infty} dv \cdot v \cdot f(v), \tag{30}$$

$$B(T) = \int_{-\infty}^{\infty} dv \cdot v^2 \cdot f(v, T) . \tag{31}$$

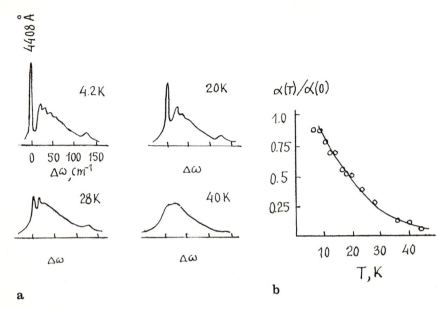

Fig. 6a, b. a Fluorescence 0–0′ band of perylene doped *n*-heptane. **b** The temperature dependence of the Debye–Waller factor for this band [22]. Curve for α(T) has been calculated by means of Eqs. (19) and (28) with the help of one-phonon function f(v) found from the optical band shown in Fig. 6a

Substituting $\varphi^{g,e}(t)$ in Eq. (3) by Eq. (29) and carrying out the integration over t we find for the homogeneous optical band the following expression:

$$I^{g,e}(\omega) \propto \left(\frac{2\pi}{B(T)}\right)^{1/2} \cdot \exp\left[-\frac{(\omega - \omega_0 \mp A)^2}{2B(T)}\right] \qquad (32)$$

The growth of both temperature and coupling constant, leads to the disappearance of ZPL. In this case, the absorption and fluorescence bands are Gaussian. The distance between their maxima equals 2A and it is called Stokes shift. It does not depend on temperature. On the contrary, the Gaussian halfwidths $\Delta\omega = (B(T) \cdot 2\ln 2)^{1/2}$ weakly depend on temperature and this can be described by a simple approximation formula:

$$\Delta\omega \simeq (2\ln 2)^{1/2} \cdot \left(B(0) + 2A \cdot \frac{kT}{h}\right)^{1/2} \qquad (33)$$

where $A = \langle v \rangle$ and $B(0) = \langle v^2 \rangle$ can be considered as the middle frequency and the averaged square frequency of phonons. Values of $\langle v \rangle$ and $\langle v^2 \rangle$ can be found from the temperature experimental data.

Equations (32) and (33) can only be applied to the homogeneous PSB. The significance of Eqs. (32) and (33) increases with the growth of temperature because ZPL disappears. However, in this situation such methods of the selective spectroscopy as FLN and HB work badly because they cannot remove inhomogeneous broadening which always exists in polymers and glasses. By now, Eqs. (32) and (33) have only been used to treat broad optical bands of F-centers in crystals [23] where the inhomogeneous broadening is negligible.

However Eqs. (32) and (33) can be applied to polymer samples with guest chromophores because PE experiments enable one to find the one-phonon function f(v) even in the case when ZPL is absent and homogeneous optical bands are described by Eq. (32). This situation will be discussed in detail in Sect. 4.8.

3.3 Mathematical Treatment of Realistic PSBs

Let us consider first the way the homogeneous optical band consisting of ZPL and PSB can be treated.

Integral intensity of whole band. The integral intensity of ZPL and PSB is simply described in Eq. (18) and f in Eq. (24). The integral intensity of ZPL decreases and the intensity of PSB increases if we heat the sample. However, the integral intensity of the whole optical band does not depend on temperature. This intensity transformation from ZPL to PSB is the peculiarity of the Franck–Condon interaction. The relevant temperature dependence of the Debye–Waller factor is shown in Fig. 6.

However, sometimes the integral intensity of the whole optical band can depend on temperature. This fact means that the so-called Herzberg–Teller (HT) electron-phonon interaction plays an appreciable role in the case under consideration. HT-interaction emerges from a modulation of the electronic dipolar moment $\vec{d}(R)$ of a guest chromophore by phonons. By expanding $\vec{d}(R)$ to a series in phonon variables R we find

$$d(R) = d(0) + \sum_q d_q R_q \ldots \ldots \tag{34}$$

The term $\sum_q d_q R_q$ describes the linear HT-interaction. It leads to appearance of one-phonon transitions in an optical band. HT-interaction describes the effect of mixing of various electronic states due to phonon modulation. This mixing is fairly small. Therefore, HT-interaction, as a rule, is smaller than the linear Frank–Condon interaction. However, HT-interaction results in the temperature dependence of the integral intensity of the whole optical band:

$$I_{ZPL} + I_{PSB} = 1 + \sum_q \left(\frac{d_q}{d(0)}\right)^2 \cdot \text{cth} \frac{h\nu_q}{2kT} \, . \tag{35}$$

This dependence is weak because $\sum_q (d_q/d(0))^2 \ll 1$.

Contribution to PSB from HT-interaction. If the electron-phonon coupling is weak, PSB consists mainly of one-phonon transitions. In this case the contribution to PSB from HT-interaction can exceed one from FC-interaction. Figure 7 shows such a situation. Here, PSB consists only of one-phonon peak which is

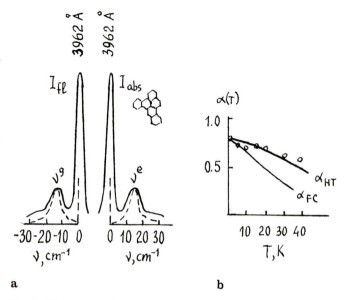

Fig. 7a. Fluorescence (left) and absorption (right) bands of the 0–0' transition of 3,4–6,7-diben-zopyrene in *n*-octane at 4.2 K [24]. The *solid lines* are the observed spectra, The *dashed lines* are the spectra after exclusion of inhomogeneous broadening; $v^g = 13.1 \, \text{cm}^{-1}$ and $v^e = 15.6 \, \text{cm}^{-1}$. **b** The temperature dependence of the Debye–Waller factor measured (circles) and calculated with the help of Eqs. (36) and (37)

related to a quasilocalized phonon mode with the frequency v_0. Different types of the electron-phonon interaction result in different formulae for the Debye–Waller factor:

$$\alpha_{FC}(T) = \exp\left(-f \cdot \text{cth} \frac{hv_0}{2kT} \right), \tag{36}$$

$$\alpha_{HT}(T) = \left[1 + \left(\frac{d_1}{d_0} \right)^2 \cdot \text{cth} \frac{hv_0}{2kT} \right]^{-1}. \tag{37}$$

By comparing Eqs. (37) and (35), we can conclude that linear HT-interaction does not cause any temperature dependence of ZPL. The ratio I_1/I_0, where I_1 and I_0 are the integral intensities of the one-phonon peak and ZPL, is equal to the parameter f or the ratio $(d_1/d_0)^2$. The solid lines in Fig. 7b have been calculated by means of Eqs. (36) and (37) with the help of I_1/I_0 found from Fig. 7a. The temperature dependence of the Debye–Waller factor proves that one-phonon peaks result from HT-interaction.

Contribution to PSB from FC-interaction. We can neglect the contribution to PSB from HT-interaction if PSB includes multiphonon transitions. The PSB of such a type is depicted in Fig. 6. In this case $\alpha(o) \simeq 0.23$. Figure 8 shows

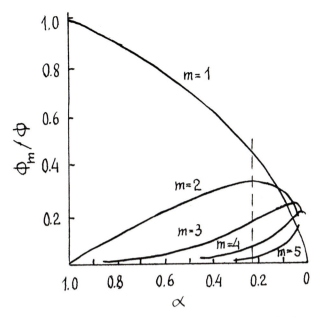

Fig. 8. The contribution of the m-phonon transitions to the integrated intensity Φ of PSB for various values of the Debye–Waller factor. The *dashed line* marks $\alpha = 0.23$ measured in the optical band shown in Fig. 6a at T = 4.2 K

a contribution to PSB from n-phonon transitions. At $\alpha = 0.23$, the phonon transition with creation of even four phonons contributes to PSB. Therefore, the problem of the extracting of one-phonon function from PSB appears. This problem can be solved with the help of the Kukushkin equation [25]. This equation uses the fact that PSB depends only on the one-phonon function $f(\nu, T)$. The Kukushkin equation can be transformed at T = 0 into the following form [22]:

$$F(n\Delta\omega) = \frac{\Delta\omega}{I_{ZPL}} \left\{ n \cdot \Phi(n\Delta\omega) - \sum_{m=1}^{n-1} \Phi[(n-m)\Delta\omega] \cdot F(m\Delta\omega) \right\} \qquad (38)$$

where $F(\omega) = \omega \cdot f(\omega)$, I_{ZPL} is integral intensity of ZPL and $\Phi(\omega)$ is the PSB shape function. The smaller the step $\Delta\omega$ along the frequency axis, the closer is Eq. (38) to the integral Kukushkin equation. The latter neglects HT-interaction. It is easy to solve Eq. (38) numerically. Since Eq. (38) is of nonlinear type with respect to I_{ZPL}, it cannot be used if I_{ZPL} is too small.

Equation (38) has been employed for treating experimental data in [22, 26–28]. If the one-phonon function $f(\nu)$ found from PSB by means of Eq. (38) is correct, the Debye–Waller factor calculated with the help of $f(\nu)$ by means of Eqs. (19) and (28) must describe the temperature dependence $\alpha(T)$ found experimentally. This verification of the function $f(\nu)$ has been realized in [22] for perylene doped n-heptane. Results are shown in Fig. 6b. Good agreement of

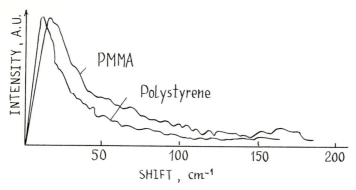

Fig. 9. Phonon density of states weighted by coupling constant f_q for Mg-octaethylporphyrine in polystyrene and PMMA [28] (see text)

curve with circles proves that the one-phonon function extracted from PSB is correct.

There are some examples of finding the one-phonon function from the homogeneous optical bands in crystals where inhomogeneous broadening is negligible. The question is: can we use this method in polymers where in-homogeneous broadening is large? Very recently the Kushida's group reported that Eq. (38) can be employed for the guest chromophores in polymers [28]. This group employed FLN to remove inhomogeneous broadening and to find the homogeneous optical band of Mg-octaethylporphyrine in a polystyrene film and in PMMA. Figure 9 shows the one-phonon function for this system found with the help of Eq. (38) from the FLN spectrum. It is the first finding of the one-phonon function for polymers from optical spectra. In order to verify this function, Kushida et al. calculated the temperature transformation of the inten-sity within the optical band. Comparison of the calculated optical band and that found from experiment is given in Fig. 10. Good agreement proves that the optical band found with the help of the selective spectroscopy methods can be treated the way the homogeneous optical bands were treated earlier in crystals and we can find a weighted density of states of polymers from optical spectra.

3.4 Role of Quadratic FC-Interaction and Temperature Broadening of ZPL

Contribution of quadratic FC-interaction into PSB. The quadratic FC-inter-action $\vec{R}\frac{W}{2}\vec{R}$ is due to the change $W = U^e - U^g$ of the force-matrix of a solvent upon the electronic excitation of a chromophore. This interaction depends on the dynamical variable $\vec{R}$ of the nearest neighbours of a chromophore because only these molecules can fill the change of the electronic state of the chromo-phore.

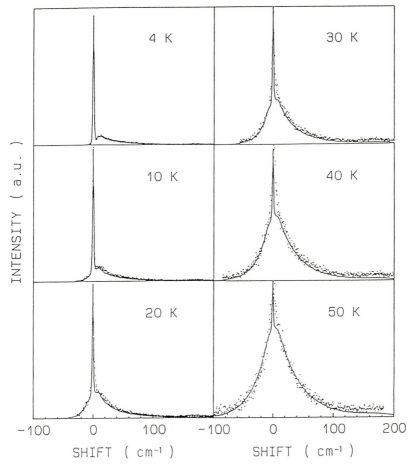

Fig. 10. Comparison of the measured band shape of Mg-octaethylporphyrine in a polystyrene film with the calculated one [28]. The calculation has been carried out in [28] with the help of the one-phonon function depicted in Fig. 9

The quadratic interaction contributes to PSB. However this interaction can never increase appreciably the intensity of PSB. Indeed, this interaction contributes only to electron-phonon transitions with creation and annihilation of even number of phonons (two, four, six, etc.) and its contribution to these transitions is proportional to

$$\left(\frac{v^e - v^g}{v^e + v^g}\right)^2 \simeq \left(\frac{\Delta}{2v}\right)^2 \tag{39}$$

where v^e and v^g are the frequencies of phonons in the excited and ground electronic states, respectively [29, 30]. It is obvious that this parameter can never exceed unity and, therefore, the contribution of the quadratic interaction to the intensity of PSB, as a rule, is negligible.

Since the force-matrices in the ground and excited electronic states differ, the phonon frequencies v^e and v^g differ as well. This fact leads to the violation of the mirror symmetry of PSBs in the absorption and fluorescence spectra. The violation of the mirror symmetry due to the quadratic interaction is shown in Fig. 7. In this case, we have $\Delta/2v \sim 0.1$.

Temperature broadening of ZPL. Although we can neglect the quadratic FC-interaction by considering PSB, we cannot do the same when we consider ZPL. We have already seen in Sect. 3.2 that the linear FC-interaction does not result in the broadening of ZPL even in the multimode case. We used Eq. (14) where ZPL is described by δ-function. If we take into account interaction of the optical electron with zero-point vibrations of the transverse electromagnetic field, the δ-function should be substituted by a Lorentzian whose halfwidth is

$$\gamma = 1/T_1 \qquad\qquad\qquad (40)$$

where T_1 is a fluorescence lifetime. If the fluorescence quantum yield is close to unity the probability of radiationless processes negligible. In this case the halfwidth $\gamma = 1/T_1$ does not depend on temperature. However experiments show that the halfwidth of ZPL is one or two order of magnitude larger than $1/T_1$ and this halfwidth depends on temperature. These facts are explained by the effect of the quadratic FC-interaction.

The quadratic interaction causes processes of creation or annihilation of two phonons. These processes do not contribute to ZPL broadening. Besides of these two-phonon processes, the quadratic interaction causes the two-phonon Raman-like processes which are characterized by the simultaneous creation and annihilation of one phonon. The probability of such two-quantum processes is proportional to $n(n + 1) = [2\,sh(hv/2kT)]^{-2}$ where n is the average phonon number.

The quadratic interaction yields the temperature broadening and shift of ZPL. The theory predicts the Lorentzian shape of ZPL with the halfwidth [8, 30, 31]:

$$\gamma(T) = \int\limits_0^\infty \frac{dv}{2\pi} \cdot \ln[1 + W^2 \cdot \Gamma^g(v) \cdot \Gamma^e(v) \cdot sh^{-2}(hv/2kT)] . \qquad (41)$$

Here, Γ^g and Γ^e are phonon spectral functions which describe phonons in the ground (g) and excited (e) electronic states of a chromophore.

Since phonons broaden ZPL, the following questions arise: will ZPLs in the absorption and fluorescence bands be in resonance with each other and will they have the same halfwidth? The answers to both questions are yes. Indeed, Eq. (41) is invariant regarding the substitution $g \rightleftharpoons e$ and $W \rightarrow -W$ which characterizes a transformation of the absorption band to the fluorescence one.

The nonperturbative theory for the halfwidth and shift of ZPL is discussed in [8, 30] in more detail.

In order to derive the perturbative formula for $\gamma(T)$, we should take into account that $z = W^2 \Gamma^g \Gamma^e sh^{-2}(h\nu/2kT) \ll 1$ and $\Gamma^g(\nu) \simeq \Gamma^e(\nu) = \Gamma(\nu)$. Substituting $\ln(1 + z)$ in Eq. (41) by z, we find the perturbative formula for $\gamma(T)$. This actual version of the nonperturbative Eq. (41) is frequently used for analysis. The function $\Gamma(\nu)$ characterizes phonons which cause the broadening of ZPL.

If acoustic phonons are responsible for the ZPL broadening, the function $\Gamma(\nu) = 5\nu^3/\nu_D^5$ where $\nu_D = k\theta$ is the Debye frequency of a solvent. In this case, Eq. (41) is transformed to the McCumber–Sturge formula [32]:

$$\gamma(T) = \frac{25\pi}{2} \cdot \frac{W^2}{\nu_D^3} \cdot \left(\frac{T}{\theta}\right)^7 \cdot \int_0^{\theta/T} dx \cdot x^6 \cdot sh^{-2}(x/2) . \qquad (42)$$

At low temperatures when $T \ll \theta$, the integral in Eq. (42) is a constant. Therefore, at the low-temperature limit we have $\gamma(T) \sim T^7$. At $T/\theta \gg 1$, the integral is proportional to $(\theta/T)^5$. Hence, at the high-temperature limit we have $\gamma(T) \sim T^2$. Equation (42) describes the temperature broadening of the R_2-line in ruby crystal [32].

A more typical situation for organic systems is when the chromophore interacts with a quasilocalized phonon mode. A guest molecule being embedded in a solvent disturbs its vicinity and creates a quasilocalized mode with $\nu_0 \ll \nu_D$. Naturally the guest chromophore interacts mainly with this very localized mode. In this case, we can describe the phonon spectral function $\Gamma(\nu)$ by a Lorentzian with a frequency ν_0 and halfwidth $\gamma = (2\tau)^{-1}$ where τ is the lifetime of this mode. Substituting $\Gamma(\nu)$ in Eq. (41) by this Lorentzian, we find the Krivoglaz formula for the half-width of ZPL [33]:

$$\gamma(T) = \frac{\Delta^2}{2\gamma} \cdot sh^{-2}(h\nu_0/kT) . \qquad (43)$$

Here, we allowed for $W \simeq \nu_0 \Delta$ where $\Delta = \nu_0^e - \nu_0^g$. At low temperature when $kT < h\nu_0$ Eq. (43) yields the temperature broadening of an Arrenius type: $\gamma(T) \sim \exp(-h\nu_0/kT)$. The temperature dependence of such types have been observed both in organic crystals [34–37], in organic glasses [38, 39] and in polyethylene [40]. At high temperatures when $kT > h\nu_0$, Eq. (43) predicts the square law of broadening: $\gamma(T) \sim T^2$. This temperature dependence has been observed in organic and inorganic glasses [41–44].

Sometimes, the frequency ν_0 of the quasilocalized mode resulting in the temperature broadening of ZPL can be identified in PSB. A good example of such a type is shown in Fig. 7. Korotaev and Kaliteevski have proved [24] that the quasilocalized mode depicted in Fig. 7 is responsible for the temperature broadening of ZPL. They have found that $W = [(\nu^e)^2 - (\nu^g)^2]/2 = [(15.6)^2 - (13.1)^2]/2 = 35$ cm^{-2}. If we substitute W in Eq. (41) by 35 cm^{-2}, a good agreement exists between the theoretical curve for $\gamma(T)$ and experimental data for the temperature broadening of ZPL [24].

However, very often we are facing another situation: a localized mode manifests itself in the temperature broadening of ZPL; however, it cannot be identified with some peak in PSB. It is not surprising because PSB is due to a linear electron-phonon coupling and the temperature broadening of ZPL is due to a quadratic electron-phonon interaction. It is possible that $v_q^e - v_q^g \neq 0$ but $a_q = 0$ for the same mode q. This very situation is shown in Fig. 7. We should remember (see Sect. 3.3) that one-phonon peaks in Fig. 7 are due to the linear HT-interaction. The linear FC-interaction equals zero ($a_q = 0$). Despite this fact the quadratic FC-interaction does not equal zero.

4 Methods of Selective Spectroscopy

In Sects. 3.2, 3.3 and 3.4, we have shown the way the information about an electron-phonon interaction and phonon dynamics can be got from the homogeneous optical bands. However, these bands are hidden under strong inhomogeneous broadening existing in polymers and glasses. Methods of selective spectroscopy enable one to remove inhomogeneous broadening. We discussed these methods briefly in Sect. 2.2. In this part, we consider methods of selective spectroscopy in more detail.

4.1 Fluorescence Line Narrowing (FLN)

Let a sample be irradiated by monochromatic light with a frequency ω_b. Then, the temporal evolution of an ensemble of two-level chromophores is described by the following equations:

$$\dot{n}_0 = - k^g n_0 + (k^e + \Gamma) n_1$$

$$\dot{n}_1 = k^g n_0 - (k^e + \Gamma) n_1 .$$

(44)

Here, n_0 and n_1 are the numbers of unexcited and excited chromophores, Γ is the rate of the spontaneous optical transitions from the excited electronic level, and the rates k^g and k^e of the induced optical transitions are related to the functions I^g and I^e which describe homogeneous absorption and fluorescence bands as follows:

$$k^{g,e}(\omega_b - \omega_0) = \Lambda^2 \cdot I^{g,e}(\omega_b - \omega_0) .$$

(45)

The constant Λ has the dimension of frequency and it depends on the intensity I of the exciting light:

$$\Lambda^2 = \frac{4\pi\omega_0}{\hbar c} \cdot (\vec{d} \cdot \vec{u})^2 \cdot I .$$

(46)

Here, $\vec{d}$ is the dipolar moment of a chromophore, $\vec{u}$ is the polarization vector of exciting light, I is the number of photons per second and cm^2 in the exciting beam.

In steady regime of the excitation, the chromophore populations do not change so that $\dot{n}_0 = \dot{n}_1 = 0$. Taking into account that $n_0 + n_1 = n(\omega_0)$, where $n(\omega_0)$ is the number of the chromophores in a sample, and solving Eq. (44) we find the population of the excited level in the steady regime:

$$n_1 = \frac{k^g}{k^g + k^e + \Gamma} \cdot n(\omega_0) . \tag{47}$$

If the light intensity I is small, the rates $k^{g,e}$ for induced transitions are much less than rate Γ for spontaneous ones, i.e. $k^{g,e}/\Gamma \ll 1$. In this case, Eq. (47) is simplified and it takes the form:

$$n_1(\omega_b, \omega_0) \simeq \frac{k^g(\omega_b - \omega_0)}{\Gamma} \cdot n(\omega_0) . \tag{48}$$

After the exciting light is off, the excited chromophores spontaneously radiate light with frequency ω_r. The FLN spectrum is described by

$$I_{FLN}(\omega_r, \omega_b) = \int_0^\infty d\omega_0 \cdot I^e(\omega_r - \omega_0) \cdot n_1(\omega_b, \omega_0)$$

$$\propto \int_0^\infty d\omega_0 \cdot n(\omega_0) \cdot I^e(\omega_r - \omega_0) \cdot I^g(\omega_b - \omega_0) . \tag{49}$$

If the homogeneous band consists solely of ZPL, we have $I^e(\omega) = I^g(\omega) = \gamma/(\omega^2 + \gamma^2)$ and Eq. (49) reads:

$$I_{FLN} \simeq n(\omega_b) \cdot \frac{2\gamma}{(\omega_r - \omega_b)^2 + (2\gamma)^2} . \tag{50}$$

Despite a strong inhomogeneous broadening, the FLN spectrum consists of a line which has the shape of the homogeneous ZPL. This line has the double homogeneous halfwidth.

The FLN spectrum lies in the spectral region of the exciting light. Therefore, the fluorescence line can be measured after excitation is off. The FLN experiment is facilitated if the chromophore has a long life-time. For instance, Sabo [13] applied the FLN methods to the inhomogeneously broadened R_2 line of ruby crystal, where the lifetime was 10^{-3} s. This situation is typical for paramagnetic ions.

In organic impurity systems, the fluorescence lifetime is too short. However, organic chromophores have vibronic lines which are shifted to the red side regarding ZPL. Therefore, if we excite in the ZPL frequency domain and record the fluorescence in the field of vibronic lines the exciting light can be cut off by a cutoff filter. Therefore, FLN is widely used to study the fine structure in vibronic and phonon spectra of organic chromophores in polymers and glasses [14, 45].

The shapes of realistic homogeneous absorption and fluorescence spectra can be described by

$$I^g(\omega_b - \omega_0) = \alpha \cdot \frac{\gamma/2}{(\omega_b - \omega_0)^2 + (\gamma/2)^2} + \Psi(\omega_b - \omega_0) \,,$$

$$I^e(\omega_r - \omega_0) = \alpha \cdot \frac{\gamma/2}{(\omega_r - \omega_0)^2 + (\gamma/2)^2} + \Psi(\omega_0 - \omega_r) \,. \tag{51}$$

Here, the Lorentzians describe ZPLs and the functions Ψ describe electron-phonon and vibronic spectra of a chromophore. Equation (51) generalizes Eqs. (14) and (17) derived for electron-phonon bands. Vibronic modes can be allowed for if we include them in the sum in Eq. (20). In this case the function $f(v)$ takes the following form:

$$f(v) = f_{ph}(v) + \sum_{s=1}^{N_0} f_s \cdot \delta(v - v_s) \tag{52}$$

where $f_{ph}(v)$ is the former one-phonon function and the sum over s includes one-phonon vibronic peaks which correspond to the intramolecular vibrations. It is obvious that the function Ψ in Eq. (51) includes both PSB and vibronic peaks.

Let us consider FLN spectrum near ZPL. In this case, we should substitute the function Ψ in Eq. (51) by the function $(1 - \alpha) \cdot \Phi$ describing PSB. Substituting $I^{g,e}$ in Eq. (49) by Eq. (51), we find FLN spectrum near ZPL

$$I_{FLN}(\omega_r, \omega_b) \sim \alpha^2 \cdot \frac{\gamma}{(\omega_r - \omega_b)^2 + \gamma^2} \cdot n(\omega_b)$$

$$+ \alpha(1 - \alpha) \cdot \Phi(\omega_b - \omega_r) \cdot (n(\omega_b) + n(\omega_r))$$

$$+ (1 - \alpha)^2 \cdot \int_0^\infty d\omega_0 \cdot n(\omega_0) \cdot \Phi(\omega_b - \omega_0) \cdot \Phi(\omega_0 - \omega_r) \,. \tag{53}$$

The FLN spectrum consists of ZPL (first term), PSB (second and third terms) and structureless spectral phon (fourth term). Sometimes, we can neglect this phon. For instance, this phon does not influence the shape of the FLN spectrum and the latter almost coincides with the homogeneous optical band shape. However, if the Debye–Waller factor is of interest, we cannot neglect the phon. By neglecting this phon, we find an effective value $\alpha_{eff}(T)$ for the Debye–Waller factor. The quantity $\alpha_{eff}(T)$ depends on the value of the frequency ω_b. In this case, the integral intensity of I_{FLN} depends on temperature although the integral intensity of the homogeneous band is temperature independent.

Let us consider now what type of information we can get from the FLN spectra if we allow for vibronic peaks. Let the function $f(v)$ in Eq. (52) have one vibronic peak, i.e. $f_s = \delta_{s0} f_0$. In this case the function Ψ in Eq. (51) can be written as follows:

$$\Psi(\omega_b - \omega_0) = \alpha \cdot f_0 \cdot \frac{\gamma_1/2}{(\omega_b - \omega_0 - \Omega^e)^2 + (\gamma_1/2)^2}$$

$$+ (1 - \alpha) \cdot \Phi(\omega_b - \omega_0) + \ldots$$

$$\Psi(\omega_0 - \omega_r) = \alpha \cdot f_0 \cdot \frac{\gamma_1/2}{(\omega_0 - \omega_r - \Omega^g)^2 + (\gamma_1/2)^2}$$

$$+ (1 - \alpha) \cdot \Phi(\omega_0 - \omega_r) + \ldots \tag{54}$$

Here, γ_1 is a halfwidth of the vibronic peak and Ω^e and Ω^g are vibronic frequencies in the excited and ground electronic states of a chromophore. In this case, the homogeneous spectrum has two narrow peaks: ZPL and the vibronic line. Therefore, two ensembles of molecules will be excited more effectively than others – see top of Fig. 11a. However, ensemble I includes many more molecules than ensemble II. Therefore, the FLN-spectrum results mainly from the fluorescence band of ensemble I. This means that the FLN-spectrum is similar to the homogeneous one if we use laser excitation in the frequency domain where the fluorescence and absorption bands are overlapped. Figure 12 proves this conclusion.

However, if we excite our sample beyond the resonant frequency domain the FLN-spectrum has a complicated character because both molecular ensembles contribute to the FLN-spectrum – Fig. 11b.

Figure 11a, b can be described mathematically. Indeed, substituting the functions Ψ in Eq. (51) by Eq. (54) we find the following expression for the four lines of the FLN-spectrum depicted in Fig. 11b:

$$I_{FLN} \propto \left[\frac{\gamma}{(\omega_b - \omega_r)^2 + \gamma^2} + f_0 \cdot \frac{(\gamma + \gamma_1)/2}{(\omega_b - \omega_r - \Omega^g)^2 + (\gamma + \gamma_1)^2/4} \right] \cdot n(\omega_b)$$

$$+ f_0 \cdot \left[\frac{(\gamma + \gamma_1)/2}{(\omega_b - \omega_r - \Omega^e)^2 + (\gamma + \gamma_1)^2/4} \right.$$

$$\left. + f_0 \cdot \frac{\gamma_1}{(\omega_b - \omega_r - \Omega^g - \Omega^e)^2 + \gamma_1^2} \right] \cdot n(\omega_b - \Omega^e) + \ldots \tag{55}$$

The situation shown in Fig. 11a corresponds to the case when $n(\omega_b) \gg n(\omega_b - \Omega^e)$. Therefore, the FLN-spectrum coincides with the homogeneous one. On the contrary, when $n(\omega_b) \simeq n(\omega_b - \Omega^e) \cdot f_0$ the FLN-spectrum has four narrow peaks (Fig. 11b). In this case, we can measure in the FLN-spectrum vibronic frequencies Ω^g and Ω^e of both electronic states.

Thus, we have seen that the FLN spectrum does not coincide exactly with the homogeneous one. However, it can be similar to the homogeneous spectra

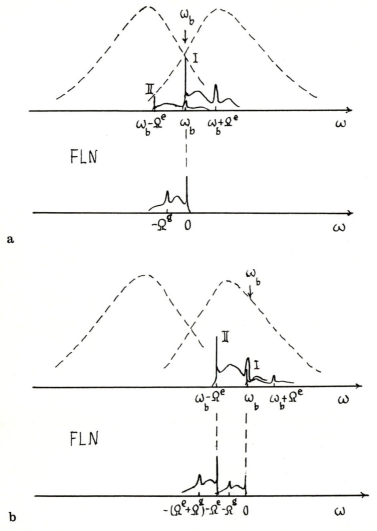

Fig. 11. The dependence of the FLN spectrum on the wave length of exciting light. *Dashed lines* show inhomogeneously broadened fluorescence (*left*) and absorption (*right*) spectra

under certain conditions (Fig. 12). The methods developed for the theoretical treatment of homogeneous spectra can be applied to the FLN spectra as well.

4.2 Spectral Hole Burning (HB): Mechanisms and Kinetics

If the chromophore has no fluorescence, the FLN method does not work. In this case, we can use the spectral hole burning (HB) method. The HB method has been discussed briefly in Sect. 2.2. Transient holes can be burnt in any impurity

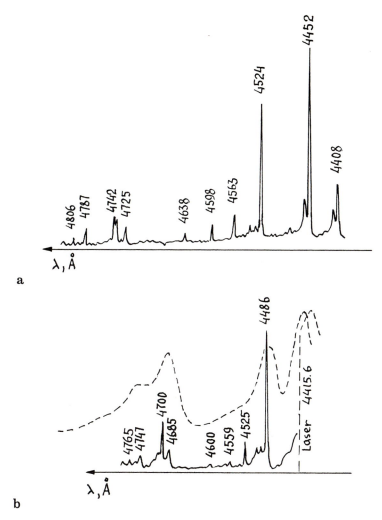

Fig. 12. a The homogeneous fluorescence spectrum of perylene in *n*-heptane at 4.2 K [22]. **b** In-homogeneously broadened spectrum of the perylene in ethanol glass at 4.2 K (*dashed line*) and FLN spectrum of the same sample (*solid line*) [45]

systems. However, these holes are inconvenient for investigations because of their short lifetimes.

The HB method became the main one for selective spectroscopy due to the existence of long life persistent holes. It turned out that persistent holes living at low temperature even for days can be burnt in many important organic and inorganic systems. The persistent holes appear because chromophores, after their excitation by light, relax to a metastable ground electronic state which is separated from the initial ground state by a potential barrier. Persistent holes can be erased if we heat our sample.

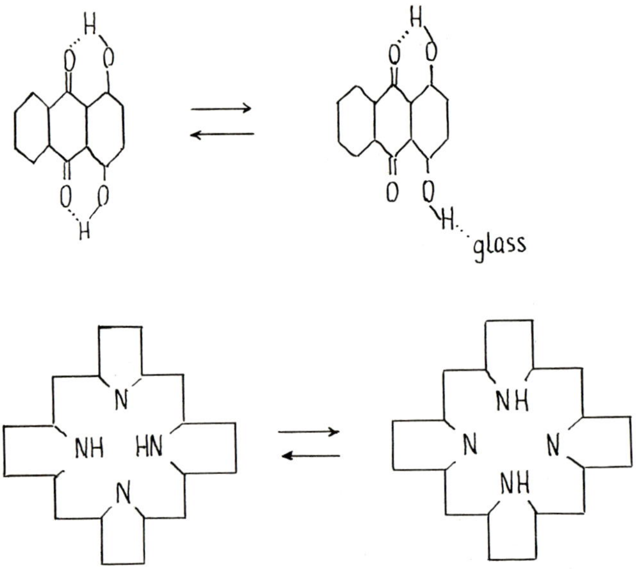

Fig. 13a, b. Examples of **a** photochemical and **b** photophysical hole-burning mechanisms

The metastable ground state can differ chemically from the initial state. In this case, we are talking about a photochemical HB. However, the metastable state can differ only by the orientation of molecules in the lattice. In this case we are talking about photophysical HB [46]. Photophysical mechanisms of HB exist mainly in disordered systems like polymers and glasses.

An example of the photochemical mechanism is shown in Fig. 13a [47]. A proton of the chromophore is bound with molecules of a solvent after electronic excitation. The metastable ground state is due to the change in a chemical bound. The boundary between photochemical and photophysical HB is frequently conditional. This situation is shown in Fig. 13b [48]. A reorientation of protons in free base porphin after electronic excitation can be considered as a change of the chemical bounds in the molecule and as a reorientation of the whole molecule in a lattice.

A mathematical description of the HB process in two-level electronic systems does not depend on the type of HB mechanism and this process can be described by the following two rate equations:

$$\dot{n}_0 = - k^g \cdot n_0 + (k^e + \Gamma) \cdot n_1 ,$$
$$\dot{n}_1 = k^g \cdot n_0 - (k^e + \Gamma + Q) \cdot n_1 . \tag{56}$$

These equations differ from Eq. (44) by rate constant Q which describes a photochemical reaction via an excited electronic state. Carrying out a summation and subtraction of Eq. (56) we find the following equations for the $n = n_1 + n_0$ and

$r = n_1 - n_0$:

$$\dot{n} = -\frac{Q}{2} \cdot (n + r) \,,$$

$$\dot{r} = \left(k^g - k^e - \Gamma - \frac{Q}{2} \right) \cdot n - \left(k^g + k^e + \Gamma + \frac{Q}{2} \right) \cdot r \,. \tag{57}$$

The total number n of chromophores decreases with the rate Q. The difference r in the population relaxes much faster with the rate Γ. Since $Q/\Gamma \ll 1$, we can take $\dot{r} = 0$ in Eq. (57). This approximation means that we consider the evolution of n at the steady value of r. Then, the evolution of n is described by the following equation:

$$n(t) = \exp(-\lambda(\omega_b - \omega_0)t) \cdot n(\omega_0) \tag{58}$$

where

$$\lambda(\omega_b - \omega_0) = Q \cdot \frac{k^g}{k^g + k^e + \Gamma + Q/2} \,. \tag{59}$$

After laser light is turned off, excited chromophores relax to the ground state for a short time. Therefore, the change in the total number of chromophores is described by $\Delta n(t) = (1 - \exp(-\lambda t)) \cdot n(\omega_0)$. The change of the optical density ΔD of a sample is proportional to $\Delta n(t)$. Therefore, by tuning the frequency ω_p of a probing laser we record a hole in the optical density. The shape of this hole is described by

$$\Delta D = h(\omega_p - \omega_b) \propto \int_0^\infty d\omega_0 \cdot k^g(\omega_p - \omega_0) \cdot (1 - e^{-\lambda(\omega_b - \omega_0)t}) \cdot n(\omega_0) \,. \tag{60}$$

If burning time t is increased, the hole becomes deeper and broader. At long burning time, the hole shape does not coincide with one of a homogeneous line. Therefore, one uses as a rule both short burning time and a weak intensity of burning light. Within this limit, we have $\lambda t \ll 1$ and $\lambda = Q \cdot k^g / \Gamma$. This simple expression for λ can be derived from Eq. (59) at $k^{g,e}/\Gamma \ll 1$. Taking into account Eq. (45), a short burning time and a weak intensity of burning light, we can transform Eq. (60) in to the following form:

$$h(\omega_p - \omega_b) \propto t \cdot \frac{Q}{\Gamma} \cdot \int_0^\infty d\omega_0 \cdot I^g(\omega_p - \omega_0) \cdot I^g(\omega_b - \omega_0) \cdot n(\omega_0) \,. \tag{61}$$

The magnitude of a hole is proportional to burning time and the photochemical quantum yield Q/Γ.

4.3 Hole Shape at Small Intensity of Burning Light

Equation (61) for hole shape is similar to Eq. (49) for the FLN spectrum. Therefore, an HB spectrum resembles a FLN spectrum and many details of our analysis made in Sect. 4.1 can be applied to an HB spectrum. However, Eq. (61) includes $I^g(\omega_p - \omega_0)$ instead of $I^e(\omega_r - \omega_0)$ which exists in Eq. (49). This difference manifests itself in the hole shape. Indeed, substituting the functions I^g in Eq. (61) by the first Eq. (51) with $\Psi = (1 - \alpha) \cdot \Phi$ we find with the help of Eq. (61) the following expression for hole shape:

$$h(\omega_p - \omega_b) \sim \alpha^2 \cdot \frac{\gamma}{(\omega_p - \omega_b)^2 + \gamma^2} \cdot n(\omega_b)$$

$$+ \alpha(1 - \alpha) \cdot \Phi(\omega_p - \omega_b) \cdot n(\omega_b)$$

$$+ \alpha(1 - \alpha) \cdot \Phi(\omega_b - \omega_p) \cdot n(\omega_p)$$

$$+ (1 - \alpha)^2 \cdot \int_0^\infty d\omega_0 \cdot \Phi(\omega_p - \omega_0) \cdot \Phi(\omega_b - \omega_0) \cdot n(\omega_0) . \quad (62)$$

The first term describes the so-called zero phonon hole (ZPH). Its halfwidth is the double homogeneous halfwidth of ZPL. The second and third terms describe the so-called phonon side hole (PSH). However, in contrast to PSB from FLN spectra the PSH is shifted to the red and blue sides with respect to ZPH (Fig. 14a). The fourth term describes a structureless phon which can be ignored if we are interested in the PSB shape. However, the phon term should be taken into account if we determine the Debye–Waller factor with the help of HB spectra. This situation resembles that existing in FLN spectra.

ZPH is a good object to investigate the temperature dependence of the ZPL halfwidth. Many of the temperature data which will be discussed in Sects. 5.2, 5.3, 5.4 and 5.5 have been obtained from the analysis of ZPH. Therefore, all details concerning ZPH is of great importance. It has been reported [49, 50] that the rate of burning of ZPH is decreased if the frequency of a burning laser is shifted from the red to the blue side of the inhomogeneously broadened optical band. Figure 15 shows the decreasing of the HB quantum efficiency with the

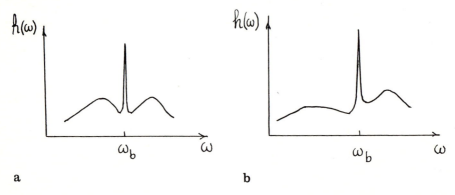

a   b

Fig. 14a, b. Zero-phonon hole and phonon-side holes at **a** small and **b** large intensities of burning light

growth of ω_b. Equation (61) enables one to find a simple explanation for this effect.

The optical density of a sample prior to HB is described by

$$D(\omega_b) \sim \int_0^\infty d\omega_0 \cdot n(\omega_0) \cdot [\alpha \cdot L(\omega_b - \omega_0) + (1 - \alpha) \cdot \Phi(\omega_b - \omega_0)] \qquad (63)$$

where $L(\omega)$ and $\Phi(\omega)$ are ZPL and PSB of a chromophore. An expression for the bottom of ZPH can be found from Eq. (61):

$$h(0) = \Delta D(\omega_b) \sim \Delta t \cdot \frac{Q}{\Gamma} \cdot \alpha^2 \cdot L(0) \cdot n(\omega_b) \qquad (64)$$

where Δt is burning time. It is obvious that the quantum efficiency of HB is described by a ratio $\Delta D(\omega_b)/\Delta t \cdot D(\omega_b)$. Inserting Eqs. (63) and (64) into this ratio, we find for the quantum efficiency of HB the following expression:

$$\frac{1}{D} \cdot \frac{\Delta D}{\Delta t} = \frac{Q}{\Gamma} \cdot L(0) \cdot \alpha^2 \cdot n(\omega_b)/[\alpha \cdot n(\omega_b) + (1 - \alpha)$$

$$\cdot \int_0^\infty d\omega_0 \cdot n(\omega_0) \cdot \Phi(\omega_b - \omega_0)] . \qquad (65)$$

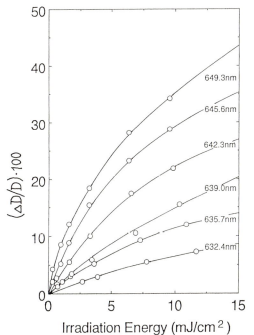

Fig. 15. Dependence of hole burning efficiency on the frequency of burning light measured in inhomogeneously broadened optical band of TPPS doped PVA [50]

The HB efficiency depends on ω_b. If the function $n(\omega_b)$ describing the in-homogeneous broadening has a maximum at $\omega_b = \Omega$ the integral in Eq. (65) has a maximum shifted to the blue side regarding Ω. Therefore, the ratio in Eq. (65) decreases if the frequency ω_b moves from the red side to the blue side of the absorption band. This decreasing takes place at $Q/\Gamma = $ const.

4.4 Power Broadening of Holes

A relationship between hole shape and homogeneous optical band shape is simple only if we use small intensity of burning light. The same conclusion can be made regarding the FLN spectrum and the relevant homogeneous band. It is important to know what value of the light intensity we can still consider as a low one. We have already seen in Sects. 4.1 and 4.3 that the physical criterion of low intensity is

$$k^{g,e}/\Gamma < 1 \ .$$

It means that the probability of the induced optical transition is smaller than one of the spontaneous transition. Equation (49) and (61) for the FLN- and HB-spectrum are correct only if the inequality $k^{g,e}/\Gamma < 1$ is fulfilled.

The ratio $k^{g,e}/\Gamma$ does not depend on the molecular dipolar moment because both $k^{g,e}$ and Γ are proportional to d^2. If we substitute $k^{g,e}$ and Γ by their microscopic expressions we can find that [51, 52]

$$\frac{k^{g,e}(\omega)}{\Gamma} = \frac{I}{I_c} \cdot \frac{\gamma}{2} \cdot I^{g,e}(\omega) \cdot \cos^2\theta \ . \tag{66}$$

Here θ is the angle between the vectors $\vec{d}$ and $\vec{u}$ in Eq. (46), γ is the ZPL halfwidth. The dimensionless function $\gamma \cdot I^{g,e}(\omega)/2$ describes a homogeneous optical band consisting of ZPL and PSB. Taking into account Eq. (51) we find that $\max(\gamma \cdot I^{g,e}(\omega)/2) = \gamma \cdot I^{g,e}(0)/2 = \alpha$ corresponds to the maximum of ZPL. Hence, Eq. (66) can be rewritten in the following form:

$$\max\left(\frac{k^{g,e}}{\Gamma}\right) = \alpha \cdot \frac{I}{I_c} < 1 \tag{67}$$

where

$$I_c = \pi \cdot \gamma/3\lambda^2 \simeq \gamma/\lambda^2 \tag{68}$$

is the critical light intensity. It depends only on the ZPL halfwidth γ and the wavelength λ of exciting light. Taking typical organic chromophores values: $\gamma = 10^{11} s^{-1}$, $\lambda = 5 \times 10^{-5}$ cm we find $I_c \simeq 4 \times 10^{19}$ phot/s cm^2. Multiplying I_c by the energy of a photon quantum $\hbar\omega_0$ we obtain $I_c \hbar\omega_0 = 16$ W/cm^2.

Equation (68) for the critical intensity is correct if the excited singlet electronic level is a doorway for the photochemical reaction. In many systems the

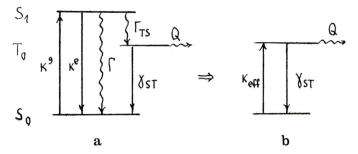

Fig. 16. Three-level scheme and relevant effective two-level scheme

photochemical reaction occurs in a triplet electronic state. An example of such type is the proton reorientation in a porphin molecule- see Fig. 13b. In this case, the chromophore is described by a tree-level scheme (Fig. 16a) HB within a tree-level scheme can be reduced to HB within an effective two-level scheme shown in Fig. 16b where $k_{eff} = k^g \Gamma_{TS}/(\Gamma + \Gamma_{TS})$ differs from k^g by the quantum yield $\Gamma_{TS}/(\Gamma + \Gamma_{TS})$ of the intersystem crossing [52]. In this case, the criterion of low intensities is

$$\max\left(\frac{k_{eff}}{\gamma_{ST}}\right) = \max\left(\frac{k^g}{\Gamma}\right) \cdot \frac{\Gamma}{\gamma_{ST}} \cdot \frac{\Gamma_{TS}}{\Gamma + \Gamma_{TS}} = \frac{I}{I'_c} < 1 \qquad (69)$$

where

$$I'_c = \frac{I_c}{\alpha} \cdot \frac{\gamma_{ST}}{\Gamma} \cdot \frac{\Gamma + \Gamma_{TS}}{\Gamma_{TS}} \qquad (70)$$

is a critical intensity for three-level system. Since $\gamma_{ST}/\Gamma \sim 10^{-4}-10^{-6}$, the new critical intensity ranges from mW/cm^2 to $\mu W/cm^2$.

The simple Eq. (61) has been derived for low intensities and for low laser fluencies. The latter means that $\lambda t < 1$. Taking, for example, free-base porphin (H_2P) we can find a burning time t which satisfied this inequality. Since the photochemical reaction in H_2P goes via the first triplet level we should substitute k^g and Γ in Eq. (59) by k_{eff} and γ_{ST}. Therefore, at low intensity we find

$$\lambda t = \max\left(\frac{k_{eff}}{\gamma_{ST}}\right) \cdot Qt = \frac{Q}{\gamma_{ST}} \cdot \frac{I}{I'_c} \cdot \gamma_{ST} t < 1 . \qquad (71)$$

Taking $I/I'_c = 10^{-1}$, $Q/\gamma_{ST} = 10^{-4}$, $\gamma_{ST} = 10^2$ s we find that $\lambda_{max} t = 10^{-1}$ if burning time is 10^2 s.

The hole shape is described by Eq. (62) only at low intensity. If the criteria at Eqs. (67) and (69) do not hold, Eq. (62) is changed and takes the following form [51]

$$h(\omega_p - \omega_b) \sim \frac{\alpha^2 \eta}{(1 + \alpha\eta)^{1/2}} \cdot \frac{\gamma_{eff}}{(\omega_p - \omega_b)^2 + \gamma_{eff}^2} \cdot n(\omega_b)$$

$$+ (1 - \alpha) \cdot \frac{\alpha\eta}{(1 + \alpha\eta)^{1/2}} \Phi(\omega_p - \omega_b) \cdot n(\omega_b)$$

$$+ \alpha \cdot \frac{(1 - \alpha)\eta \cdot \Phi(\omega_b - \omega_p)}{1 + (1 - \alpha)\eta \cdot \Phi(\omega_b - \omega_p) \cdot \gamma/2} \cdot n(\omega_p)$$

$$+ \text{the phon term.} \tag{72}$$

where

$$\gamma_{eff} = \frac{\gamma}{2} \cdot (1 + (1 + \alpha\eta)^{1/2}) . \tag{73}$$

Equations (72) and (73) are correct at an arbitrary value of the dimensionless intensity $\eta = I/I_c$ or I/I'_c. The most important peculiarities of Eqs. (72) and (73) are:

(i) ZPH described by the first term in Eq. (72) broadens at the large intensity as $\sqrt{\eta}$ [53, 54];

(ii) The mirror symmetry of the blue and red PSHs disappears. The intensity of the blue PSH and ZPH increases as $\alpha\eta/(1 + \alpha\eta)^{1/2}$;

(iii) The red PSH broadens when the intensity increases.

This behaviour of the red PSH can easily be explained if we take into account that this PSH is due to the burning of chromophores via their PSB. The efficiency of such nonselective hole burning is increased with the growth of the laser intensity. The hole shape at large intensity is schematically shown in Fig. 14b.

4.5 Two-Quantum Photochemical HB

Persistent holes can serve not only as a tool for scientific research but also for information storage. By burning a hole, we write a bit of information. We can read this information via light from the scanning laser. Although scanning laser light is weak, it causes changes to the holes. At repeated usage of the scanning laser, the original information will be erased. This is inherent in every molecular system involving hole burning which is due to the absorption of one quantum of light.

However, there are impurity systems where holes can be burnt if the sample is irradiated by light of two lasers with different wavelengths. In this case a photochemical reaction occurs via a highly excited electron level which can be reached only after absorption of two light quanta. Therefore, HB of such a type

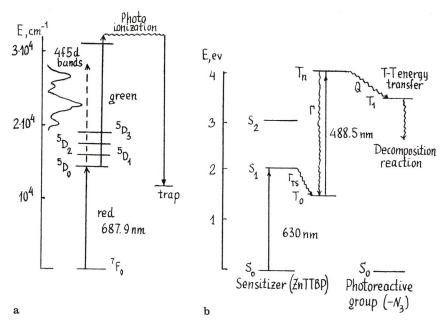

Fig. 17. a Energy levels of the systems undergone two-colour photochemistry Sm^{2+} ions doped BaClF [55]. **b** ZnTTBP doped glycidyl azide polymer crosslinked with trimethylol-propane and isophorone diisocyanate [62]

is called photon gated HB. In photon gated HB, as a rule, light quanta of different colours are used. Therefore, photochemistry which provides photon gated HB is called two-colour photochemistry. Photon gated HB has been discovered both in inorganic [55–57] and organic [58–63] systems.

A scheme of the energy levels of systems undergoing two-color photochemistry is shown in Fig. 17a, b. The highly excited electron levels are a doorway for a photochemical reaction. These levels cannot be reached by absorption of one light quantum. After irradiation of the sample by two-colour light, a hole in the red band appears. This hole cannot be erased by the reading red laser. The photochemical quantum yield η_p of two-colour photochemistry is two or three orders of magnitude larger than that in one-quantum photochemical reactions. For instance, $\eta_p = 10^{-4}$ for H_2P which exhibits one-quantum photochemistry and $\eta_p = 0.4$ for Zn-tetrabenzoporphin undergoing two-colour photochemistry [59].

Photon gated HB can be used for finding the rate constants of higher electronic levels which serve as a gateway to photochemical reactions. Although these rate constants lie in the picosecond time region, they can be found by recording slow hole-burning kinetics in the $S_1 \leftarrow S_0$ absorption band. The relationship between high rate constants Γ and Q (see Fig. 17b) and slow HB kinetics has been found [52]. It has been considered a regime of the photon-gated HB which is usually used in practice [59]. The red CW He–Ne laser populates the lowest triplet level with the help of the fast intersystem transitions with the rate Γ_{TLS} (Fig. 17b). The blue pulses of N_2-laser simultaneously populate an upper triplet level T_n with the help of the $T_n \leftarrow T_0$ transitions. A photochemical reaction goes via the populated T_n level but a persistent hole appears in the red optical band.

The relationship between the optical density D in the bottom of hole and the rate constants Γ and Q has a complicated form. However, the theoretical expression for $D(t)$ is simplified if the CW laser is able to depopulate the ground electronic level completely and the powerful blue laser pulses are able to transfer all the chromophores from T_0 to the T_n level. In order to satisfy the first condition, the intensity of the CW laser should violate the inequality Eq. (69). It can be fulfilled at several mW/cm^2. The power of the pulse blue laser should be 100 MW/cm^2 [52]. In this case, the kinetics of the hole bottom is described by

$$D(N) \simeq \frac{\Gamma + Q/2}{\Gamma + Q} \cdot \exp(-N\tau Q/2)$$

$$\cdot \left(\frac{\Gamma + Q/2}{\Gamma + Q} + \frac{Q/2}{\Gamma + Q} \cdot \exp[-(\Gamma + Q)T_p] \right)^{N-1}. \qquad (74)$$

Here, N is the number of blue laser pulses, τ is the pulse duration and T_p is a pause between pulses. If this pause T_p is longer than $(\Gamma + Q)^{-1}$, Eq. (74) is simplified and it takes the form:

$$D(N, \tau) \simeq \left(1 - \frac{\varphi}{2} \right)^N \cdot \exp(-N\tau Q/2) \qquad (75)$$

where $\varphi = Q/(\Gamma + Q)$ is the photochemical quantum yield. Equation (75) can be used to find the unknown constants φ and Q from the slow kinetics of the hole bottom. For this purpose we should measure the function $D(N, \tau)$ for two values of τ.

4.6 Polarization Aspects of HB-Spectroscopy

The light absorption by a chromophore has an anisotropy because the absorption coefficient k^g is proportional to $(\vec{d} \cdot \vec{u})^2 = d^2 \cos^2 \theta$ where $\vec{d}$ is the chromophore dipole and $\vec{u}$ is the vector of light polarization. Polymers and glasses in contrast to crystals are isotropic solids. Various orientations of the embedded impurity centers have equal probabilities. Therefore, the absorption coefficient of polymers and glasses is an average of the molecular absorption coefficient:

$$\langle k \rangle = \frac{1}{8\pi^2} \cdot \int_0^{2\pi} d\alpha \cdot \int_0^{2\pi} d\gamma \cdot \int_{-1}^1 d(\cos\beta) \cdot k^g(\alpha, \beta, \gamma; \omega) \qquad (76)$$

where α, β and γ are the Euler angles which describe the orientation of an impurity center regarding a laboratory coordinate system. Taking the axis oz along the polarization vector $\vec{u}$ we find that $\beta = \theta$ and $\langle k \rangle = k(\omega)/3$, where $k(\omega)$ is the chromophore coefficient $k^g(\theta, \omega)$ at $\theta = 0$. The absorption coefficient of a polymer or glass is an isotropic function.

However, an anisotropy emerges after we burn hole by a polarized light. Indeed, the molecules whose dipole lies along the polarization vector $\vec{u}$ absorbs light more effectively than others. This anisotropy can be measured if we change the polarization of the scanning laser. Taking into account orientational factors in Eq. (61) we find after averaging over orientations the following expression for a hole [64]

$$h(\Delta, \chi) \sim (2\cos^2\chi + 1) \cdot h(\Delta) \tag{77}$$

where $\Delta = \omega_p - \omega_b$ is detuning and χ is the angle between the polarization of the probing and burning lasers.

More interesting information can be found from the polarization measurements in the photoproduct optical band. If the photoproduct optical band lies outside the optical band of chromophores we can find the angle between the dipolar moments of a chromophore and its photoproduct form. Sometimes this situation takes place [65]. We can consider the photoproduct band as an antihole $h_-(\Delta)$. The dipole d_{ph} of a photoproduct can differ from the dipole d of a chromophore. It is possible to find a theoretical expression for the antihole band. Carrying out an averaging over different orientations of chromophores we find the following expression for the antihole [64]:

$$h_-(\Delta, \chi, \varphi) = [(2\cos^2\chi + 1) \cdot \cos^2\varphi + (\sin^2\chi + 1) \cdot \sin^2\varphi] \cdot h_-(\Delta) . \tag{78}$$

Here, φ is the angle between the dipole $\vec{d}_{ph}$ of a photoproduct and the dipole $\vec{d}$ of the chromophores. χ is the angle between the polarization of the probing and burning laser. By carrying out a measurement of the ratio $h_-(\Delta 0\varphi)/h_-(\Delta\frac{\pi}{2}\varphi)$ we find an equation for the angle φ.

In the case of the photon gated hole burning, we can find from the polarization measurements an angle between the dipoles d_1 and d_2 of both absorption optical bands. Since all the chromophores, independent of their orientation in a solvent, are burnt at the high intensity of burning light, the holes burnt under this condition will not exhibit any anisotropy. Therefore, for polarization measurements we should use only holes burnt in the regime of low intensity. This condition is in contrast to that which has been used to derive Eqs. (74) and (75) for the Γ and Q finding.

If the inequalities, Eqs. (67) and (69), for the intensities of both burning lasers which take part in the photon-gated HB are fulfilled the angle ψ between dipoles d_1 and d_2 can be found from the following equation [52]

$$\cos^2\psi = \frac{2h_\parallel - h_\perp}{h_\parallel - 2h_\perp} \tag{79}$$

where $h_\parallel$ and $h_\perp$ are the optical density of the bottom of a hole measured at the parallel and perpendicular orientation of the polarization of the probing and burning *red* laser which depopulates the ground electronic state.

4.7 Elementary Theory for Photon Echoes (PE)

FLN and HB methods are based on the frequency selection of chromophore doped polymers and glasses. They are methods of spectral type. PE, in contrast to FLN and HB, is a kinetic type of method. The PE method is based on the measurement of a temporal evolution of a chromophore dipolar moment. We shall see in Sect. 4.8 that the PE method has a great advantage over FLN and HB. Therefore, PE deserves more detailed consideration than has been made in Sect. 2.2.

Let the functions $\varphi_0(r)$ and $\varphi_1(r)$ describe the ground and excited electron stationary states of a chromophore. They satisfy the Schrödinger equations:

$$H\varphi_0 = \varepsilon_0\varphi_0, \quad H\varphi_1 = \varepsilon_1\varphi_1 \tag{80}$$

where H is the chromophore Hamiltonian; ε_0 and ε_1 are energies of the chromophore. A temporal evolution of the chromophore is determined by the operator $\exp(-iHt/\hbar)$. Therefore, we find

$$\varphi_{0,1}(t) = e^{-i\frac{H}{\hbar}(t-t_0)} \cdot \varphi_{0,1}(t_0) = e^{-i\Omega_{0,1}(t-t_0)} \cdot \varphi_{0,1}(t_0) \tag{81}$$

where $\Omega_{0,1} = \varepsilon_{0,1}/\hbar$.

The interaction between a chromophore and the transverse electromagnetic field is described by the operator $\hbar\hat{\Lambda} = (\hat{\vec{d}} \cdot \vec{E})$ where $\hat{\vec{d}} = e\hat{\vec{r}}$ is the operator of the electronic dipolar moment of a chromophore and $\vec{E}$ is the electric field in a classical electromagnetic wave. The chromophore can absorb and radiate light by changing electronic state. These transitions are due to the operator $\hat{\Lambda}$:

$$\hat{\Lambda}\varphi_0 = \Lambda\varphi_1, \quad \hat{\Lambda}\varphi_1 = \Lambda\varphi_0 \tag{82}$$

where Λ is a matrix element of the operator $\hat{\Lambda}$.

Let us consider a process schematically depicted in Fig. 3. At the initial moment the chromophore is described by φ_0. It is irradiated by two laser pulses with a duration Δt and a pause τ. At time τ after the second laser pulse ends, the chromophores spontaneously radiate a light pulse which is called a photon echo.

This process can be described mathematically. After applying the first laser pulse to the sample, the chromophore gets into nonstationary quantum state which is described by

$$\psi(\Delta t) = c_1\varphi_1 + c_0\varphi_0 . \tag{83}$$

Here, $|c_0|^2$ and $|c_1|^2$ are probabilities of finding the chromophore in the ground and excited electronic state. During the pause τ, there are no electronic transitions. Therefore, the temporal evolution of the chromophore wave function is described by Eq. (81). However, now the initial state is described by $\psi(\Delta t)$:

$$\psi(\tau + \Delta t) = e^{-i\frac{H}{\hbar}\tau} \cdot \psi(\Delta t) = c_0 e^{-i\Omega_0\tau} \cdot \varphi_0 + c_1 e^{-i\Omega_1\tau} \cdot \varphi_1 . \tag{84}$$

After application of the second pulse to the chromophore the latter changes its own state:

$$\psi(\Delta t + \tau + \Delta t) = \hat{\Lambda} \cdot \Delta t \cdot \psi(\tau + \Delta t)$$

$$= \Lambda \cdot \Delta t \cdot (c_0 e^{-i\Omega_0\tau} \cdot \varphi_1 + c_1 e^{-i\Omega_1\tau} \cdot \varphi_0) . \tag{85}$$

During the second pause, the temporal evolution of the chromophore wave function is given by

$$\psi(t) = e^{-i\frac{H}{\hbar}\tau'} \cdot \psi(\Delta t + \tau + \Delta t)$$

$$= \Lambda \cdot \Delta t \cdot (c_0 e^{-i\Omega_0\tau - i\Omega_1\tau'} \varphi_1 + c_1 e^{-i\Omega_1\tau - i\Omega_0\tau'} \varphi_0) \tag{86}$$

where $t = \tau' + \Delta t + \tau + \Delta t$.

The chromophore in the quantum state $\psi(t)$ will radiate light at $\tau' = \tau$. Indeed, classical electrodynamics relates light radiation to the temporal change of classical dipolar moment. The classical value of the chromophore dipolar moment in a quantum state $\psi(t)$ is given by

$$\langle \hat{d} \rangle = \langle \psi(t)| \hat{d} | \psi(t) \rangle = 2d \cdot (\Lambda \cdot \Delta t)^2 \cdot c_0 c_1 \cdot \cos(\omega_0(\tau - \tau')) \tag{87}$$

where $\omega_0 = \Omega_1 - \Omega_0$ is the resonant frequency; $d = \langle \varphi_1 | \hat{d} | \varphi_0 \rangle$ is the dipolar matrix element. The classical dipolar moment $\langle \hat{d} \rangle$ reaches a maximum at $\tau' = \tau$. At this very moment, its radiation will achieve a maximum as well. Hence, the echo signal emerges in the sample at $\tau' = \tau$.

The echo signal is more pronounced in the ensemble of chromophores. Indeed, by carrying out the sum over the molecules from this ensemble we find for the dipolar moment of the ensemble the following expression:

$$D(\tau - \tau') = \int_0^\infty d\omega_0 \cdot N(\omega_0) \langle d(\omega_0) \rangle$$

$$\propto d \cdot \int_0^\infty d\omega_0 \cdot N(\omega_0) \cdot \cos(\omega_0(\tau - \tau'))$$

$$= N(\tau - \tau') \cdot d . \tag{88}$$

Here, the function $N(\omega_0)$ describes a distribution of chromophores over their resonant frequencies. We know from Fourier analysis that the broader function $N(\omega_0)$ becomes the narrower function $N(\tau - \tau')$. The function $N(\omega_0)$ describes an inhomogeneous broadening and it is broad. Therefore, the echo signal described by $N(\tau - \tau')$ will be fairly short. An amplitude $E(2\tau)$ of the echo signal is proportional to $D(\tau - \tau')$ at $\tau' = \tau$: $E(2\tau) \sim D(0)$. This signal has a coherent character because its amplitude is proportional to $d \cdot N(0) = d \cdot N$ where N is the number of chromophores in the ensemble.

Until now, we have neglected the influence of phonons on the electronic transition of a chromophore. This influence is taken into account by means of the substitution.

$$e^{-i\omega_0\tau} \rightarrow \exp\left(-i \cdot \int_0^\tau dx \cdot \omega_0(x) \right). \tag{89}$$

Carrying out this substitution in Eq. (88), we find for the amplitude of the echo signal the following expression:

$$E(2\tau) \sim \int_0^\infty d\omega_0 \cdot N(\omega_0) \cdot |I(\tau, \omega_0)|^2 \tag{90}$$

where

$$I(\tau, \omega_0) = \exp\left[-i \cdot \int_0^\tau dx \cdot \omega_0(x) \right]. \tag{91}$$

If the resonant frequency does not fluctuate, the echo signal does not depend on pause time τ. The fluctuations due to an interaction with phonons decrease the echo signal. Hence, we can get information about the electron-phonon interaction from the dependence of the echo signal on the pause time τ. Comparing Eq. (91) with Eq. (2) enables us to conclude that the homogeneous optical band can be expressed via the echo-decay function $E(2\tau)$. Therefore, PE is one of the selective spectroscopy methods.

4.8 Relation Between Homogeneous Optical Bands and PE Signals

The substitution Eq. (89) leading to Eq. (90) describes the phonon influence on a resonant frequency in a stochastic way. Any stochastic theory is unable to explain two experimental facts: why the homogeneous band consists of ZPL and PSB and why the PSBs of the absorption and fluorescence bands are not in resonance. We have already seen that the dynamic theory successfully explains both experimental facts. Therefore, we should use a dynamic theory for PE.

The dynamic theory for three-pulse PE (see Fig. 29) yields the following expression for the echo signal [66, 67]:

$$E(2\tau) \sim e^{-t_w/T_1} \int_0^\infty d\Delta \cdot N(\Delta) \cdot (I^g(\tau, \Delta) + I^e(\tau, \Delta)) \cdot I^g(\tau, \Delta) \tag{92}$$

where

$$I^{g,e}(\tau, \Delta) = 2 \operatorname{Re} \exp[i\Delta\tau - |\tau|/2T_1 + \varphi^{g,e}(\tau)]. \tag{93}$$

Here, T_1 is the fluorescence life time; $\Delta = \omega_b - \omega_0$ is a frequency detuning and $\varphi^{g,e}(\tau)$ is the cumulant function which describes a phonon influence on the electronic transition. The cumulant function has already been discussed in Sect. 3.2. Although Eq. (92) resembles Eq. (90), there are some differences between these equations. First of all Eq. (92) takes into account that the optical band consists of ZPL and PSB. Besides this fact, Eq. (92) allows for the difference between absorption and fluorescence bands, so that $I^g(\tau, \Delta) \neq I^e(\tau, \Delta)$.

Equations (22) and (23) for the cumulant function φ^g take into account only the linear electron-phonon interaction. This cumulant function cannot explain temperature broadening and shift of ZPL. We have already seen in Sect. 3.4 that the quadratic electron-phonon interaction is responsible for the line broadening and shift. Therefore, the cumulant function which allows for both linear and quadratic interaction is given by

$$\varphi^g(\tau) = -i \cdot \delta(T) \cdot t - \frac{\gamma(T)}{2}|t| + f(\tau, T) - f(0, T),$$

$$(94)$$

$$\varphi^e(\tau) = -i \cdot \delta(T) \cdot t - \frac{\gamma(T)}{2}|t| + \overset{*}{f}(\tau, T) - f(0, T).$$

Here, $\delta(T)$ and $\gamma(T)$ are the ZPL shift and halfwidth due to the quadratic FC interaction The function $f(\tau, T)$ is described by Eq. (23). It yields PSB, due to the linear FC interaction. The effect of the quadratic FC interaction on the function $f(\tau, T)$ is small and we neglect it. This approximation is discussed in Sect. 3.4. Equation (23) can be rewritten as follows:

$$f(\tau, T) = \text{Re} f(\tau, T) + i \, \text{Im} f(\tau)$$

where

$$\text{Re} f = \int_0^\infty dv \cdot f(v) \cdot \text{cth} \frac{h v}{2kT} \cdot \cos(vt),$$

$$(95)$$

$$\text{Im} f = \int_0^\infty dv \cdot f(v) \cdot \sin(vt).$$

The function $f(v)$ describes the weighted density of phonon states and we have already discussed the way this function can be found from the homogeneous PSB (Sect. 3.3). However the function $f(v)$ can be found from PE experiments as well.

In order to prove this statement, we should express PE signal via $\text{Re} f$ and $\text{Im} f$. For this purpose, we rewrite Eq. (93) with the help of Eq. (94) in the following form:

$$I^{g,e}(\tau, \Delta) = 2e^{-|\tau|/T_2 + \text{Re} f(\tau,T)} \cdot e^{-f(0, T)} \cdot \cos[(\Delta - \delta)\tau \pm \text{Im} f(\tau)] \qquad (96)$$

where

$$\frac{1}{T_2} = \frac{1}{2T_1} + \frac{\gamma(T)}{2}. \qquad (97)$$

T_2 is usually called the optical dephasing time [15, 18]. Substituting $I^{g,e}$ in Eq. (92) by Eq. (96) and taking into account that

$$\int_{-\infty}^{\infty} d\Delta \cdot N(\Delta) \cdot \cos(\Delta - \delta)\tau \propto \int_{-\infty}^{\infty} d\Delta \cdot N(\Delta) \cdot \sin(\Delta - \delta)\tau \propto 0 \qquad (98)$$

because of the large halfwidth of the function $N(\Delta)$, we find the following final expression for the echo signal:

$$E(2\tau) \propto e^{-t_w/T_1 - 2\tau/T_2 + 2\mathrm{Re}\,f(\tau,\,T)} \cdot e^{-2f(0,\,T)} \cdot [1 + \cos(2\mathrm{Im}\,f(\tau))] \,. \qquad (99)$$

It is obvious that the measurement of the echo-decay function $E(2\tau)$ enables one to find T_2, $\mathrm{Re}\,f(\tau, T)$ and $\mathrm{Im}\,f(\tau)$ which are characteristics of a homogeneous optical band.

The pure exponential term $\exp(-2\tau/T_2)$ in Eq. (99) describes a contribution to the echo signal from ZPL. The function $f(\tau, T)$ describes a contribution from PSB. Since the one-phonon function $f(v)$ has a width of the order of v_D the width of the function $f(\tau, T)$ is of the order of v_D^{-1}. The frequency v_D characterizes phonons which manifest themselves in PSB. For instance, $v_D \simeq 4\ \mathrm{cm}^{-1}$ and $v_D \simeq 50\ \mathrm{cm}^{-1}$ for PSBs shown in Fig. 7 and Fig. 6, respectively. As v_D exceeds the halfwidth $\gamma(T)$ of ZPL, we have $v_D^{-1} \ll T_2$. This inequality means that the phonon function $f(\tau, T)$ determines the temporal behaviour of the echo signal at $\tau \ll T_2$. However, the function $f(\tau, T)$ at $\tau > T_2$ equals zero and the echo signal falls off exponentially. This behaviour of the echo signal is schematically shown in Fig. 18.

Figure 18 shows the decay of the phase of a dipole. Indeed, Eq. (87) shows that the echo signal depends on the phase $\omega_0(\tau - \tau')$ of the dipolar moment. The maximum of the signal is achieved at $\tau' = \tau$ and its value does not depend on the pause time τ if we neglect the influence of phonons. By carrying out the substitution Eq. (89), we take into account the effect of phonons on the phase of dipoles. An interaction with phonons leads to "jumps" of the phase i.e. to the loss of a phase memory. This process is called *optical dephasing*. Optical dephasing leads to the dependence of the echo amplitude on the pause time τ.

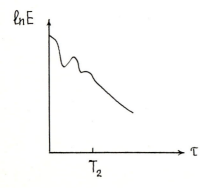

Fig. 18. Nonexponential (ultrafast) and exponential (fast) echo decay revealing a complicated character of optical dephasing in a chromophore ensemble

This decreasing of the echo signal is depicted in Fig. 18. Optical dephasing has an ultrafast (nonexponential) part and a fast (exponential) part. At $\tau \gg T_2$, the dipoles lose the memory phase completely and the echo signal disappears. Therefore, T_2 is called dephasing time.

The PE method has an important advantage compared to the FLN and HB methods because it enables one to get information about a homogeneous optical band in the case when this optical band consists only of PSB. In this case, the frequency selection underlying the FLN and HB methods is impossible. ZPL disappears when we raise the temperature (see Fig. 6). We are dealing with optical bands of such type when investigating proteins at $T > 77$ K. The FLN and HB methods do not work in this case. However, a theoretical treatment of the ultrafast echo decay with the help of Eq. (99) enables us to find the one-phonon function $f(\tau, T)$ even when ZPL is absent.

5 Tunneling Systems of Polymers and Glasses Probed by Selective Spectroscopy

Phonons are not the only type of low-frequency excitations existing in polymers and glasses. It has been known for the last two decades that so-called tunneling excitations exist in polymers and glasses in addition to the conventional vibrational excitations which we call phonons. Today this fact is beyond doubt. The tunneling excitations have been discovered in glasses and they have been extensively studied there by means of various physical methods. The tunneling excitations of polymers were discovered only ten years ago but they have been little investigated. In the following sections, new information concerning tunneling excitations of polymers and glasses is discussed.

5.1 Tunneling Two-Level Systems (TLS) of Polymers and Glasses

Polymers and glasses are metastable systems because spontaneous changes of their atomic structure are possible. This fact can be attributed to the change of the atomic equilibrium positions separated by a potential barrier (Fig. 19). A coordinate X can be both of the angular and translational type. It can be directed along any chemical bound or can describe a group of atoms, i.e. it can be of a delocalized type.

The rate of transition between two equilibrium positions can be described by the following simplified expression:

$$R = R_0 \cdot e^{-\lambda} + R_1 \cdot e^{-\frac{V}{kT}} . \tag{100}$$

The first term describes the rate of tunneling transition through a potential barrier. This term does not depend on temperature. A constant λ depends on the

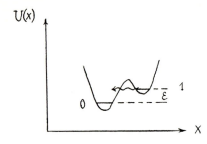

Fig. 19. Dependence of the adiabatic potential on tunneling coordinate

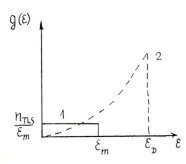

Fig. 20. Densities of states for phonons (*dashed line*) and for excitations of TLS (*solid line*)

mass m of a tunneling group and on the height V and width d of a potential barrier:

$$\lambda = \frac{\sqrt{2mV}}{\hbar} \cdot d \ . \tag{101}$$

The second term in Eq. (100) describes the rate of transition above the barrier. These transitions are accompanied by the absorption of phonons with a frequency $\omega = V/\hbar$. Therefore, these transitions of Arrenius type depend strongly on temperature and they dominate Eq. (100) at $kT \geqslant V$.

It was established twenty years ago [68] that the physical properties of glasses reveal a number of anomalies at low temperatures. Anderson, Halperin and Varma [69] and Phillips [70] supposed that all these anomalies are due to the tunneling degrees of freedom of a glass. Since all the levels, except the two lowest ones depicted in Fig. 19, are depopulated at low temperature, authors of [69, 70] described these tunneling systems by taking into account only the two lowest levels. These two level systems (TLS) are responsible for all the low temperature anomalies in glasses. Excitations in TLS are of the Fermi-type in contrast to conventional phonons which are excitations of the Bose-type. It means that the average number of excitation in TLS is described by the Fermi-function $f(\varepsilon) = [\exp(\varepsilon/kT) + 1]^{-1}$ where ε is the energy of TLS (Fig. 19). The density of states of TLS exceeds the density of states of phonons at small ε (Fig. 20). This fact and the Fermi-type of the excitations of TLS lead to the appearance of the low temperature anomalies in glasses and polymers.

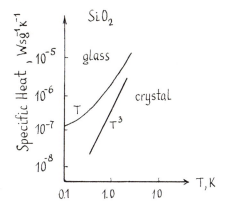

Fig. 21. Low-temperature dependence of specific heat [71]

Let us consider, for example, the temperature behaviour of heat capacity. If we assume that there are two types of low frequency excitation in a glass, the heat capacity is the sum of two terms:

$$C = \frac{d\bar{E}_{ph}}{dT} + \frac{d\bar{E}_{TLS}}{dT} \qquad (102)$$

where

$$\bar{E}_{ph} = \int_0^{\varepsilon_D} d\varepsilon \cdot \varepsilon \cdot n(\varepsilon) \cdot g_{ph}(\varepsilon), \quad \bar{E}_{TLS} = \int_0^{\varepsilon_D} d\varepsilon \cdot \varepsilon \cdot f(\varepsilon) \cdot g_{TLS}(\varepsilon) \qquad (103)$$

are the average energies of phonons and excitations of TLS at given temperature. Here, ε_D and ε_m are the upper edges of phonon and TLS spectra; $n(\varepsilon) = [\exp(\varepsilon/kT) - 1]^{-1}$ is the average number of phonons. The phonon density of states is given by $g_{ph}(\varepsilon) = 9N\varepsilon^2/\varepsilon_D^3$, where N is the number of atoms in a sample. The function $g_{TLS}(\varepsilon)$ is shown in Fig. 20. By using a dimensionless variable ε/kT we find that $\bar{E}_{ph} = (kT)^4 \cdot I_{ph}(T)$ and $\bar{E}_{TLS} = (kT)^2 \cdot I_{TLS}(T)$. At low temperature when $\varepsilon_D/kT \gg 1$ and $\varepsilon_m/kT \gg 1$ we find that integrals do not depend on temperature: $I_{ph}(T) \simeq I_{ph}(0)$ and $I_{TLS}(T) \simeq I_{TLS}(0)$. Therefore, Eq. (102) takes the following form:

$$C = 36Nk \cdot I_{ph}(0) \cdot \left(\frac{kT}{\varepsilon_D}\right)^3 + 2n_{TLS}k \cdot I_{TLS}(0) \cdot \left(\frac{kT}{\varepsilon_m}\right). \qquad (104)$$

The first term describes the contribution to the heat capacity by phonons. The second term describes the contribution by TLS. In crystals, there are no TLS. Therefore, $C \sim T^3$ as shown in Fig. 21. However, in glasses there are both phonons and excitations of TLS. Therefore, C is described by Eq. (104) and $C \sim T$ at low temperature (Fig. 21).

TLS is the reason for anomalies in other physical properties of glasses at low temperature: thermal conductivity [72], ultrasonic attenuation [73], Raman scattering [74], optical dephasing [2, 6, 8, 75], etc. Unfortunately, the low temperature properties of polymers have been investigated much less than those of glasses where TLS have been discovered. It would be very interesting to show the same low temperature anomalies in polymers. This task seems realistic because low temperature spectroscopic properties of glasses and polymers are similar to each other and they reveal the existence of TLS both in glasses and polymers. This statement is proved in the next Sections.

5.2 Cycling Experiments. Distribution Over Potential Barriers

All the TLS existing in a polymer with impurity centres can be divided into two groups: extrinsic TLS and intrinsic TLS. The *intrinsic* TLS are spread throughout the whole bulk of a solvent. The distance between intrinsic TLS and an impurity chromophore, as a rule, is large. Chromophores do not affect these TLS and they determine bulk properties of a polymer at low temperature. These very TLS are responsible for the low temperature anomalies in heat capacity.

The *extrinsic* TLS are created in polymers by an embedded chromophore. The extrinsic TLS is closely related to the adiabatic potential of a chromophore.These TLS are characteristic of a couple: chromophore + polymer. It is obvious that the number of the extrinsic TLS is proportional to the chromophore concentration. These very TLS are responsible for the burning of persistent holes in a chromophore optical band (Fig. 22).

Due to potential barriers the chromophore exists in an educt- and a photoproduct-form (Fig. 22). An electronically unexcited chromophore exists mainly in the educt-form. After absorption of light the chromophore easily overcomes the low barrier in the excited state and, therefore, the chromophore gets into product-form. It immediately relaxes to the ground state of this product-form. The chromophore lives in this ground state for a long time because the potential barrier in the ground state is high. The lack of chromophores in the educt-form manifests itself as a persistent hole in the chromophore absorption band. It is obvious that this hole can be erased by heating the sample.

The experimental data prove that there is a distribution over the barrier heights in the ground electronic state. This distribution can be measured in

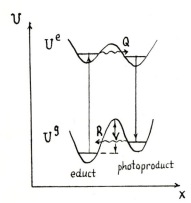

Fig. 22. Adiabatic potentials of a chromophore in the ground and excited electron states resulting in a hole-burning process

so-called cycling experiments [76]. The experimental data of the Friedrich group [76] which confirm the existence of this distribution is shown in Fig. 23. Hole 1 has been burnt at T_b. After increasing the temperature up to T_c and returning it back to T_b the hole area decreases. Curve 2 in Fig. 23 reveals this fact. However, after repetition of the temperature cycle $T_b \rightarrow T_c \rightarrow T_b$ the hole area does not change its value. Curve 3 coincides in fact with curve 2.

These results can be explained as follows: when we raise the temperature from T_b to T_c in the first cycle, chromophores with low barriers existing in the ground state of the product-form cross the barriers and come back to the educt-form. Therefore, the hole area decreases after coming back to the initial temperature T_b. After the second temperature cycle, there are no additional transitions through the barrier from the product- to the educt-form. Therefore, the number of chromophores in the product-form is not changed and curve 3 coincides with curve 2 in Fig. 23.

Cycling experiments enable one to find a distribution of the extrinsic TLS over the barrier heights V. Since cycling experiments use a temperature variation, we can neglect the first term in Eq. (100) and take

$$R(T) = R_1 \cdot e^{-\frac{V}{kT}} . \qquad (105)$$

$P(R)dR$ is the number of chromophores in the product-form with a rate constant $R(T)$. The number ΔA of chromophores which make a transition from product to educt during a time τ of the experiment is given by

$$\Delta A = \int_{R_{min}}^{R_{max}} dR \cdot P(R) \cdot (1 - e^{-R\tau}) \simeq \int_{1/\tau}^{R_{max}} dR \cdot P(R) . \qquad (106)$$

Here, we simplify the situation and assumed that all the chromophores in a photoproduct form with $R \geqslant 1/\tau$ change their form. By introducing a distribu-

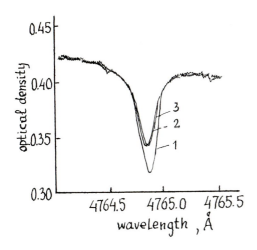

Fig. 23. Reduction of the hole area in the course of the temperature cycling in tetra-cene doped alcohol glass (*curves 1 and 2*). If the same cycle is repeated twice, no further reduction occurs (*curve 3*) [76]

tion function $P_1(V)$ over the potential barriers we can write:

$$P(R)dR = - P_1(V)dV .$$ (107)

We can find with the help of Eq.(105) that R_{max} and $1/\tau$ are related to $V_{min} = 0$ and $V_{max} = \alpha T$, respectively. Here $\alpha = k \cdot \ln R_1 \tau$. By carrying out a substitution of the variable R in Eq. (106), we find

$$\Delta A = \int_0^{\alpha T} dV \cdot P_1(V) .$$ (108)

Equation (108) describes the recovery of a hole area as a function of cycling temperature. It is obvious that the derivative of ΔA with respect to cycling temperature T describes a distribution $P_1(V)$ over barrier heights:

$$\frac{d}{dT}\Delta A = \alpha \cdot P_1(V) .$$ (109)

Equation (109) has been used by the Friedrich group [77] to find $P_1(V)$ from the temperature behaviour of ΔA. Experimental data for $A(T)/A(0) = 1 - \Delta A(T)$ are depicted in Fig. 24. The function

$$\Delta A(T) = \sqrt{\frac{\alpha T}{V_1}}$$ (110)

where V_1 is a constant fits these experimental data well – Fig. 24. Substituting $\Delta A(T)$ in Eq. (107) by Eq. (110), we find a distribution function over the barrier heights:

$$P_1(V) = \frac{1}{2\sqrt{VV_1}} .$$ (111)

Equations (109) and (111) can also be used to find a distribution function $P(R)$ over tunneling rates R. Tunneling processes dominate at low temperature and,

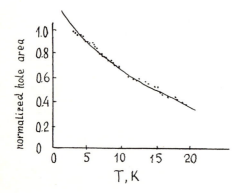

Fig. 24. Hole area in tetracene doped alcohol glass as a function of cycling temperature [76]. The fit curve has been calculated with the help of $P_1 \propto V^{-1/2}$

therefore, we can neglect the second term in Eq. (100). The first term we can rewrite as follows:

$$R = R_0 \cdot \exp(-\sqrt{V/V_2}) .\qquad(112)$$

Substituting the function $P_1(V)$ in Eq. (107) by Eq. (111) and obtaining $dV/dR = -2\sqrt{VV_2}/R$ with the help of Eq. (112), we can easily find from Eq. (107) the following distribution over tunneling rates R:

$$P(R) = \sqrt{\frac{V_2}{V_1}} \cdot \frac{1}{R} .\qquad(113)$$

The distribution functions $P_1(V)$ and $P(R)$ described by Eqs. (111) and (113) have been found from experimental data which characterize the extrinsic TLS. However, they seem to be correct for the intrinsic TLS which characterize a solvent. For instance, authors of [69] used Eq. (113) to explain anomalies in a thermal conductivity of inorganic glasses without impurities. Anomalies in such systems can be due solely to intrinsic TLS.

5.3 Spectral Diffusion

The temperature cycling experiments not only reveal the recovery of the hole area. They also reveal the broadening of zero-phonon holes. After a temperature cycle $T_b \rightarrow T_c \rightarrow T_b$, the hole width exceeds that the initial value. This effect is shown in Fig. 25. It can be supposed that the additional so-called irreversible hole width is due to changes in the glass structure which emerge during a temperature cycle. These changes in atomic configuration of a glass affect the resonant frequency of a chromophore. Every chromophore has its own frequency shift of ZPL. It means that a homogeneous ensemble of chromophores acquires an additional inhomogeneous halfwidth as shown schematically in

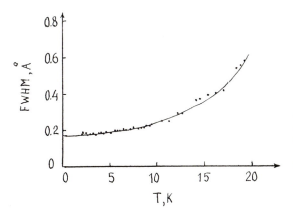

Fig. 25. Hole width measured in the tetracene + alcohol glass system at the burning temperature T_b as a function of cycling temperature T [76]

Fig. 26. The emergence in the homogeneous ensemble of chromophores of an inhomogeneous broadening is called a *spectral diffusion* (SD).

SD is due to the interaction of a chromophore with an ensemble of the intrinsic TLS. Indeed, a change of the molecular structure in a polymer or glass can be considered as a transition in intrinsic TLS. SD can emerge spontaneously and can be triggered by the heating of a sample (thermally-induced SD) and by light irradiation (light-induced SD). For instance, Fig. 25 shows a thermally-induced SD.

SD has been discovered in the magnetic resonance of inorganic systems [78, 79]. For instance, Mims et al. [79] found that the decay of the spin-echo-signal over microsecond time scale is described by the expression

$$E(2\tau) = \exp(-m\tau^2) \, . \tag{114}$$

This means that the dephasing rate constant $1/T_2$ depends on the duration τ of a pause between exciting pulses: $1/T_2 = m\tau$. Klauder and Anderson [19] have offered a simple explanation of this fact. They assumed that a homogeneous ensemble of atomic spins spontaneously acquires features of an inhomogeneous ensemble because the function $N(\Delta)$, which describes a distribution over resonant frequencies in a hole burnt in the spin population, depends on time:

$$N(\Delta) = \frac{m\tau}{\Delta^2 + (m\tau)^2} \, . \tag{115}$$

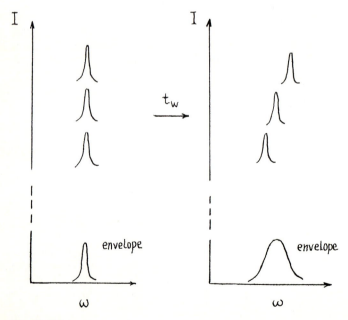

Fig. 26. Inhomogeneous character of the spectral diffusion broadening. t_w is waiting time

Here Δ is the detuning between frequencies of spin and the electromagnetic field. At $\tau = 0$ the function $N(\Delta)$ equals δ-function. It describes a homogeneous ensemble of excited spins.

Effects of spin- and photon-echoes are similar with each others. Therefore, we can derive Eq. (114) from Eq. (92) for photon-echo and Eqs. (93) and (94) for the functions $I^{g,e}$ and $\varphi^{g,e}$. Since a magnetic line consists only of ZPL we should neglect PSB by taking $f(\tau, T) = f(0, T) = 0$ and $I^g(\tau, \Delta) = I^e(\tau, \Delta)$. Then, substituting $N(\Delta)$ in Eq. (92) by Eq. (115) we find after integration over Δ:

$$E(2\tau) \sim \exp[-(m\tau + 2/T_2)\tau] \tag{116}$$

where $1/T_2 = 1/2T_1 + \gamma(T)/2$. SD increases dephasing rate with time.

SD manifests itself in the temporal broadening of holes as well. The problem of great interest is finding the time scale of SD processes. Indeed, all the experiments can be attributed to fast or slow ones as compared to the SD time scale. Certainly, a photochemical HB experiment taking no less than several hundred seconds is an example of the slow experiments. On the contrary, the measurements in a two-pulse PE experiment take nanoseconds or even picoseconds. This is an example of the fast experiments. The slower the experiment, the more SD is pronounced.

SD in optical spectra has been discovered when a comparison of experimental data of slow and fast experiments has been made. Figure 27 shows experimental data for the dephasing rate $1/T_2$ found in slow (HB) and fast (PE) experiments. There is a difference between experimental data in the low temperature range. The slower the experiment the larger the dephasing rate $1/T_2$ found from this experiment. This effect has been attributed to SD.

When comparing experimental data from slow and fast experiments, we should take into account that some effects can distort experimental data. Völker et al. [80, 81] have shown that holes recorded in the luminescence spectra, as a rule, are narrower than holes recorded in the absorption spectra. This effect is attributed to the thermo- or light-induced SD. The sample heating existing in PE experiments can lead to an artificial increase in the dephasing rate. This rate

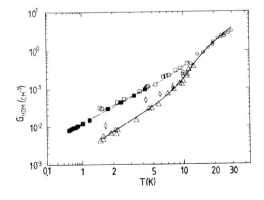

Fig. 27. Temperature dependence of the "homogeneous" line width of resorufin in d-ethanol glass measured in "fast" and "slow" experiments. □ – hole burning [38], ■ – hole burning [80], ○ – incoherent nano-second photon echo [39], △ – two-pulse picosecond photon echo [38]

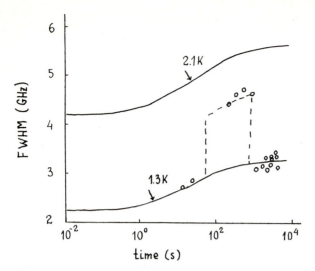

Fig. 28. Plot of the hole width vs waiting time for cresyl violet in ethanol glass at two temperatures. *Circles* show results of the temperature cycling experiment [83]

can be even larger than that found in slow experiments (HB). Wiersma et al. [83] have shown that the difference between PE and HB experiments can change its sign and correspond to the picture presented in Fig. 27 if the effect of a heating by laser pulses is removed.

Since the temperature cycling experiments take a long time the irreversible hole width measured in these experiments can be due both to thermal-induced SD and to spontaneous SD. The latter depends on time. Experiments by the Fayer group [83] prove this statement. Their experimental data plotted in Fig. 28 show the SD effect which manifests itself in the time-related increase in the width of holes burnt at various temperatures. However, if the hole burnt at $T_b = 1.3$ K is subjected to a temperature cycle $T_b \rightarrow T_c \rightarrow T_b$ the hole acquires an additional halfwidth. This excess of halfwidth is due mainly to spontaneous SD which depends only on the duration of the experiment. The thermally-induced SD does not contribute to the result of this temperature cycling experiment. In the light of these results, we can assume that some of the irreversible halfwidth plotted in Fig. 25 is due to a spontaneous SD and does not depend on the value of a temperature excursion.

The duration of PE experiments is ten or twelve orders of magnitude shorter than HB experiments. It is obvious that the time scale of SD lies inside this widest time interval. Therefore, the comparison of PE and HB experimental data is not the best way to investigate the distribution over rates of the SD processes. The best way to find this distribution is to use the so-called three-pulse photon echo (3PE). This echo signal emerges in a sample after three laser pulses are applied (Fig. 29). 3PE can be considered the way the two-pulse (2PE) echo was considered in Sect. 4.7. A 3PE signal depends both on pauses τ and

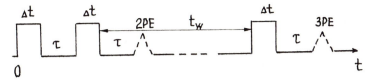

Fig. 29. Train of laser pulses yielding 3PE signal

t_w (see Fig. 29). If we can neglect SD, the 3PE signal is described by [18]:

$$E_{3PE}(2\tau, t_w) \sim e^{-\Gamma t_w} \cdot e^{-2\tau/T_2} \tag{117}$$

where Γ is a rate of the excited state depopulation. If the chromophore has a triplet level the rate Γ is determined by a bottleneck in the scheme depicted in Fig. 16a, i.e. $\Gamma = \gamma_{ST}$. By varying chromophores we can vary γ_{ST}^{-1} from nanoseconds to seconds.

Bai and Fayer [84] have shown that the allowance for SD changes Eq. (117). They used the stochastic approach developed in [19, 85, 86] and derived the following expression for the 3PE signal:

$$E_{3PE} \sim \exp[-2\tau/T_2(t_w)] , \tag{118}$$

where

$$\frac{1}{T_2(t_w)} = \int_{R_{min}}^{R_{max}} dR \cdot P(R) \cdot (1 - \exp(-Rt_w)) . \tag{119}$$

Here, $P(R)$ is a rate distribution function for intrinsic TLS.

Meijers and Wiersma [87] tested the Bai–Fayer theory. They measured the dependence of the rate $1/T_2$ on waiting time t_w (Fig. 30) and used this dependence to find, with the help of Eq. (119), the distribution function $P(R)$. This function is given by $P(R) = \begin{cases} a/R, & \log R > 6 \quad \text{and} \quad \log R < 2.5 \\ 0 & 2.5 < \log R < 6 \end{cases}$. It differs considerably from the distribution $1/R$ found in Sect. 5.2 for the extrinsic TLS. Using the function $P(R)$ we can calculate $E_{3PE}(t_w)$ using Eq. (118) and also measure the function $E_{3PE}(t_w)$ by an experiment. The calculated and measured function $E_{3PE}(t_w)$ is shown in Fig. 31. It reveals a discrepancy between theory and experiment. Experimental data by Meijers and Wiersma for $E_{3PE}(t_w)$ could be described by the function

$$E_{3PE}(t_w) = \exp[-a(t_w)] \cdot \exp[-2\tau/T(t_w)] \tag{120}$$

which, however, differs from Eq. (118). This means that the Bai-Fayer theory should be modified.

However, it has also been reported [83, 88] that SD does not manifest itself in some amorphous systems. For instance, there is no temporal dependence of hole width for the system resorufin in glycerol glass at 2.13 K although burning time was varied from 10^{-2} to 10^4 seconds [83]. There is no visible SD in the

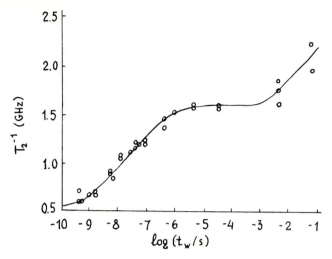

Fig. 30. Plot of the dephasing rate measured in 3PE experiment as a function of waiting time t_w. Zn-porphin doped ethanol glass at T = 1.5 K [87]

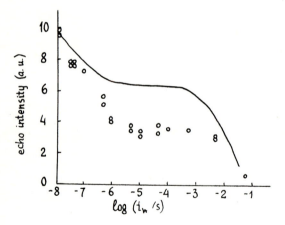

Fig. 31. Intensity of 3PE as a function of waiting time for a fixed time decay τ measured (*circles*) and calculated with the help of Eq. (118) and experimental data for $1/T_2$ depicted in Fig. 30 (*solid line*) [87]

millisecond time scale for the system porphin in polyethylene [88]. Perhaps SD in these systems occurs over other time scales.

5.4 Effect of TLS on Homogeneous Halfwidth of ZPL

Experimental data for HB and PE presented in Fig. 27 lead to two conclusions. *First*, a contribution from SD hole width has an appreciable value only at low temperature. When temperature rises, the temperature broadening of homogeneous ZPL exceeds the SD effect and all the data coincide at T > 15 K.

Second, a contribution from SD to hole width does not change the quasilinear temperature behaviour of the hole width over the low-temperature range (Fig. 27). The temperature behaviour of such a type can be described by $\gamma(t, T) = a(t) \cdot T$ or by $\gamma(t, T) = b(t) + c \cdot T$. The first variant means that low-temperature hole broadening results solely from SD and this inhomogeneous broadening depends on temperature. A mechanism of the temperature broadening does not depend on the time scale of the experiment. Varying this time scale only affects scaling along the rate axes. The second variant for $\gamma(t, T)$ means that there are two mechanisms of hole broadening. The first mechanism resulting from SD yields an inhomogeneous halfwidth $b(t)$ which does not depend on temperature. The second mechanism yields a homogeneous halfwidth cT. This mechanism results from the chromophore – TLS interaction and it resembles that existing in electron-phonon systems. Let us consider this mechanism of homogeneous temperature broadening.

We have already seen in Sect. 3.1 that a change of the adiabatic potential of the electronic excitation of a chromophore is the electron-phonon interaction. This very interaction transforms a δ-like optical line into an optical band. In polymers and glasses, the number of the potential minima exceeds the number of atoms. Therefore, a change of the adiabatic potential of such a type upon the electronic excitation of a chromophore can be written as follows:

$$V = V_{ph} + V_{TLS} . \tag{121}$$

Here, V_{ph} is the operator of the electron-phonon interaction which is described by Eq. (12). The operator V_{TLS} describes changes in TLS upon the electronic excitation of a chromophore.

If a system has only two quantum states, its Hamiltonian can be expressed via the Pauli matrices σ_z and σ_x:

$$H^{g,e} = \tfrac{1}{2}\varepsilon^{g,e} \cdot \sigma_z + \tfrac{1}{2}(\vec{\lambda}^{g,e} \cdot \vec{R}) \cdot \sigma_x . \tag{122}$$

Here, $\varepsilon^{g,e}$ is the energy of TLS in the ground (g) and excited (e) electron states of a chromophore. The second term in Eq. (122) describes an operator yielding tunneling transitions in TLS. These transitions result from the phonon modulation of a barrier. An excitation of TLS with an energy ε we shall call *a tunnelon*. It is obvious that tunnelons in contrast to phonons are the excitations of Fermi-type, i.e. the operators C^+ and C of the creation and annihilation of a tunnelon obey the Fermi-relation: $CC^+ + C^+C = 1$. These operators relate to the Pauli matrices as follows:

$$C = \frac{\sigma_x - i\sigma_y}{2}, \quad C^+ = \frac{\sigma_x + i\sigma_y}{2}, \quad C^+C - CC^+ = \sigma_z . \tag{123}$$

Taking into account Eq. (123), we can write the operator $V_{TLS} = H^e - H^g$ of the electron-tunnelon interaction as follows:

$$V_{TLS} = \Delta C^+C + (\vec{\lambda} \cdot \vec{R}) \cdot (C^+ + C) - \Delta/2 \tag{124}$$

where $\Delta = \varepsilon^e - \varepsilon^g$ and $\vec{\lambda} = (\vec{\lambda}^e - \vec{\lambda}^g)/2$. The operator V_{TLS} describes the interaction with one tunnelon (TLS). Substituting Δ, $\vec{\lambda}$, C^+ and C by Δ_j, $\vec{\lambda}_j$, C_j^+ and C_j where j is the tunnelon index, and carrying out summation over j, we obtain the electron-tunnelon interaction with many tunnelons. Only the immediate environment of a chromophore is affected by the change of its electronic state. Therefore, only one or couple parameters Δ_j and λ_j are not equal to zero. This means that the electron-tunnelon interaction has a localized character.

We have already seen in Sect. 3.4 that the homogeneous halfwidth of ZPL is due to two-phonon Raman-like processes: one phonon is created and the other phonon is annihilated. It is obvious that the quadratic interaction $\Delta C^+ C$ yields two-tunnelon processes of such a type in the electron-tunnelon system and the probability of a simultaneous creation and annihilation of a tunnelon is proportional to $(1-f)f = (2ch(\varepsilon/2kT))^{-2}$. Therefore, the effect of the quadratic electron-tunnelon interaction on the homogeneous halfwidth of ZPL is given by the following formula [89, 90]

$$\gamma_{TLS}(T) = -\int_0^\infty \frac{d\nu}{2\pi} \cdot \ln[1 - \Delta^2 \cdot \Gamma_{TLS}^g(\nu) \cdot \Gamma_{TLS}^e(\nu) \cdot ch^{-2}(h\nu/2kT)] \quad (125)$$

where Γ_{TLS}^g and Γ_{TLS}^e are tunnelon spectral functions in the ground and excited electron state, respectively (Fig. 32). Equation (125) resembles Eq. (41) which describes a line halfwidth resulting from the quadratic interaction with phonons. The substitution of sh^{-2} by ch^{-2} is due to the fact that Eqs. (41) and (125) take

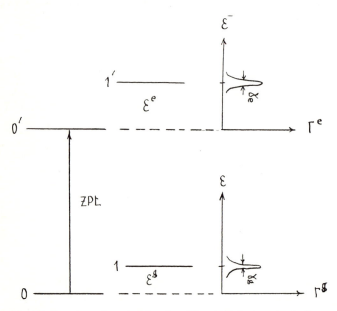

Fig. 32. Energies and spectral functions Γ^g and Γ^e of a tunnelon in the ground and excited electron states

into account interaction with bosons (phonons) and fermions (tunnelons), respectively. It is possible to prove that $z = \Delta^2 \Gamma_{TLS}^g \Gamma_{TLS}^e \mathrm{ch}^{-2}(h\nu/2kT)$ is always less than unity. Therefore, $-\ln(1 - z)$ in Eq. (125) can be substituted by z. Taking the tunnelon spectral functions in a Lorentzian form, $\Gamma_{TLS}^{g,e}(\nu) = \gamma_{g,e}/[(\nu - \varepsilon^{g,e})^2 + \gamma_{g,e}^2]$, we can carry out the integration in Eq. (125). For the weak and strong electron-tunnelon coupling we obtain:

$$\gamma_{TLS}(T) = \left(\frac{\Delta}{\gamma}\right)^2 \cdot \frac{\gamma}{2} \cdot \mathrm{ch}^{-2}\left(\frac{\varepsilon}{2kT}\right) \quad \text{for } \Delta/\gamma \ll 1 \tag{126}$$

$$\gamma_{TLS}(T) = \frac{\gamma_g}{4} \cdot \mathrm{ch}^{-2}\left(\frac{\varepsilon^g}{2kT}\right) + \frac{\gamma_e}{4} \cdot \mathrm{ch}^{-2}\left(\frac{\varepsilon^e}{2kT}\right) \quad \text{for } \frac{\Delta}{\gamma_g} \sim \frac{\Delta}{\gamma_e} \gg 1 . \tag{127}$$

Equation (126) for a weak electron-tunnelon coupling takes into account that $\varepsilon^e \simeq \varepsilon^g = \varepsilon$ and $\gamma_g \simeq \gamma_e = \gamma$. It is easy to find from Eq. (127) that long living tunnelons do not cause line broadening. Indeed, $\gamma_{TLS}(T)$ disappears if γ_g and γ_e are fairly small. The telling feature of a two-tunnelon mechanism of line broadening is a saturation of the broadening over a high temperature range because ch^{-2} tends to unity if 2kT exceeds ε^g and ε^e.

The result of a change in tunneling upon the electron excitation of a chromophore is given by the second term in Eq. (124). This bilinear interaction allows for the processes of a simultaneous creation (annihilation) of a tunnelon and annihilation (creation) of a phonon. The probability of these processes is proportional to $(1 + n)f = n(1 - f) = [2 \mathrm{sh}(\hbar\nu/kT)]^{-1}$. Therefore, this bilinear interaction yields the following expression for the homogeneous halfwidth of ZPL [46, 91, 92]:

$$\gamma_{TLS}(T) = 2\lambda^2 \cdot \int_0^\infty \frac{d\nu}{2\pi} \cdot \Gamma(\nu) \cdot \Gamma_{TLS}(\nu) \cdot \mathrm{sh}^{-1}(\hbar\nu/kT) . \tag{128}$$

Here, $\Gamma(\nu)$ and $\Gamma_{TLS}(\nu)$ are the spectral functions of phonons and tunnelons which are affected by the change in the electron state of a chromophore.

If we consider the interaction with a single tunnelon, we should substitute Γ_{TLS} in Eq. (128) by the Lorentzian with a tunnelon energy ε. In this case, Eq. (128) takes a simple form:

$$\gamma_{TLS} = 2\lambda^2 \cdot \Gamma(\varepsilon) \cdot \mathrm{sh}^{-1}(\varepsilon/kT) . \tag{129}$$

It is obvious that γ_{TLS} is a linear temperature function at $kT \gg \varepsilon$ and it has the behaviour of the Arrhenius type at $kT \ll \varepsilon$.

The temperature dependence of a homogeneous linewidth resulting from the two-phonon, two-tunnelon and tunnelon-phonon processes are schematically depicted in Fig. 33. It should emphasize that only two-tunnelon processes are able to cause the saturation effects in the temperature dependence of homogeneous halfwidths. We shall use this fact when interpretating experimental data.

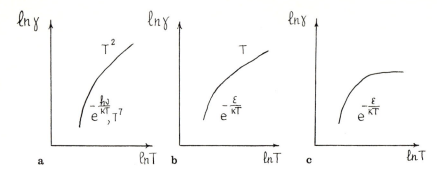

Fig. 33a–c. Temperature dependence of a homogeneous halfwidth $\gamma = 1/T_2$ due to **a** two-phonon, **b** tunnelon-phonon and **c** two-tunnelon Raman-like dephasing processes. Only the two-tunnelon process leads to the plateau in the high temperature range

By now, there are a number of facts showing that polymers can have the so-called fractal structure [93, 94]. In this case, we have to substitute the one-phonon function $\Gamma(\nu)$ in Eq. (128) by the fracton density of states. However, this substitution can be of importance only if the electron-tunnelon interaction is of the long-range type. Indeed, in this case the spectral tunnelon function $\Gamma_{TLS}(\nu)$ is a sum over all tunnelons affected by the electronic state of a chromophore and, therefore, the integral in Eq. (128) depends on the specific feature of the function $\Gamma(\nu)$. In this very fashion, Lyo and Orbach [92] tried to explain the temperature broadening of ZPL which reveals the T^α dependence with a noninteger exponent α. Perhaps the situation with the long-ranged electron-tunnelon interaction can occur if the chromophore is an ion because in this case the electron-tunnelon interaction can be of long-range type. However, the uncharged chromophores have the electron-tunnelon interaction of short-range type and, therefore, the substitution of phonons by fractons does not change Eq. (129).

5.5 Low Temperature Anomalous Hole Broadening in Polymers and Glasses

It has been shown in Sect. 4.3 that the halfwidth of zero-phonon holes equals the double homogeneous halfwidth of ZPL. However, this statement requires correction because it does not allow for the effect of SD and the possibility that chromophores with a single resonant frequency can have ZPLs with different homogeneous halfwidths. However, if the distribution over homogeneous halfwidths is of importance the hole has a non-Lorentzian shape. Experimental data reveal the Lorentzian hole shape [38]. Therefore, we can neglect this distribution and take the theoretical expression for hole halfwidth as a sum of three terms:

$$\gamma(t, T) = \gamma_{SD}(t, T) + \gamma_{TLS}(T) + \gamma_{ph}(T) . \tag{130}$$

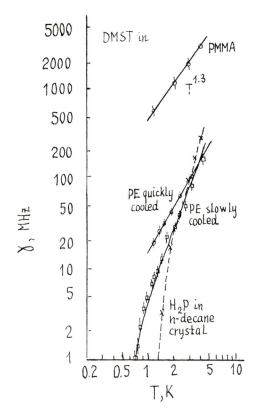

Fig. 34. Slow temperature hole broadening in polymers and a fast one in crystal [95]

Here, γ_{SD} describes inhomogeneous hole broadening due to SD and $\gamma_{TLS} + \gamma_{ph}$ is the homogeneous halfwidth resulting from the electron-tunnelon and electron-phonon interactions.

Anomalies in the temperature behaviour of halfwidth in polymers and glasses are of two types: first, the holes broaden over a low temperature domain slower than ZPLs in crytalline matrices and, second, the hole width is one or two orders of magnitude larger than the line width in crystals. Both anomalies are shown in Fig. 34 [95]. The closer the structure of a solvent to that of a crystal, the narrower the hole and the steeper its temperature dependence. This excess of halfwidth in polymers and glasses can be due to γ_{SD} or γ_{TLS}. The hole width behaves as T^{α} over the low temperature range. Here, α is close to unity. Theoretical work on the SD effect predicts a quasilinear temperature dependence for γ_{SD} [96, 97]. Therefore, it is commonly accepted that the quasilinear low temperature hole broadening is due to SD [98, 99]. However, if this statement is correct how can we explain the quasilinear temperature dependence measured in PE experiments (Fig. 27)? Two pulse picosecond PE cannot be undergone by means of SD effect. Therefore, the temperature dependence of the dephasing rate found by means of picosecond PE in Fig. 27 can be attributed to the homogeneous broadening.

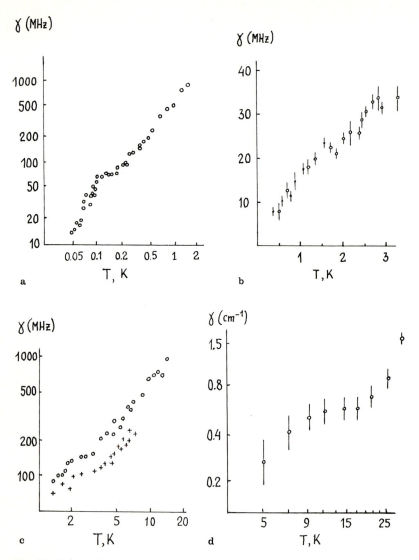

Fig. 35a–d. Anomalous temperature broadening with the bend in the slope: **a** octaethylporphin in polystyrene [99]; **b** Eu^{3+} doped silicate glass [100]. Time scale of the hole-burning > 100 s (*dots*) and < 5 s (*circles*); **c** Nd doped BIGaZYTZ heavy metal fluoride glass (*crosses*) and Nd doped ED-2 silicate glass (*circles*) [101]; **d** perylene doped *n*-octane crystal [102, 103]

Indeed, Eq. (129) allowing for the bilinear electron-phonon-tunnelon inter-action can easily explain a linear temperature broadening at $kT > \varepsilon$. Hence, the temperature dependence of dephasing time in Fig. 27 results from the interaction with a tunnelon having $\varepsilon < 1\,cm^{-1}$ and the difference between experimental data found over various timescales is due to the temperature independent SD effect. At $T > 15$ K, the third term γ_{ph} in Eq. (130) exceeds others. This is the reason for the coalescence of the experimental data found in slow and fast experiments.

There are some facts which unambiguously prove that the two-tunnelon mechanism plays an important role in low-temperature dephasing in amorphous and even in crystalline solids. The contribution from this mechanism is described by Eq. (126) at the weak electron-tunnelon coupling and by Eq. (127) at the strong coupling. Two-tunnelon processes are characterized by a very small slope in the temperature curve at $kT > \varepsilon$. None of the other conceivable mechanisms can cause such saturation effect in the temperature broadening. This saturation is shown in Fig. 33.

Experimental data for the temperature hole broadening which displays a bend in the slope is shown in Fig. 35. The plateau in the temperature dependence can be explained only with the help of the two-tunnelon mechanism causing the saturation. What is the nature of this TLS? It cannot be TLS that is responsible for the burning mechanism because hole burning is due to long-living extrinsic TLS but optical dephasing results from the interaction with short-living intrinsic TLS. The existence of these TLS in polymers and glasses is not surprising. However the experimental data by Korotaev group (Fig. 35d) prove that TLS of such type can exist in polycrystalline matrices as well.

The theoretical analysis of the experimental data for the temperature behaviour of hole widths and dephasing rates reveals that they can be explained by means of physical mechanisms of various types. For instance, the quasilinear temperature dependence depicted in Fig. 27, 34 and 35 can be explained with the help of inhomogeneous SD mechanism or homogeneous bilinear and two-tunnelon mechanisms. This uncertainty results from the fact that all existing theories include some adjustable parameters which facilitate their fitting. It seems that only the temperature dependence with a bend in the slope like that depicted in Fig. 35 enables one to make a definite conclusion. It unambiguously proves the two-tunnelon character of homogeneous line broadening in polymers and glasses.

6 Conclusion

We have seen that impurity optical bands are a good tool for investigating low-frequency excitations in polymers and glasses. In fact, just the spectroscopic methods have enabled one to discover tunneling systems in polymers. Upto now these systems have been investigated solely by means of spectroscopic methods. This situation is in contrast to that existing in glasses where tunneling systems were investigated by various physical methods. It is thought that investigation of polymer tunneling systems by various physical methods is a task for the very near future. This task could be carried out in super-low temperature experiments (T < 4 K) for supersonic attenuation, thermal conductivity, heat capacity and so on.

However, it should be emphasized that selective spectra of chromophores embedded in polymers or glasses have their own specific properties and they are the best tool to investigate chromophore-solvent interaction and the related molecular dynamics. Today there are many unanswered questions in this field. Some of these are: is hole broadening in low-temperature domain due to the inhomogeneous spectral diffusion effect or the homogeneous line broadening mechanism manifests itself? In what time scale does spectral diffusion exist? Can the distribution over barriers and relaxation rates found for the extrinsic tunneling systems be extended for the intrinsic ones? Certainly, this list of questions is far from exhaustive. I hope the answers will be found in the not too distant future.

Acknowledgement. I am grateful to Professor K. Horie who initiated my work with this review and Dr. Ludmila Bykovskaya for her comprehensive help in the course of the preparation of this manuscript.

7 References

1. Agranovich V, Hochstrasser R (eds) (1983) Spectroscopy and excitation dynamics of condensed molecular systems. North-Holland, Amsterdam
2. Weber M (ed) (1987) Optical linewidths in glasses. J Lum 36: N4 and N5
3. Moerner WE (ed) (1988) Persistent spectral hole-burning: science and applications. Springer, Berlin Heidelberg New York
4. Rebane L, Gorokhovski AA, Kikas JV (1982) Appl Phys B29: 235
5. Friedrich J, Haarer D (1984) Angewante Chemie 23: 113
6. Narasimhan LR, Littau KA, Pack DW, Bai YS, Elshner A, Fayer MD (1990) Chem Rev 90: 439
7. Horie K, Furusawa A (1990) In: Rabek JF (ed) Progress in photochemistry and photophysics. CRC, Boston 5: 50
8. Osad'ko IS (1991) Phys Rep 206: 45
9. Kubo R (1962) In: ter Haar D (ed) Fluctuations relaxation and resonance in magnetic systems. Plenum, N-Y
10. Anderson PW (1954) J Phys Soc Japan 9: 316
11. Kun Huang, Rhys A (1950) Proc Roy Soc 204: 406
12. Lax M (1952) J Chem Phys 20: 1752
13. Szabo A (1970) Phys Rev Lett 25: 924
14. Personov RI, Al'shits EI, Bykovskaya LA (1972) Opt Comm 6: 169
15. Allen L, Eberly JH (1975) Optical resonance and two-level atoms. John Wiley and Sons, N-Y
16. Kharlamov BM, Personov RI, Bykovskaya LA (1974) Opt Comm 12: 191
17. Gorokhovski AA, Kaarli RK, Rebane LA (1974) Zh Eksp Teor Fiz Pis'ma 20: 474
18. Hesselink WH, Wiersma DA (1983) In: Agranovich V, Hochstrasser R (eds) Spectroscopy and excitation dynamics of condensed molecular systems. North-Holland, Amsterdam
19. Klauder R, Anderson PW (1962) Phys Rev 125: 912
20. Davydov AS (1964) Theory of molecular excitons. McGraw-Hill
21. Born M, Kun Huang (1954) Dynamical theory of crystalline lattices. Clareadon, Oxford
22. Personov RI, Osad'ko IS, Godyaev ED, Al'shits EI (1971) Fiz Tverd Tela 13: 2653 [(1971) Sov Phys Solid State 13: 2224]
23. Markham JJ (1959) Rev Mod Phys 31: 956
24. Korotaev ON, Kaliteevski MJu (1980) Zh Eksp Teor Fiz 79: 439
25. Kukushkin LS (1963) Fiz Tverd Tela 5: 2170 [(1964) Sov Phys Solid State 5: 1581]
26. Osad'ko IS, Al'shits EI, Personov RI (1974) Fiz Tverd Tela 16: 1974 [(1974) Sov Phys Solid State 16: 1286]
27. Ranson P, Peretti P, Laporte J-L, Rousset Y (1976) J de Chim Phys 73: 545
28. Kanematsu Y, Ahn JS, Kushida T (1992) J Lum 53: 235

29. Osad'ko IS (1973) Fiz Tverd Tela 15: 2429 [(1974) Sov Phys Solid State 15: 1614]
30. Osad'ko IS (1983) In: Agranovich V, Hochstrasser R (eds) Spectroscopy and excitation dynamics of condensed molecular systems. North-Holland, Amsterdam
31. Osad'ko IS, Zaitsev NN (1985) Chem Phys Lett 121: 209
32. McCumber DE, Sturge MD (1963) J Appl Phys 34: 1682
33. Krivoglaz MA (1964) Fiz Tverd Tela 6: 1707 [(1964) Sov Phys Solid State 6: 1340]
34. Aartsma TJ, Morsink J, Wiersma DA (1977) Chem Phys Lett 47: 425
35. Dicker AIM, Dobkowski J, Völker S (1981) Chem Phys Lett 84: 415
36. Duppen K, Molenkamp LW, Morsink JBW, Wiersma DA, Trommsdorff HP (1981) Chem Phys Lett 84: 421
37. Lee HWH, Patterson FG, Olson RW, Wiersma DA, Fayer MD (1982) Chem Phys Lett 90: 172
38. Berg M, Walsh CA, Narasimhan LR, Littau KA, Fayer MD (1988) J Chem Phys 88: 1565
39. Gruzdev NV, Sil'kis EG, Titov VD, Vanier YuG (1992) J Opt Soc Am B9: 941
40. Gutierrez A, Castro G, Schulte G, Haarer D (1986) in: Zschokke I (ed) Optical Spectroscopy of Glasses. Reidel Publ Comp
41. Selzer PM, Huber DL, Hamilton DS, Yen WM, Weber MJ (1976) Phys Rev Lett 36: 813
42. Hegarty J, Yen WM (1979) Phys Rev Lett 43: 1126
43. Lee HWH, Huston AL, Gehrtz M, Moerner WE (1985) Chem Phys Lett 14: 491
44. Mariotto G, Montagna M, Rossy F (1988) Phys Rev B38: 1072
45. Personov RI (1983) In: Agranovich V, Hochstrasser R (eds) Spectroscopy and excitation dynamics of condensed molecular systems. North-Holland Amsterdam
46. Small GJ (1983) In: Agranovich V, Hochstrasser R (eds) Spectroscopy and excitation dynamics of condensed molecular systems. North-Holland Amsterdam
47. Friedrich J, Haarer D (1986) In: Zschokke I (ed) Optical spectroscopy of glasses. Reidel Publishing Comp.
48. Völker S, van der Waals (1976) Mol Phys 32: 1703
49. Horie K, Murase N, Ikemoto M (1992) Spectral hole burning and luminescence line narrowing: science and applications. Technical Digest ser, Ascona, Switzerland, 22: 253
50. Horie K, Ikemoto M, Suzuki T, Machida S, Yamashita T, Murase N (1992) Chem Phys Lett 195: 563
51. Jalmukhambetov AU, Osad'ko IS (1983) Chem Phys 77: 247
52. Osad'ko IS, Soldatov SL (1988) Chem Phys 128: 125
53. Gorokhovski A, Kikas YV (1978) Zh Priklad Spektr 28: 824
54. de Vries H, Wiersma DA (1980) J Chem Phys 72: 1851
55. Winnacker A, Shelby RM, Macfarlane RM (1985) Opt Lett 10: 350
56. Macfarlane R, Vial J-C (1986) Phys Rev B34: 1
57. Macfarlane R (1987) J Lum 38: 20
58. Lee HWH, Gehrtz M, Marinero E, Moerner WE (1985) Chem Phys Lett 118: 118, 611
59. Korotaev ON, Donskoi EI, Glyadkovski VI (1985) Opt i Spektrosk 59: 492 [(1985) Opt Spectrosc (USSR)]
60. Iannone M, Scott GW, Brinza D, Coulter DR (1986) J Chem Phys 85: 4863
61. Carter TP, Braenchle C, Lee VV, Manavi M, Moerner WE (1987) J Phys Chem 91: 3998
62. Mashida S, Horie K, Yamashita T (1992) Appl Phys Lett 60: 286
63. Mashida S, Kyono O, Horie K, Yamashita T (1992) Spectral hole burning and luminescence line narrowing: science and applications. Technical Digest ser, Ascona, Switzerland, 22: 107
64. Osad'ko IS, Soldatov SL, Jalmukhambetov AU (1985) Chem Phys Lett 118: 97
65. Olson RW, Lee HWH, Patterson FG, Fayer MD, Shelby RM, Burum DP, Macfarlane RM (1982) J Chem Phys 77: 2283
66. Osad'ko IS, Gladenkova SN (1992) Chem Phys Lett 198: 531
67. Osad'ko IS, Gladenkova SN (1992) Spectral hole burning and luminescence line narrowing: science and applications. Technical Digest ser, Ascona, Switzerland, 22: 111
68. Phillips WA (ed) (1981) Amorphous solids: low temperature properties. Springer-Verlag, Heidelberg
69. Anderson PW, Halperin BI, Varma CM (1972) Phil Mag 25: 1
70. Phillips WA (1972) J Low Temp Phys 7: 351
71. Zeller RC, Pohl RO (1971) Phys Rev B4: 2029
72. Zaithin RC, Anderson AC (1975) Phys Rev B12: 4475
73. Hunklinger S, Schickfus M in [68]
74. Winterling G (1975) Phys Rev B12: 2432
75. Völker S (1989) Annu Rev Phys Chem 40: 499
76. Köler W, Meiler J, Friedrich J (1987) Phys Rev B35: 403

77. Köler W, Friedrich J (1987) Phys Rev Lett 59: 2199
78. Portis AM (1953) Phys Rev 91: 1070
79. Mims WB, Nassau K, McGee JD (1961) Phys Rev 123: 2059
80. van der Berg R, Völker S (1987) Chem Phys Lett 137: 201
81. van der Berg R, Visser A, Völker S (1988) Chem Phys Lett 144: 105
82. Fidder H, de Boer S, Wiersma DA (1989) Chem Phys 139: 317
83. Littau K, Bai YS, Fayer MD (1990) J Chem Phys 92: 4145
84. Bai YS, Fayer MD (1989) Phys Rev B39: 11066
85. Hu P, Hartmann SR (1974) Phys Rev B9: 1
86. Hu P, Walker LR (1978) Phys Rev B18: 1300
87. Meijers H, Wiersma DA (1992) Phys Rev Lett 68: 381
88. van der Zaag PV, Galaup JP, Völker S (1990) Chem Phys Lett 166: 263
89. Osad'ko IS (1986) Zh Eks Teor Fiz 90: 1453 [(1986) Sov Phys JETP 63: 851]
90. Osad'ko IS, Shtygashev AA (1987) J Lum 37: 255
91. Hayes JM, Stout RP, Small GJ (1981) J Chem Phys 74: 4226
92. Lyo SK, Orbach R (1984) Phys Rev B29: 2300
93. Landau DL, Family F (eds) (1984) Proceedings of the International Conference on Kinetics of Aggregation and Gellation. North-Holland, Amsterdam
94. Kaufman JH, Nazzal AI, Melroy OR, Kapitulnik A (1987) Phys Rev B35: 1881
95. Thijssen HPH, Völker S (1986) J Chem Phys 85: 785
96. Bai YS, Fayer MD (1988) Chem Phys 128: 135
97. Putikka WO, Huber DL (1987) Phys Rev B36: 3436
98. Müller KP, Haarer D (1991) Phys Rev Lett 66: 2344
99. Gorokhovski AA, Korrovits V, Palm V, Trummal M (1986) Chem Phys Lett 125: 355
100. van der Zaag PJ, Schokker BC, Schmidt T, Macfarlane RM, Völker S (1990) J Lum 45: 80
101. Jacquier B, Macfarlane R (1992) Spectral hole burning and luminescence line narrowing: science and applications. Technical Digest ser, Ascona, Switzerland, 22: 160
102. Korotaev ON, Kolmakov IP, Karpov VP, Shchanov MF (1992) Spectral hole burning and luminescence line narrowing: science and applications. Technical Digest ser, Ascona, Switzerland, 22: 95
103. Korotaev ON, Kolmakov IP, Shchanov MF, Karpov VP, Godyaev ED (1992) Pis'ma Zh Eksp Teor Fiz 55: 417

Editor: Prof. K. Dušek
Received: March 1993

Small-Angle and Ultra-Small-Angle Scattering Study of the Ordered Structure in Polyelectrolyte Solutions and Colloidal Dispersions

Hideki Matsuoka[1], Norio Ise[2]
[1] Department of Polymer Chemistry, Kyoto University, Kyoto 606-01, Japan
[2] Fukui Research Laboratory, RENGO Co. LTD., 10-8-1 Jiyugaoka, Kanazu-cho, Sakai-gun, Fukui 919-06, Japan

Small-angle and ultra-small-angle X-ray scattering studies of ionic polymer solutions and colloidal dispersions are reviewed. A single, broad peak was observed in the small-angle region, attributed to ordering of the ionic polymers. From the Bragg spacing being smaller than the average spacing, a two-state structure was suggested. A paracrystal theory was developed for 3-D lattice systems, by which the observed single, broad peak was shown to be due to highly distorted ordered structures. The Bonse-Hart type ultra-small-angle X-ray scattering apparatus, by which a micron-size density fluctuation can be investigated, was applied to the structure study of latex dispersions and others. A distinct upturn was observed at low angles by neutron scattering method for polystyrenesulfonate solutions, justifying the two-state structure, and the size of the localized structure was thereby determined. Attention has been drawn to a direct microscopic study of latex dispersions, showing structural inhomogeneities in dispersions such as the two-state structure and voids. Various counterinterpretations are critically discussed.

Advances in Polymer Science, Vol. 114
© Springer Verlag Berlin Heidelberg 1994

"In NaCl solutions, sodium ions gather around negatively charged chloride ions and chloride ions around sodium ions by electrostatic attraction and repulsion so that a certain type of ionic distribution is established. A similar phenomenon will take place in solutions of dissociated polyacrylates. In other words, the polyacrylate ions are surrounded by sodium ions and a group of sodium ions interact in turn with carboxyl groups of polyions . . . . **The relative positions of the polyanions are fixed by the ionic charges. Such a situation in the solutions corresponds to a kind of structure formation (Strukturierung)".**

(Hermann Staudinger, 1930)

1 Introduction: Historical Review

Polyelectrolytes are high molecular weight compounds carrying dissociable groups. When introduced into dissociating solvents, ionic dissociation takes place so that electrically charged macromolecules are suspended in solutions. The electrostatic interaction is operative between ionized groups inside the macromolecules, between the macromolecules, between the counterions and between the macromolecules and counterions. Since this interaction is characterized by its long-range nature, the solutions exhibit various properties very different from solutions of neutral polymers. Because this interaction is shielded by coexisting salt ions, the unique properties are magnified and strong under low salt conditions. This fact had been noticed very early in the study of macromolecules. For example, Staudinger noticed the very high viscosity of salt-free polyacrylate aqueous solutions in 1930 [1]. In connection with the main theme of the present article, it is important to mention Staudinger's opinion that the high viscosities are caused by "Strukturierung" in the solutions. As can be inferred from his statement quoted in the very beginning of this article, it seems that he knew the truth before he had proof.

In the present article, we discuss the structure formation in polyelectrolyte solutions as studied by scattering techniques. Our main interest will be in dilute solutions; Highly concentrated solutions would not be given too much emphasis for the following reason; monodisperse spherical particles would show space-filling close-packed hexagonal structure if confined in a small space (one-state structure). This is contrasted with the "two-state structure" to be discussed later, in which non-space-filling, localized structure coexists with unstructured particles. There seems to be no points to discuss how and why the ordered structure is formed and maintained at very high particle concentration. The main factor for such regular packing is simply an exclusion volume effect (purely repulsive interaction) of the particles. Although concentrated solutions of ionic polymers are of interest from practical points of view, the study of structural aspects of these systems is not highly exciting from fundamental points of view.

To our knowledge, Bernal and Funkuchen were the first to perform detailed X-ray analysis of plant viruses (such as tobacco mosaic virus) in solution [2]. Four strong reflections were observed from dry gels (virus concentration greater than 80%), whereas three and two reflection lines were distinguished for wet gels (80–40%) and solutions (40–10%), respectively. These lines were attributed to a 2D-hexagonal packing of virus particles with particle spacing changing in inverse proportion to the square root of the virus concentration. Bernal and Funkuchen took this spacing-concentration relation as implying that "the particles distribute themselves in a hexagonal array so as to fill the available space as uniformly as possible".

X-ray diffraction study was then extended by Riley and Oster to bovine serum albumin (BSA), human hemoglobin, and nucleic acid [3]. All aqueous solutions of these biopolymers without added salt gave one or more reasonably well-defined small-angle diffraction bands between 30 and 50%wt vol^{-1}. When the Bragg spacing is plotted against the concentration on a double logarithmic scale, the points lay reasonably well on straight lines of slopes -0.5 and -0.33 for BSA and hemoglobin, respectively. Thus, these authors suggested that the molecules took up mean positions equidistant from each other throughout the whole volume in the 2D- and 3D-regularities, respectively.

Other ionic systems were also studied: for example, solutions of soap micelles such as potassium oleate were found to give sharp small-angle X-ray diffraction bands [4]. Brady carried out a Fourier analysis of the X-ray diffraction of solutions of sodium dodecyl sulfate to calculate the radial distribution function g(r) and the number of nearest neighbors (N_n) [5]. At about 30%, the N_n was about 12, indicating that the spherical micelles tended to assume a close-packed hexagonal arrangement.

Guinand et al. [6] found that the scattering intensity of salt-free polyacrylate solutions drastically decreased to one-fiftieth of that of the corresponding unionized acid. Doty and Steiner pointed out that, unlike uncharged polymers, concentration fluctuation cannot be independent in polyelectrolyte solutions and scattering intensity is lowered by a non-randomness of solute distribution [7]. The strong diminution of the scattering intensity was furthermore noticed for poly-4-vinyl-N-n-butylpyridinium bromide and arabic acid by Fuoss and Edelson [8] and Veis and Eggenberger [9] in the early 1950s, who also noticed a distinct minimum in the dissymmetry-concentration curve.

Fuoss inferred that "the long-range of Coulomb forces tends to establish an ordered distribution". However, the technical difficulty of performing the light scattering experiments for salt-free polyelectrolyte solutions hindered quantitative consideration of the structure formation of macroions.

The introduction of dynamic light scattering techniques unveiled an interesting aspect of the diffusion coefficient of ionic polymers. Schurr and his associates [10, 11] noticed that the diffusion coefficient of poly-L-lysine hydrobromide (PLL-HBr) first monotonically increased with decreasing NaBr concentration and then dropped at about 10^{-3} M by over an order of magnitude. Although readers are referred to a recent book by Schmitz [12] for more detail, this bizarre behavior was named the "ordinary-extraordinary transition" and accounted for in terms of the formation of temporal aggregates of macroions by Schmitz and Parthasarathy [13], in which the dynamics of all of the flexible coils in the solution became highly correlated. Independently from this work, French groups started small-angle neutron scattering (SANS) studies of polyelectrolyte solutions and noticed a single broad peak in the SANS profiles under low salt conditions [14], which according to our interpretation is definitely reminiscent of an ordered arrangement of ionic polymers in solutions.

Argument supporting the presence of ordered structure in ionic solutions was also advanced, albeit in an indirect manner, from thermodynamic properties. In 1959, Frank and Thompson [15] pointed out on the basis of statistical-mechanical considerations that the Debye–Hückel formalism [16] of simple electrolyte solutions breaks down at a concentration m_0, where the Debye length $(1/\kappa)$ equals an average distance apart of nearest-neighboring ions $(\bar{l})$. As an interesting support for the failure of the Debye–Hückel theory, they pointed out for 1:1 type alkali halides that the logarithm of the (mean) activity coefficient $(\log f_{\pm})$ shows a cube-root dependence on concentration m (mol l^{-1}) above m_0, while a square-root dependence is valid below m_0, as follows:

$$\log f_{\pm} = a - b\, m^{1/3} \quad \text{for } m > m_0 \tag{1-1}$$

$$\log f_{\pm} = 1 - b'\, m^{1/2} \quad \text{for } m < m_0 \,. \tag{1-2}$$

They took $\bar{l}$ as "lattice length", namely 9.4 m$^{-1/3}$ for symmetrical-valence type electrolytes. They suggested that, although in solution, the ions form a "disordered lattice" (rather than the Debye ion cloud) at higher concentrations than m_0, which was about 10^{-3} moles l^{-1} for 1:1 electrolytes in water at 25 °C. A similar lattice model was also proposed by Robinson and Stokes [17], who predicted an exponent of 0.29 for electrolytes with the "sodium chloride lattice". The cube-root dependence of the activity coefficient was also noted for other alkali halides [18] in a wide concentration range between 10^{-3} and 1 mol l^{-1}. Bahe proceeded further: for 1:1 electrolytes he proposed a "loose" face-centered cubic (fcc) structure for a concentration range between 10^{-2} and 3 moles l^{-1} and noted that existing X-ray scattering curves are compatible with this structure [19]. For 2:1 electrolytes up to concentrations as high as 2 or 3 mol l^{-1}, a loose fluorite structure was proposed [20].

The mean activity and activity coefficient of polyelectrolytes were measured for the first time by Ise et al. and the activity coefficient was generally found to obey the cube-root dependence in a much wider concentration range than simple electrolytes [21]. From this concentration dependence, it was inferred that ordered arrangements were maintained in polyelectrolyte solutions.

Dusek and Ise and their associates then started small-angle X-ray scattering (SAXS) studies of polyelectrolyte solutions [22, 23, 24]. In the following we discuss further development in this area. Analyses of ordered structure and cluster size by small-angle and ultra-small-angle X-ray scattering and small-angle neutron scattering techniques are described. The ordered structure of polyelectrolyte solutions, particularly their distorted nature, is briefly discussed with relation to a 3D paracrystal theory.

2 Small-Angle X-ray Scattering (SAXS) of Polyelectrolyte Solutions

2.1 Synthetic Polyelectrolytes

Polyacrylate (PAA) and Polymethacrylate (PMA)

We begin with early SAXS curves obtained for polyelectrolytes. In Fig. 2-1 are plotted SAXS data of (fractionated) sodium polyacrylate (NaPAA) [22, 24].

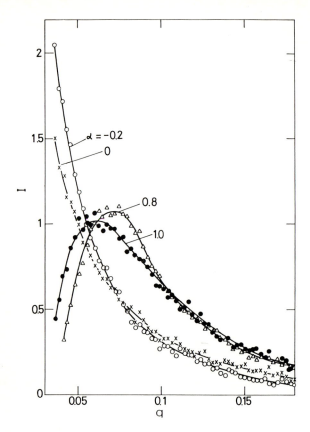

Fig. 2-1. Scattering intensity (I) vs scattering vector (q) plots for sodium polyacrylate (NaPAA) solutions at 22 °C: influence of degree of neutralization (α) at a weight-average degree of polymerization $Pw = 1470$ and $[NaPAA] = 0.02 \text{ g ml}^{-1}$. Taken from [24] with the permission of the American Chemical Society

Clearly, a single broad peak was observed for electrically charged polyacrylate, whereas the intensity decreased monotonically when PAA was uncharged. The same trend was reported for polymethacrylate by Dusek et al. [23]. These authors noticed a peak even at a low degree of neutralization of 0.05. This seems to be true for other polyelectrolytes and reflects the long-range nature of the Coulombic interaction.

Figure 2-2 presents the curves of the salt concentration influence on the SAXS data for PAA. As the salt concentration was increased, the peak position (q_m) shifted toward lower scattering vectors ($q = 4\pi \sin\theta/\lambda$, where 2θ is the scattering angle and λ is the wavelength of X-ray) and the peak was not observed in the presence of a large amount of salt. Although various counterinterpretations were presented, our interpretation was and is that the scattering peak reflects some sort of ordered arrangement of ionic polymers in solutions. The

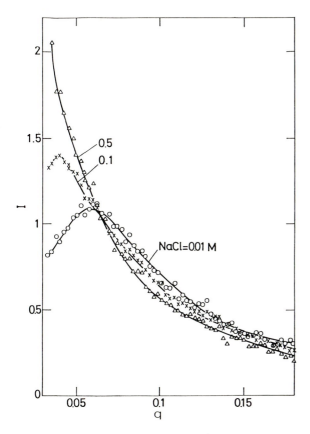

Fig. 2-2. Scattering intensity (I) vs scattering vector (q) plots for NaPAA solutions at 22 °C: influence of NaCl added at $\alpha = 0.5$, Pw = 1470, and [NaPAA] = 0.02 g ml^{-1}. Taken from [24] with the permission of the American Chemical Society

salt concentration dependence indicates that the ordered arrangement is of an electrostatic origin.

Sodium Polystyrenesulfonate (NaPSS)

Figure 2-3 shows the curves of the polymer concentration influence on the scattering data for sodium polystyrenesulfonate (NaPSS) aqueous solutions. The measurements were carried out independently for a sample of NaPSS in Prague and in Kyoto; the agreement (in the peak position, q_m) between the two groups is most satisfactory: With increasing polymer concentration, the q_m shifted toward higher q, indicating that the interparticle distance became smaller. This concentration dependence was also observed for other polyelectrolytes and colloidal systems in the concentration range covered.

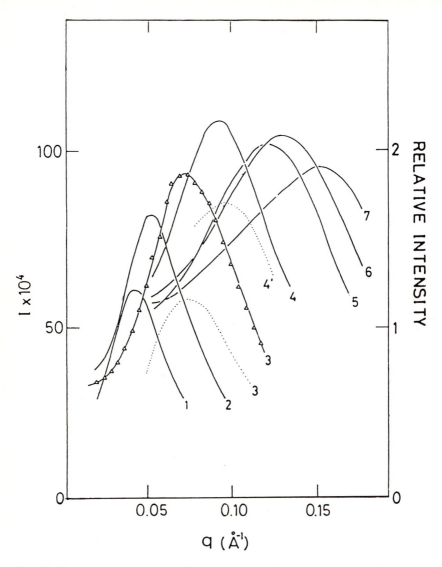

Fig. 2-3. Concentration dependence of scattering curves of sodium polystyrenesulfonate (NaPSS) solutions ($M_w = 780\,000$). The curves show the data observed by Dušek's group (coordinate: left) whereas the dotted curves were obtained by us (coordinate: right). Polymer concentration; *curve 1*: 0.0101 g ml^{-1}, *curve 2*: 0.0204, *curve 3*: 0.0399, *curve 3'*: 0.04, *curve 4*: 0.0696, *curve 4'*: 0.08, *curve 5*: 0.1053, *curve 6*: 0.1425, *curve 7*: 0.1796

Figure 2-4 presents the influence of four kinds of salts [25]. At a fixed concentration (0.1 M), $CaCl_2$ and $BaCl_2$ destroyed the ordered structure more effectively than NaCl or KCl. This is quite understandable in view of the more efficient shielding effects of the higher valence cations. The difference between Ca^{2+} and Ba^{2+} would be due to the smaller Pauling radius and hence the

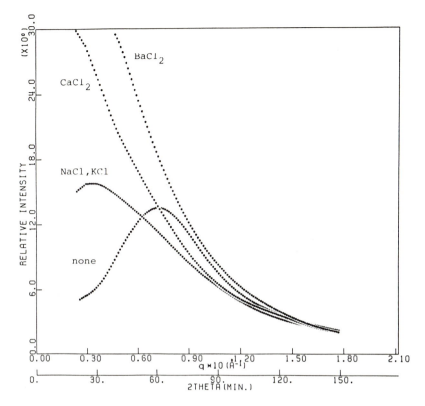

Fig. 2-4. Scattering profile for NaPSS solutions in the presence of NaCl, KCl, CaCl$_2$, and BaCl$_2$. $M_w = 74\,000$, [NaPSS] $= 0.04$ g ml^{-1}, [salt] $= 0.1$ M. Taken from [25] with the permission of the American Institute of Physics

stronger hydration of the former, which would cause a larger hydration sphere and a lower shielding efficiency.

The molecular weight dependence is shown in Fig. 2-5 for $M_w = 4600$, 18 000, 74 000 and 780 000 at 0.04 g ml^{-1}. For the three samples of lower M_w, the peak position clearly depended on the M_w, indicating that the peak reflects the distribution of the centers of mass of the polymers. If the peak is due to intersegmental correlation, the peak position should be independent of M_w because the concentration (in g ml^{-1}) is same for all solutions.

Figure 2-6 shows the effect of the addition of NaPAA in NaPSS solution. Only one scattering peak is observed even in binary solutions, indicating that the SAXS peak reflected an intermolecular ordering.

Polyallylamine Hydrochloride (PAA-HCl)

The dependencies of the scattering curves on polymer concentration and salt concentration of PAA-HCl were basically the same as those discussed above for

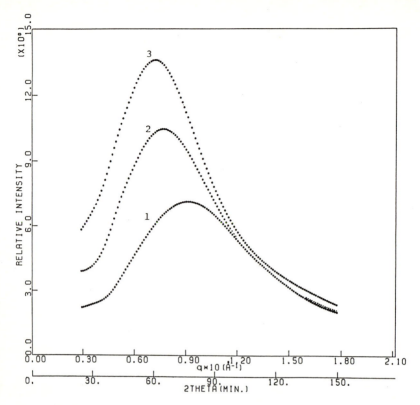

Fig. 2-5. Molecular weight dependence of scattering of NaPSS solutions. Polymer concentration: 0.04 g ml^{-1}, M_w *curve 1*: 4600, *curve 2*: 18 000, *curve 3*: 74 000. The scattering curve for M_w = 780 000 coincides with curve 3. Taken from [25] with the permission of the American Institute of Physics

other ionic polymers [26]. Its pH profile was superficially different from other anionic polymers in that the peak became more distinct at higher pH's. These observations indicate that the ordering phenomenon is general whether the polymers are cationic or anionic.

Figure 2-7 shows the variation of the scattering curves with time. The measurements were repeated three times. The first one was made immediately after solution preparation, the second on the fifth day, and the third on the 21st day. The solution (0.04 g ml^{-1}) was kept in the capillary in the meantime. The absolute intensity of the incident X-ray was determined by using Lupolen as standard before and after the measurements, and the scattering intensity of the solution was corrected for the time change of the intensity thus determined. The scattering curves shown here are "raw" data, i.e. not smoothed, nor desmeared. Practically no change was observed in the period investigated, which indicates that the ordered structure was stable and did not change in this time scale. However, it has been reported that the scattering peak became higher and

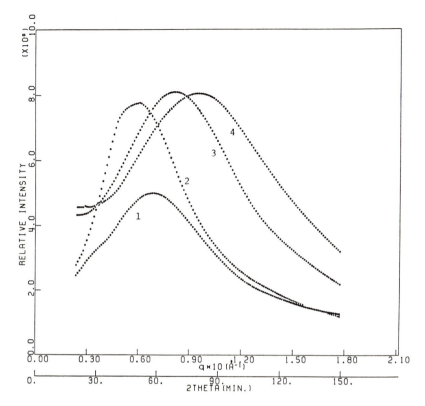

Fig. 2-6. Scattering curves of a NaPSS ($M_w = 18\,000$) solution, NaPAA ($M_w = 97\,000$) and their mixtures. *Curve 1*: NaPAA $0.02\,\mathrm{g\,ml^{-1}}$, *curve 2*: NaPSS $0.02\,\mathrm{g\,ml^{-1}}$, *curve 3*: NaPAA ($0.02\,\mathrm{g\,ml^{-1}}$) + NaPSS ($0.02\,\mathrm{g\,ml^{-1}}$), *curve 4*: NaPAA ($0.04\,\mathrm{g^{-1}\,ml}$) + NaPSS ($0.02\,\mathrm{g\,ml^{-1}}$). Taken from [25] with the permission of the American Institute of Physics

sharper about one year after sample preparation [27]. There may be a possibility that in such a long time scale, aging or growth of the structure occured.

2.2 Bragg Spacing and Average Spacing: The Two-State Structure

From the experimental results mentioned above, it is suggested that the scattering peak observed is due to an interparticle interference. If so, the interparticle distance ($2D_{exp}$) can be estimated from the peak position by the Bragg equation. $2D_{exp}$ is given by

$$2D_{exp} = 2\pi/q_m \tag{2-1}'$$

On the other hand, by assuming a uniform distribution of polymer molecules in the solution, an average interparticle spacing ($2D_0$) can be calculated from

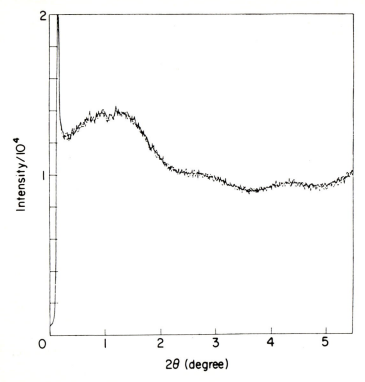

Fig. 2-7. Three 'raw' scattering curves of polyallylamine hydrochloride observed immediately after the solution preparation, on the fifth day and on the 21st day. Taken from [26] with the permission of the Society of Chemical Industry

concentration as follows,

$$2D_0 = (cN_A/M_w)^{1/3} \, 10^8 \, [\text{Å}] \tag{2-2}$$

where c is the concentration in $g \, ml^{-1}$, and N_A the Avogadro number. This equation assumes a simple cubic (sc) lattice for solute distribution. As will be described later, for face-centered cubic (fcc) and body-centered cubic (bcc) lattices, other equations should be used. However, since the peak observed is single and broad under our experimental conditions, too much emphasis should not and cannot be placed on the lattice system. The evaluation by Eq. (2-2) is enough for our purpose.

It is necessary to comment on the reliability of the peak position and $2D_{exp}$. As will be discussed in Sect. 3, the scattering profile is dependent on three factors, namely paracrystalline distortion, the Debye–Waller effect and the size of the ordered structure. When the first two factors become stronger, and when the size becomes smaller, the scattering intensity is lowered and higher order scattering peaks become indiscernible. Fortunately, the peak position itself is

Table 2-1. Small-angle X-ray scattering data of sodium poly-styrenesulfonates in salt-free aqueous solutions at 25 °C

$M_w \times 10^{-3}$	Concn (g ml^{-1})	$2D_{exp}$ (Å)	$2D_0$ (Å)
4.6	0.04	69	58
18	0.01	136	144
	0.02	119	114
	0.04	81	91
	0.08	60	72
74	0.01	157	231
	0.02	122	183
	0.04	87	145
	0.08	65	115
	0.16	47	92
780	0.04	87	319

almost independent of these factors. Thus, the determination of the $2D_{exp}$ from the peak top of the single, broad peak is valid.

Table 2-1 summarizes the $2D_{exp}$ and $2D_0$ values with relevant information for NaPSS. It should be noted that $2D_{exp}$ values are smaller than $2D_0$ values except for very low molecular weight samples ($M_w = 4600$). The inequality relation suggests that the ordered structures are maintained not throughout the solution, but they exist as non-space-filling "clusters" of limited sizes and of a high polymer density in coexistence with disordered regions of a lower polymer density, although the cluster size is not known at this stage. (This problem will be described in Sect. 5). Such a structure has been called the "two-state structure" [24] which is schematically shown in Fig. 2-8. Although this structure is quite surprising, we note that a similar structure has been directly confirmed for ionic polymer latices in dispersions by ultramicroscopy [28]. Another type of structural inhomogeneity, namely void structure, has also been observed by Ito et al. by metallurgical microscopy and confocal laser scanning microscopy [29]. The presence of these structures indicates that the overall interaction between ionic polymers is attractive, although it has been widely believed that the electrostatic interaction between macroions is purely repulsive [28].

This attraction is generated through the intermediary of counterions [28, 30], and is purely electrostatic. As discussed in great detail previously [28, 31], the attraction arises as an immediate consequence of the electrostatic forces between the ions that, "considering any one ion, we shall find on the average more dissimilar than similar ions in its surroundings," as Debye and Hückel stated [16]. Since the Coulombic potential is inversely proportional to the interionic distance, such an ionic configuration in solutions would lead to an attraction-induced stabilization of the solutions. It seems to be a matter of course that colloidal systems and macroionic solutions must also satisfy this

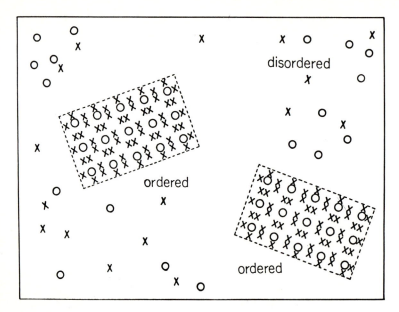

Fig. 2-8. Highly simplified schematic representation of the "two-state structure". The *circles* indicate macroions and the *crosses* counterions. Taken from [24] with the permission of the American Chemical Society

axiomatic condition. In other words, any correct theory of such systems must predict a short-range electrostatic repulsion between charged colloidal particles (and between macroions) and, in addition, a similar long-range electrostatic attraction. This problem and the related matters will be discussed in Sect. 6.

2.3 Ionic Biopolymers

The ordering phenomena were also found for ionic biopolymers in solutions. The first X-ray study for biopolymers was done by Bernal and Funkuchen in 1941, as mentioned above [2]. However, the polymer concentration was very high in their experiments. Thanks to recent technical developments in the diffraction methods, it became feasible to perform X-ray scattering experiments at relatively low concentrations and at low scattering angles.

Bovine Serum Albumin (BSA)

Bovine serum albumin is a typical globular protein. It has a prolate ellipsoidal shape whose long axis is about 140 Å and short axis is about 40 Å. The surface charge depends on the solution pH, and its isoelectric point (pI) is reported to be about 4.8.

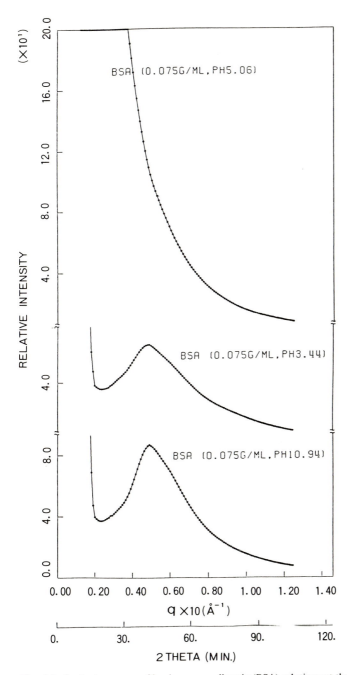

Fig. 2-9. Scattering curves of bovine serum albumin (BSA) solutions at three different pH's. Taken from [32] with the permission of the American Institute of Physics

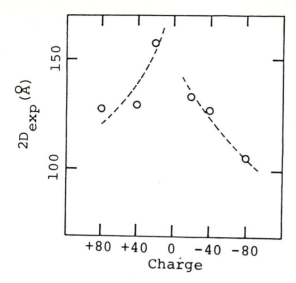

Fig. 2-10. Charge number dependence of $2D_{exp}$ for BSA. [BSA] $\simeq 0.075\ g\,ml^{-1}$. Taken from [32] with the permission of the American Institute of Physics

Figure 2-9 shows scattering curves of BSA solutions at three different pH's [32]. At pH = 5.06, which is very close to pI, the scattering intensity decreased monotonically with increasing q, that is, no peak was observed, as was the case with unneutralized weak polyacids (e.g., PAA in Fig. 2-1) and polybases. On the other hand, under both acidic and alkaline conditions, in which BSA molecules are positively and negatively charged, respectively, the scattering curve showed a clear, distinct peak although it was single and broad. These results again indicate that the ordering phenomena have an electrostatic origin.

Figure 2-10 shows the $2D_{exp}$ as a function of the number of analytical charges on BSA molecules. With increasing surface charge number, $2D_{exp}$ clearly decreased both for positive and negative charges. This tendency indicates that the net electrostatic interaction between BSA molecules is attractive and is intensified with increasing charges. It may sound strange but is quite reasonable if we recall that the attraction in question is generated through the intermediary of counterions. In other words, as the charge number increases, the counterions in between the macroions will increase, giving rise to enhanced attraction between the macroions.

Other Ionic Biopolymers

Besides BSA, a single, broad peak was observed also for lysozyme, transfer ribonucleic acid (tRNA), and chondroitin sulfate [32, 33]. It was concluded that an ordered structure was formed in these solutions at low salt conditions. For lysozyme and tRNA, the relation $2D_{exp} \approx 2D_0$ was found, indicating that the structure is maintained throughout the solution (the one-state structure). This can be easily accepted if we realize that these macroions have relatively low charge number (in comparison with PSS), which would result in a weaker electrostatic attraction.

2.4 Ionic Micelles

For ionic micelle solutions, also, a SAXS peak was found as mentioned in Sect. 1. Figure 2-11 shows scattering curves of dodecyltrimethylammonium chloride (DTAC), a cationic micelle [34]. From this behavior, it is clear that ordering occurred in these systems. A spherical scatterer such as DTAC micelles has an advantage with regard to the detailed analysis of the solution structure, because its scattering intensity I(q) can be represented in a simple form as follows:

$$I(q) = P(q)S(q) \tag{2-3}$$

where P(q) is the intraparticle scattering function which depends only on the size of the spherical scatterer and S(q) is the interparticle structure factor which

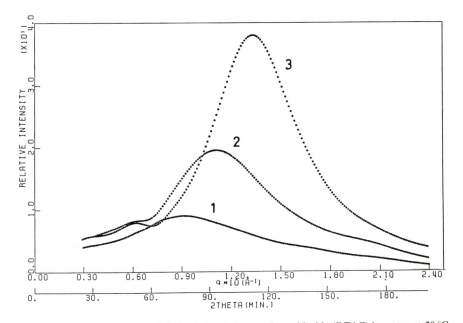

Fig. 2-11. Scattering curves of dodecyltrimethylammonium chloride (DTAC) in water at 28 °C. [DTAC]; *curve 1*: 0.05 g ml^{-1}, *curve 2*: 0.10 g ml^{-1}, *curve 3*: 0.20 g ml^{-1}. Taken from [34] with the permission of the Bunsengesellschaft

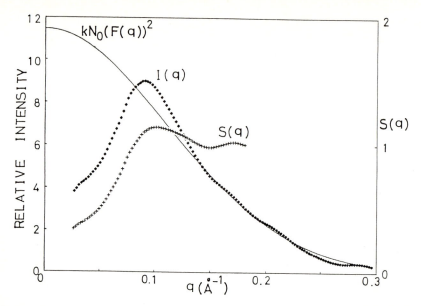

Fig. 2-12. I(q), P(q) and S(q) for 0.05 g ml^{-1} DTAC in water. Taken from [34] with the permission of the Bunsengesellschaft. $kN_0(F(q))^2$ is the intraparticle scattering function P(q), where k is a constant, N_0 the number of spherical scatters and $(F(q))^2$ the form factor

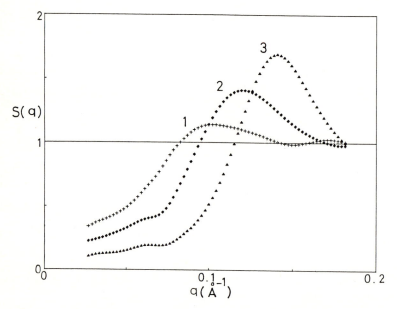

Fig. 2-13. S(q) for DTAC in water. [DTAC]; *curve 1*: 0.05 g ml^{-1}, *curve 2*: 0.10 g ml^{-1}, *curve 3*: 0.20 g ml^{-1}. Taken from [34] with the permission of the Bunsengesellschaft

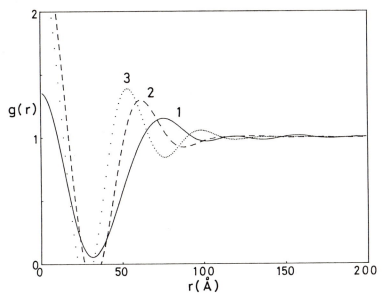

Fig. 2-14. Radial distribution function for DTAC in water obtained from S(q). [DTAC]; *curve 1*: 0.05 g ml^{-1}, *curve 2*: 0.10 g ml^{-1}, *curve 3*: 0.20 g ml^{-1}. Taken from [34] with the permission of the Bunsengesellschaft

reflects the spatial distribution of the center of mass of the scatterer. By Fourier transformation, S(q) is related to the radial distribution function g(r) by:

$$g(r) = 1 + \frac{1}{2\pi^2 nr} \int_0^\infty [S(q) - 1] \, q \sin qr \, dq \, , \qquad (2\text{-}4)$$

where n is the number density of the scatterer. If the P(q) function is known, S(q) and hence g(r) can be estimated from Eqs. (2-3) and (2-4). Figure 2-12 shows I(q), P(q) and S(q) for DTAC solution, and Figs. 2-13 and 2-14 are S(q) and g(r) functions thereby estimated. The S(q) function also showed a single broad peak, and g(r) displayed a peak and somewhat oscillating behavior at higher concentrations. Concerning S(q) and g(r) functions, it has been claimed [35] that g(r) is a more exact function than S(q) and that g(r) should be used to get more exact information about the structure. However, this seems not to be correct: as is easily seen from Eq. (2-4), g(r) and S(q) are interrelated by Fourier transformation, i.e. these two functions contain exactly the same information on the structure. The only difference is that g(r) is in real space while S(q) is in reciprocal space. Certainly, in this sense, g(r) is easy to understand. However, if a termination effect and statistical error contained in the data would cause some uncertainty in the integral in Eq. (2-4) [36], the S(q) function could often give more exact information about the structure.

2.5 Colloidal Dispersions

Colloidal Silica

For colloidal silica dispersion, the same analysis as for spherical micelles can be used to estimate the S(q) function. Figure 2-15 shows S(q) functions of Ludox SM (DuPont Co.) dispersions at various concentrations [37]. A clear peak and the second shoulder could be observed at high concentrations. Values for $2D_{exp}$ can be calculated from the peak position by assuming an fcc lattice from the following equation:

$$2D_{exp} = (\sqrt{3}/\sqrt{2})(2\pi/q_m) . \tag{2-5}$$

When the particle density (d) was assumed to be 2.2 g cm^{-3} [38], we obtained $2D_{exp} < 2D_0$, implying the two-state structure [39]. For example, at 3.93 vol%, the $2D_{exp}$'s were 225, 275 and 275 Å while the $2D_0$'s were 288, 324 and 315 Å for sc, fcc and bcc, respectively.

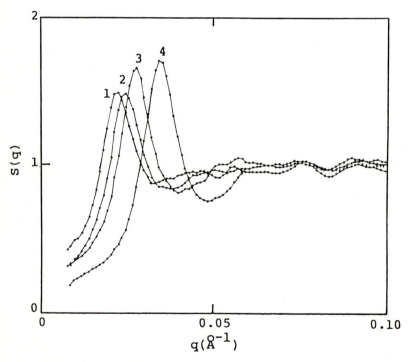

Fig. 2-15. Concentration dependence of S(q) of Ludox SM. *Curve 1*: [Ludox] = 2.75 vol.%, *curve 2*: 3.66, *curve 3*: 5.49, *curve 4*: 10.98. Taken from [37] with the permission of the American Physical Society

3 Three-Dimensional Paracrystal Theory and its Applications to the Structure in Solutions and Dispersions

As was described in Sect. 2, SAXS curves of polyelectrolytes at very low ionic strengths showed an interparticle interference peak, indicating an ordered arrangement of ionic solutes in solution. Since the structure is formed in solution, the degree of order would not be so high. In fact, the scattering peak is single and broad. Sometimes it is claimed that such a peak does not imply the existence of "ordering". However, it is clear that without some kind of ordering, a peak should not be observed. The shape and height of the peak are in principle very sensitive to the degree of order and also to defects in ordering. There have been some attempts to interpret the scattering peak by application of liquid theory [40, 41], and a rather good agreement with observation has been obtained. However, it is quite reasonable to infer that the true nature of the structure in solution lies between those of liquids and solid crystals. Approaches both from liquid theory and from theory of solid crystal by taking account of lattice distortions should be possible. In this Section, it will be shown that a single, broad peak, which was often observed for polyelectrolyte solutions and colloidal dispersions, can be interpreted as due to highly distorted crystals if distortions of ordering are taken into account. For this purpose, we apply a paracrystal theory. (Further discussion on the liquid theory approach is given in Sect. 3.2.)

3.1 Theory

A paracrystal theory was first proposed by Hosemann in 1965 [42] for calculating scattering profiles of distorted crystals. The general equation of the lattice factor $Z(q)$ is introduced for 1D, 2D, and 3D cases, and the total scattering intensity $I(q)$ is written as follows for spherical and monodisperse scatterers;

$$I(q) = P(q)Z(q) \tag{3-1}$$

For $P(q)$, we have

$$P(q) = [3(\cos(qR) - qR\sin(qR))/(qR)^3]^2 \tag{3-2}$$

with R, the sphere radius, and $Z(q)$, the lattice factor calculated from the paracrystal theory.

For the 1D case, the lattice factor $Z(q)$ is calculated numerically, and has been compared with experimental results. Cooper et al. calculated the $Z(q)$ for 3D cases [43] applied to sc lattices. The general equation of $Z(q)$ for 3D cases was given by us as [44, 45];

$$Z(\mathbf{q}) = \prod_{k=1}^{3} Z_k(\mathbf{q})$$

$$Z_k = \mathrm{Re}\left[\frac{1 + F_k(\mathbf{q})}{1 - F_k(\mathbf{q})}\right] = \frac{1 - |F_k|^2}{1 - 2|F_k|\cos(\mathbf{a}_k \cdot \mathbf{q}) + |F_k|^2}$$

$$F_k = |F_k(\mathbf{q})| \exp(-i\mathbf{q} \cdot \mathbf{a}_k) \tag{3-3}$$

Although our original papers should be referred to for the details, we formulated the theory for fcc and bcc cases. Their Z(q)'s are given by using unit vectors defined previously [44] and

$$|F_k(\mathbf{q})| = \exp\{-\tfrac{1}{2}(\Delta a^2/a^2)[(\mathbf{q}\cdot\mathbf{a}_1)^2 + (\mathbf{q}\cdot\mathbf{a}_2)^2 + (\mathbf{q}\cdot\mathbf{a}_3)^2]\}$$

$$= \exp\{-\tfrac{1}{8}\Delta a^2 q^2[(\sin\theta\sin\phi + \cos\theta)^2 + (-\sin\theta\cos\phi + \cos\theta)^2$$

$$+ (-\sin\theta\cos\phi + \sin\theta\sin\phi)^2]\} \tag{3-4}$$

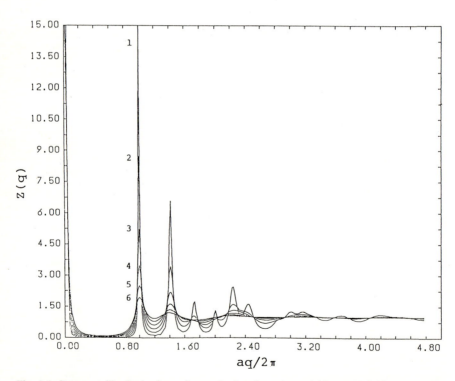

Fig. 3-1. Paracrystalline lattice factor for a sc lattice. *Curve 1*: g = 0.05, *curve 2*: 0.07, *curve 3*: 0.09, *curve 4*: 0.11, *curve 5*: 0.13, *curve 6*: 0.15. Taken from [44] with the permission of the American Physical Society

for fcc, and

$$|F_k(\mathbf{q})| = \exp\{-\tfrac{1}{8}\Delta a^2 q^2 [(\sin\theta\cos\phi + \sin\theta\sin\phi + \cos\theta)^2$$
$$+ (-\sin\theta\cos\phi - \sin\theta\sin\phi + \cos\theta)^2$$
$$+ (-\sin\theta\cos\phi + \sin\theta\sin\phi - \cos\theta)^2]\} \qquad (3\text{-}5)$$

for bcc.

Figures 3-1 – 3-3 show Z(q)'s for sc, fcc, and bcc lattices at various distortion factors, g, which represent the degree of distortion of the lattice. For the ideal (perfect) lattice, g = 0, and for the completely distorted one, g = 1. (Strictly speaking, g = Δa/a, where a is the distance between neighbouring lattice points and Δa is the standard deviation of the probability function of the lattice points, which is assumed to be Gaussian in this theory.) As seen from these figures, the interference peak is well reproduced at the characteristic positions for each lattice. It is observed that with increasing g-factor, in other words with increasing degree of distortion of the lattice, the peak becomes lower and broader, the

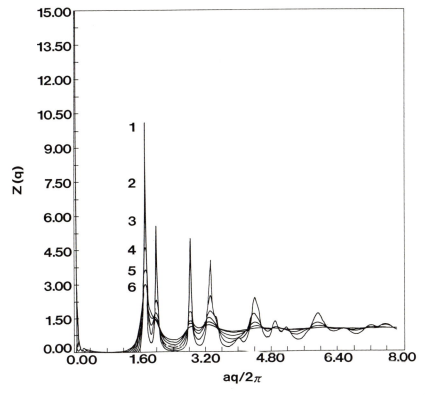

Fig. 3-2. Paracrystalline lattice factor for a fcc lattice. *Curve 1*: g = 0.05, *curve 2*: 0.07, *curve 3*: 0.09, *curve 4*: 0.11, *curve 5*: 0.13, *curve 6*: 0.15. Taken from [45] with the permission of the American Physical Society

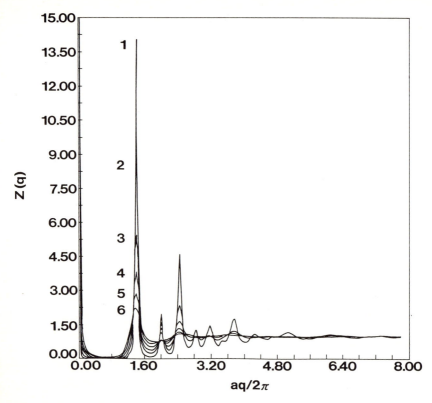

Fig. 3-3. Paracrystalline lattice factor for a bcc lattice. *Curve 1*: g = 0.05, *curve 2*: 0.07, *curve 3*: 0.09, *curve 4*: 0.11, *curve 5*: 0.13, *curve 6*: 0.15. Taken from [45] with the permission of the American Physical Society

higher order peak disappears, and at about g > 0.3 Z(q) shows only a single, broad peak. The observed single, broad peak is thus reproduced by this approach. It should be noted that the peak position is not largely influenced by the g-factor, indicating that the interparticle distance calculated by the Bragg equation from the position of the single broad peak is correct as an "average" one, although, strictly speaking, the peak position shifts to smaller q very slightly with increasing g.

3.2 Comparison with Experiments

Figure 3-4 shows the comparison of experimental S(q) function [46] (exactly corresponding to Z(q) in the paracrystal theory) for latex dispersion with theoretical Z(q)'s for three lattice systems. Clearly the sc lattice should be excluded, and both fcc and bcc can well reproduce the experimental S(q). It can be concluded that the arrangement of latex particles in dispersion is fcc- or

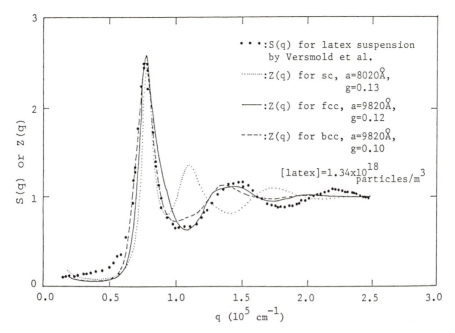

Fig. 3-4. Comparison of experimental S(q) for latex dispersion and theoretical Z(q)'s. The data are taken from Versmold et al. [46]

bcc-like, not sc-like, and that the fcc structure is slightly more realistic than the bcc under the present experimental conditions. Similarly, a less satisfactory agreement was obtained with a sc lattice symmetry for silica particle solutions than with fcc and bcc systems. Figures 3-5 and 3-6 show excellent agreement for latex dispersions between experimental S(q) [47, 48] and theoretical Z(q) calculated with an fcc lattice and g-factors of around 0.15–0.20.

In conclusion, a single, broad peak which has been observed for polyelectrolyte solutions and colloidal dispersions at low ionic strengths is well explained by the formation of some kind of order containing rather large distortion (say, g = 0.15 to 0.3). Concerning the fact that a modified liquid theory also shows good agreement, it must be remembered that the ordered structure in solutions or dispersions would be between those of liquids and solid crystals. It is a mistake to imagine a perfect lattice like solid crystals in solutions and dispersions. The 3D paracrystal theory reproduces experimental results well, and can be used to analyze the distortion of the structure quantitatively. On the other hand, it is well known that the thermal vibration and the size of the crystal also affect the scattering profile. Although the details are not mentioned here, it can be said that the reason why we obtain only a single, broad peak is due to the fairly large lattice distortion and thermal vibration, and to the small size of the crystal.

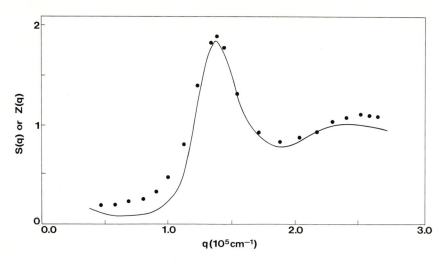

Fig. 3-5. Comparison of the experimental interference function S(q) with the theoretical paracrystalline lattice factor Z(q). The *filled circles* denote S(q) for a latex dispersion by Ottewill [47] with a volume fraction 10^{-3} and a particle radius of 256 Å. The *line* denotes Z(q) for a fcc lattice, a = 5597 Å and g = 0.22. Taken from [45] with the permission of the American Physical Society

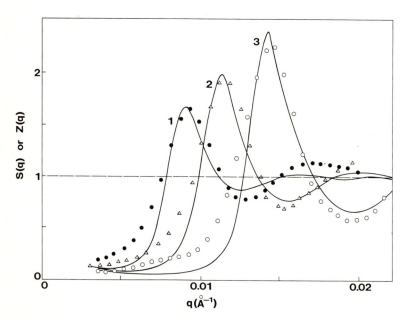

Fig. 3-6. Comparison of the experimental interference function S(q) obtained by neutron scattering for latex dispersions by Cebula et al. [48] with theoretical lattice factors Z(q). *The filled circles, triangles* and *open circles* represent S(q) observed for a latex particle of R = 157 Å at volume fractions 0.04, 0.08, and 0.13, respectively. *Curves 1, 2,* and 3 represent Z(q) for fcc structures with [a = 830 Å, g = 0.24], [a = 679 Å, g = 0.21], [a = 539 Å, g = 0.18], respectively. Taken from [45] with the permission of the American Physical Society

A comment may be worthwhile on the application of liquid theory. As mentioned above, the so-called Mean Spherical Approximation or Rescaled Mean Spherical Approximation methods, in which only an electrostatic repulsion is assumed between polyions or ionic particles, have been applied to interpret the $S(q)$ functions. Good agreement has been obtained between observed and calculated values. From this agreement, it has often been claimed that the interaction in the real systems is purely repulsive. However, recently Sood [49] demonstrated by the Monte Carlo simulation that both the (repulsive) DLVO or Yukawa potential and the (repulsive and attractive) Sogami potential [50] provide $S(q)$ in excellent agreement with experimental values. Rajagopalan et al. [51] examined the structure of charged dispersions by using the Ornstein–Zernike formalism and concluded that fitting the experimental $S(q)$ in the neighborhood and beyond the primary peak in $S(q)$ could lead to inaccurate conclusions concerning the nature of interparticle forces. In addition, the paracrystal theory, in which only statistical fluctuation of lattice points is considered, could reproduce experimental $S(q)$ functions very well without assuming any specific potentials. These considerations tell us that it is difficult to determine the potential form from experimental $S(q)$ functions, and the previous argument in favor of the repulsion must be viewed with caution.

It should be pointed out that Overbeek criticized the Sogami theory and contended that, if solvent contribution is taken into account, the attraction in the Sogami theory disappears [52]. This conclusion cannot be correct, however, because his argument violated the Gibbs–Duhem relation, as was discussed recently by us [31].

4 Ultra-Small-Angle X-ray Scattering (USAXS) and its Application to the Structure of Colloidal Dispersions

A conventional SAXS apparatus whose optical systems consist usually of some kind of slit system has an upper limit of resolution: about 1000 Å. This limit is caused by the requirement to remove the influence of strong direct beam. To get a much better small-angle resolution for investigation of micron-size structures, it is necessary to make the beam width very narrow or extend the camera length to several hundred meters. The former technique reduces the intensity dramatically and is mechanically very difficult, and the latter is impossible on a laboratory scale.

On the other hand, there have frequently been requests to make the small-angle resolution of SAXS instruments much higher for study of the structure of various systems. Our direct incentive for the higher resolution was related to, among other things, the questions:

(1) is the ordered structure in latex dispersions, which has been confirmed by ultramicroscopy, also maintained within the dispersion?

(2) how large is the ordered "cluster" in "the two-state structure"?

As was described in Sect. 2, the ordered structures in latex dispersions have been well investigated by optical method [28]. However, by this technique, we could study only the structure of the dispersion which is close to (usually 2–3 μm, at most 10 μm away from) the cell (glass) wall as far as the combination of polystyrene-based latex particles and water (solvent) is concerned: the internal structure of the dispersion could not be studied, although the difficulty was mitigated by confocal laser scanning microscopy, to great extents [29, 31]. The electron microscope (EM) is only for solid samples. A light scattering (LS) technique is limited to transparent samples. For these reasons, we constructed ultra-small-angle X-ray scattering (USAXS) instruments [53–56] following the principle of Bonse and Hart [57]. By this technique, the problems mentioned above can be eliminated. In this Section, we will describe the principle of USAXS and discuss some applications to the structure study of colloidal systems. For the second question, i.c. the size of the cluster, we so far obtained information by small-angle-neutron scattering (SANS) at low angles and by dynamic light scattering (DLS) which will be discussed in the next Section.

4.1 Principle of Bonse–Hart Type Ultra-Small-Angle X-Ray Scattering

In 1966, Bonse and Hart proposed a new type of SAXS instrument [57]. Their optical system is shown in Fig. 4-1. It has two grooved single crystals (instead of slits) for both sides of the sample. The groove is just parallel to a lattice plane (220) and the position of the first crystal is adjusted to satisfy the Bragg condition for the incident X-ray beam. Hence, the X-ray beam is Bragg-diffracted in the groove, and then hits the sample. By this procedure, the X-ray beam becomes highly monochromatized and parallel. The X-rays transmitted and scattered by the sample go into the groove of the second crystal which is similar to the first crystal. However, only the X-ray which satisfies exactly the Bragg condition for the crystal plane (parallel to the groove surface) of the second crystal, can repeat the Bragg reflection and hence can reach the detector.

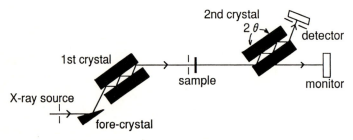

Fig. 4-1. The optical system of the Ultra-Small-Angle X-ray Scattering (USAXS) apparatus. Taken from [53], Proc. of the National Academy of Sciences, USA

According to this principle, by rotating the second crystal, we can detect weak scattered X-ray very close to the direct beam without influence by the latter. This idea was important, but no systematic structure studies by this apparatus have been done; only some applications to neutron scattering were documented [58].

In our system, Si single crystals were used as the first and second crystals, because of their high quality. The FWHM of the rocking curve was about 4 s, implying that the small-angle resolution of our systems was about 8 μm. From a simple calculation of optical geometry, the lower limit was estimated to be about 300 Å. This measurable range is much wider than that of light scattering and the lower limit overlaps that of SAXS. Although the details of the characteristics of our USAXS instrument should be referred to elsewhere [54], it should be noted that, in this system, the beam size and the camera length do not determine the small-angle resolution: the preciseness of the Bragg condition in the groove, i.e., the perfectness of the two grooved crystals, is the most important factor for resolution.

4.2 Calibration of the USAXS Apparatus

Figure 4-2 shows the USAXS patterns of polystyrene latex spheres in powder form and in a dispersion, together with the theoretical scattering curve of a sphere. For the experimental scattering curve, distinct maxima were observed both for powder and dispersion. This is due to intraparticle scattering of the sphere. Since the position of the maximum depends on the size of the sphere, the radius of the sphere can be estimated from its position. For example, the position of the first maximum q_m is related to the radius of the sphere R by the following relation;

$$q_m R = 5.76 \qquad\qquad\qquad (4\text{-}1)$$

For the scattering curve from the powder sample, an interparticle interference peak (designated A) due to the close packing of the particles was observed at very small scattering angles in addition to the (intraparticle) maxima (B). The positions of the maxima of the dispersed sample showed a satisfactory agreement with those of the theoretical one for R = 3000 Å. On the other hand, the positions for the powder sample were located at slightly larger angle values, indicating a swelling effect of the particles in ethanol dispersion. It is remarkable that USAXS can detect such a small change in size, and gives us the correct size in dispersion, which is impossible by EM technique. Figure 4-3 shows a USAXS pattern of a much larger latex sphere nearly of the order of micrometers. The characteristic maxima were also clearly observed for this sample, indicating that the micron-size structure can be investigated by the USAXS instrument. The diameter of this latex sphere could be estimated to be 8800 Å from the positions of the maxima.

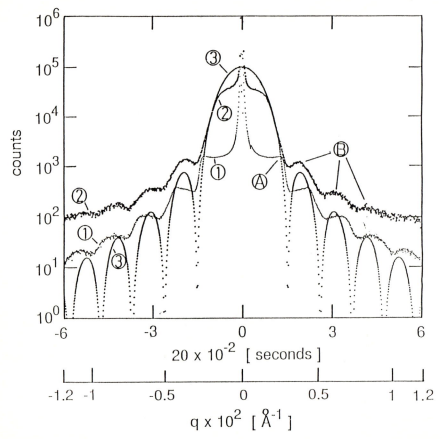

Fig. 4-2. USAXS curves of SS-80 latex (stated diameter = 3000 Å) and theoretical curve of a sphere (diameter = 3000 Å). The peak (A) results from an interparticle interference, and maxima (B) from the shape of the latex. *Curve 1*: powder sample, cps, *curve 2*: dispersion (in ethanol) sample, cps × 100, *curve 3*: theoretical curve of a sphere. Taken from [53], Proc. of the National Academy of Sciences, USA

4.3 Applications to Ordered Structures in Latex Dispersions

As already mentioned, latex particles in dispersions at very low ionic strength form an ordered structure. This phenomenon has been confirmed by ultra-microscopy techniques [28, 59]. Figure 4-4 shows USAXS patterns of latex dispersions at various concentrations. In the profile, clear and distinct Bragg peaks were observed, reflecting the formation of the ordered structure. Hence, we contend that the ordered structure was formed in the interior portion of the dispersion in addition to the location close to the cell wall. From the peak position, the interparticle distance $2D_{exp}$ could be estimated to be 7200 Å for a 6 vol.% dispersion, which is very close to the calculated value, $2D_0$. The

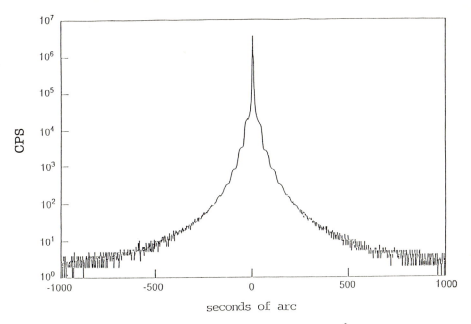

Fig. 4-3. USAXS curve of SS-83 latex powder. Stated diameter = 12 000 Å. From the positions of the maxima, the diameter was found to be 8800 Å

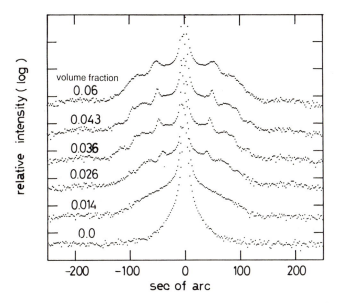

Fig. 4-4. Concentration dependence of USAXS curve of latex dispersions. Latex: MC-8 (methyl-methacrylate based). Stated diameter: 3000 Å

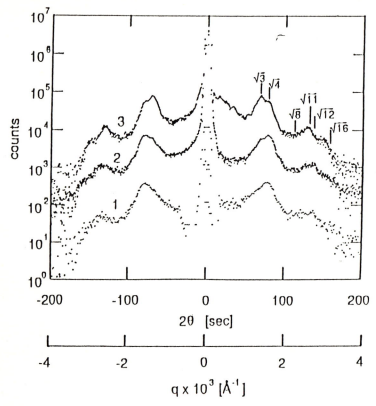

Fig. 4-5. Time evolution of the scattering curves of SS-121 latex dispersed in a water-ethanol mixture. The curves 1, 2, and 3 were taken at 1 week, 2 weeks, and 4 weeks after simple preparation, respectively. Curves 2 and 3 were shifted vertically by an order of 10. Latex concentration = 3.7 vol.%. Water:ethanol = 3:2 in volume. Accumulation time = 30 s for each point. Taken from [53], Proc. of the National Academy of Sciences, USA

relation $2D_{exp} = 2D_0$ indicates that the one-state structure was maintained at this concentration. It has been confirmed that the two-state structure is formed at much lower concentrations, below about 2 vol.% [60]. Figure 4-5 shows the time evolution of USAXS curve for a latex dispersion in a water-ethanol mixture. An addition of ethanol was necessary for polystyrene-based latices to get a sufficient density contrast: At 1 week after sample preparation, only a broad peak was observed. However, at 2 and 4 weeks, the peak grew up and higher order peaks appeared, indicating that an ordered structure of a higher degree was formed. An interesting feature is the splitting of the first peak into two peaks. Such a time evolution of scattering profiles deserves more careful investigation. The ratio of the positions of the first and second peaks of the final curve was about $3^{1/2}:4^{1/2}$, which is characteristic for an fcc structure. It is interesting to note that the fcc symmetry could not be detected by the scattering method for three weeks after sample preparation.

From analyses of the peak height and width by the 3D paracrystal theory, it may be feasible to estimate the distortion of the lattice and the size of the microcrystal, which is our next goal.

Although the details are not discussed here, the USAXS technique can be applied to a variety of systems. In fact, we are now investigating the long-range density fluctuations in polymer alloys, polymer films, fibers, metal alloys etc., which have been impossible to determine by conventional scattering techniques. Some of the preliminary results have already been described [54].

It is necessary to note that most recently four groups constructed Bose–Hart type USAXS apparatuses [61–64].

5 Small-Angle Neutron Scattering Study of the Cluster Size in Polyelectrolyte Solutions

For latex dispersions, the two-state structure has been confirmed by microscopy. In polyelectrolyte solutions it has been inferred from the fact that $2D_{exp} < 2D_0$ as described in Sect. 2. However, the size of the ordered "cluster" was not estimated. For polyelectrolyte solutions, the intermolecular distance calculated from the Bragg peak position was of the order of 100 Å. Hence, it may be a reasonable estimate that the size of the polyelectrolyte cluster would be of the order of several thousand Å to several micrometers. Because the USAXS intensity was not sufficiently high enough for solutions of ionic polymers such as NaPSS, we determined the size of the cluster by using neutron scattering [65]. In the following the SANS measurements for NaPSS-D_2O solutions are described.

The existence of the cluster has also been proposed from dynamic light scattering experiments: the time correlation function obtained for solutions of some ionic polymers at low salt concentrations showed two or three dynamic modes, fast and slow. The very slow mode was first reported by Lin, Lee, and Schurr, who coined the term "ordinary-extraordinary transition" for its appearance at low salt concentrations [11]. Several authors have also observed essentially the same phenomena by DLS [66–70]. The slow mode has been attributed to the translational diffusion of (temporal) cluster by Schmitz [66], or of multi-chain domains (clusters) by Sedlak [68, 70]. We also observed two modes for polystyrenesulfonate solutions; we attributed the slow mode to the translational diffusion of the localized ordered structures [71].

5.1 SANS Apparatus

The SANS apparatuses used were KWS I and II systems in Forschungszentrum Jülich, Germany [72]. Its maximum camera length was 20 m, which enabled us to investigate the q-range down to 10^{-3} Å^{-1}. The detector position could be changed from 2 m to 20 m to cover a wide q-range. The scattering profiles at different camera lengths were brought together to one scattering profile in a wide q-range.

5.2 Cluster Size in Sodium Polystyrenesulfonate Solutions

Figure 5-1 shows SANS patterns of NaPSS in D$_2$O at four different polymer concentrations without added salt [65]. The solvent was D$_2$O to get a high scattering contrast and to eliminate incoherent scattering. At larger q regions, a single broad but distinct peak was observed as was the case in SAXS curves (see Fig. 2-3). The peak positions of the SANS curves were nearly the same as in the SAXS cases, indicating that almost the same ordered structure was formed in

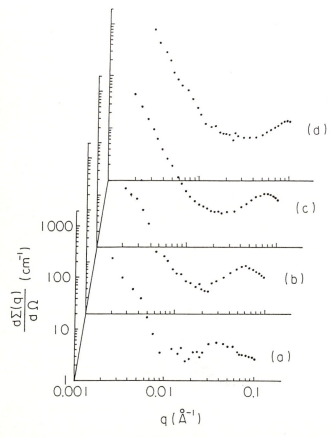

Fig. 5-1. SANS curves of an NaPSS-D$_2$O solution at different polymer concentrations without added salt. [NaPSS]: *curve (a)*: 0.01, *curve (b)*: 0.02, *curve (c)*: 0.04, *curve (d)*: 0.08 g ml^{-1}. Taken from [65] with the permission of the American Chemical Society

D_2O as in H_2O. On the other hand, in the q regions which could not be investigated by the SAXS technique, i.e. $q < 0.01$ Å^{-1}, a distinct upturn behavior was observed. The scattered intensity did not monotonically decrease with decreasing q. This outstandingly strong scattering at very small-angle regions means that there certainly exists a large scale density fluctuation in the system.

By the Guinier method [73], we conveniently estimated the size of the cluster, although we were fully aware of its approximate nature for such an inhomogeneous entity as the cluster. A typical example of the Guinier plot is shown in Fig. 5-2. From the slope, the radius of gyration, Rg, of the cluster can be calculated. The Rg values thus estimated are summarized in Table 5-1 with relevant data. The Rg value increased with increasing polymer concentrations. Because the radius (R) and hence the volume (V) of the cluster can be calculated by assuming a spherical shape for the cluster, and because the intermolecular distance ($2D_{exp}$) can be obtained from the peak position, the number of polymer ions in one cluster, N, can be found from $N = V/(2D_{exp})^3$. With increasing polymer concentration, the N value increased dramatically. The forward scattering intensity I(0) was obtained by extrapolation of the Guinier plot to $q = 0$.

Now the cluster size could be obtained. However, the number of the cluster in unit volume of solution, in other words, the number density of the cluster (n) is still unknown. To evaluate it, analysis with absolute scattering intensity and model calculation are necessary. The intensity of the forward scattering I(0) is given by

$$I(0) = nV^2(\Delta\rho)^2 \qquad\qquad (5\text{-}1)$$

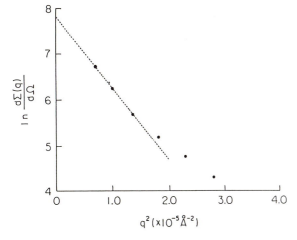

Fig. 5-2. Guinier plot of the scattering intensity at [NaPSS] = 0.04 g ml^{-1}. Taken from [65] with the permission of the American Chemical Society

Table 5-1. Small-angle neutron scattering data of salt-free sodium polystyrenesulfonate – D_2O solutions

[polymer] (g ml^{-1})	R_g [a] (A)	R [b] (A)	$2D_{exp}$ [c] (A)	$2D_0$ [d] (A)	N [e] (10^{12} cm^{-3})	n [f]	$d\Sigma(0)/d\Omega$ [g] (cm^{-1})
0.01	407	525	166	255	130	346	194
0.02	516	666	103	202	1100	14.9	532
0.04	686	886	85	160	4700	4.1	2590
0.08	630[h]	813	64	128	8600	1.5	3043

[a] The radius of gyration of the cluster determined by the Guinier method
[b] The radius of the spherical cluster
[c] Bragg spacing between macroions in the cluster
[d] Average intermacroion spacing obtained from the concentration
[e] The number of macroions in one cluster
[f] The number of clusters in unit volume
[g] The scattered intensity extrapolated to q = 0
[h] The linearity of the Guinier plot was not so good as those for the other three concentrations

where $\Delta\rho$ is the difference in the scattering length density between the inside and the outside of the cluster. Since I(0) and V are experimentally found as described above, if the $\Delta\rho$ value can be obtained, we can estimate the n value. As regards the density difference, a model of the density fluctuation shown in Fig. 5-3 was assumed. In the figure, ρ_{polym} and ρ_{solv} are the scattering length densities of NaPSS and solvent, respectively. $\rho_{cluster}$ and ρ_{dis} are the average scattering length densities of the cluster and disordered region, respectively. Although the details are to be fully described elsewhere, the n value can be calculated by

$$n^{1/2} = \frac{-\alpha^{1/2}\beta \pm \sqrt{\alpha\beta^2 + 4VI(0)}}{2I(0)^{1/2}V} \qquad (5\text{-}2)$$

where $\alpha = (M_w/N_A d)^2 (\rho_{polym} - \rho_{solv})^2$, $\beta = N - VN_{tot}$, N_{tot} is the number of polymers in the system, and d is the density of polymer. Although this model calculation cannot be applied to high salt conditions where no peak in the scattering curve was found, it can be claimed that, with increasing polymer concentration, the ordered cluster becomes larger and the polymer density in the cluster goes up but the number of the clusters decreases.

It is repeated that the numerical data in Table 5-1 were obtained from measurements in the q range shown in Fig. 5-1. More reliable data would be obtained after future measurements at much lower q values.

Following the Babinet principle, we cannot exclude the possibility that Rg is a size of void (polymer-poor region) instead of cluster (polymer-rich region), since the scattering intensity is proportional to the square of the $\Delta\rho$: whether $\Delta\rho$ is positive or negative cannot be judged by the present scattering experiments. A similar calculation with the void model is also possible, and will be described elsewhere.

It has often been taken for granted that, for salt-free cases, the scattering intensity decreases monotonically with decreasing q in the q range which could

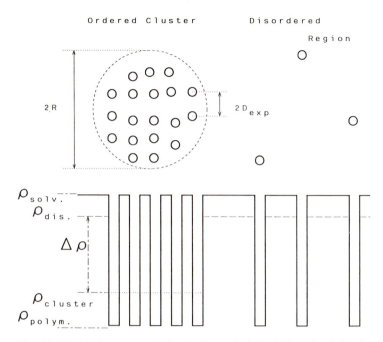

Fig. 5-3. Model of structure and density fluctuation in NaPSS-D_2O solution for the estimation of the number density of the cluster from SANS data. The circles denote macroions

be covered in previous measurements [74] and that the $I(0)$ value determined by this extrapolation is related to the osmotic compressibility. In light of the upturn observed above, this extrapolation seems unwarranted. To obtain the correct osmotic compressibility, scattering information at much lower q's than in the previous studies is required.

A further comment is necessary about the upturn. Several years ago, Boue et al. reported an upturn behavior at $(1 \sim 5) \times 10^{-2} \text{ Å}^{-1}$, which is much higher than in our measurements (between 3×10^{-2} and $3 \times 10^{-3} \text{ Å}^{-1}$), and they attributed its origin to some dusts or colloidal aggregates formed by impurities in D_2O [75]. However, according to our study, the upturn behavior is dependent of polymer and salt concentrations, which is not compatible with their interpretation. If some polymer aggregates are formed in solution, the number of polymers in solution should be decreased, which would cause a change in the scattering peak at larger q regions. However, this was not the case: the peak positions in our SAXS measurements (in H_2O) (Sect. 2) and SANS experiments (in D_2O) were almost the same. Furthermore, it is difficult to accept their interpretation in terms of the ionic impurities for the following reasons. Their curve assuredly shows an upturn in the absence of salt whereas that in the presence of salt simply increased with decreasing q. However they did not provide scattering profiles at higher q, where a Bragg peak could be expected in salt-free condition and no peak would be found in high salt condition. Thus, we find it difficult to accept the claim of Boue et al. that the observed upturn was due to ionic impurities.

It is important to refer to the work by Förster et al. [69], who observed the upturn at low angles in the SANS profile for salt-free aqueous solutions of quaternized poly(2-vinyl pyridine). By DLS experiments, they also obtained fast and slow diffusion modes and a striking q dependence of the total scattering function I(q), from which an apparent radius of gyration $R_{G, app}$ was estimated to be between 400 and 1000 Å. The slow diffusion coefficient was attributed to the motion of large domains, whose existence was linked to the SANS upturn. As mentioned above, Sedlak and Amis [70] observed two diffusion modes and evaluated the size of what they called the "multi-chain domain". From their results the radius of gyration of the "domain" is found to be in the range of 400 to 800 Å at 0.045 g ml^{-1} for a NaPSS sample of $M_W = 74\,000$. The R_G value that we obtained at 0.04 g ml^{-1} was 686 Å (Table 5-1). The numerical disagreement appears to be much narrower than the gap between the namings (multi-chain domain and localized ordered structure).

6 Remarks on Counterinterpretations

Our interpretation of the peak in the SAXS region and the upturn at low q's for ionic polymer solutions and for colloidal dispersions is reiterated as follows: the peak reflects a largely distorted ordered structure of the solute ionic polymers, though in solutions, and the upturn at low q's, testifying to the presence of a large-scale density fluctuation, is reminiscent of non-space-filling, localized ordered structures (clusters). The existence of localized ordered structures has been and is our claim since the first observation of the peak for polyacrylates [24]. On the other hand, there were proposed counterinterpretations by several research groups. It seems to be pertinent to dwell here on the pros and cons of the existing interpretations.

Our confidence in the translational ordering of ionic polymers in solutions is based on the direct observation of localized ordered structures of latex particles in dilute dispersions by microscopy [28, 31]. The microscopic study clearly showed not only ordering of the latex particles but also the two-state structure, namely coexistence of the ordered structure and free particles. The computer-assisted Fourier transformation provided a limited number of rings in the reciprocal space from the microscopic image (the real space density function) showing the two-state structure [76]. This analysis corroborates our interpretation of the single, broad peak in the SAXS study of invisible ionic polymers, although the Fourier transformation is 2-dimensional whereas the SAXS patterns are obtained from 3D real density functions. The inequality relation, $2D_{exp} < 2D_0$, was also substantiated by the microscopic investigation of dilute dispersions.

The most representative counterinterpretation is the so-called correlation hole concept proposed by Benmouna et al. [77]. According to their calculation based on the purely repulsive potential, the structure factor was shown to display a maximum. Accordingly, these authors judged our claim as misinterpretation and contended that the observed maximum is not "reminiscent of an order in the particle distribution". It is most difficult to accept the correlation hole concept for the following reasons:

(1) as discussed in our previous review article [28], this theory predicts a negative scattering intensity for bovine serum albumin under some of our experimental conditions, which is in contradiction to experimental results;

(2) the solute distribution was assumed to be homogeneous in theory, which is contrary to experimental observations at very low angles. As mentioned above, the microscopic observations also demonstrate microscopic inhomogeneity of solute distribution in macroscopically homogeneous dispersions. As an example of such inhomogeneities, the two-state structure must first be pointed out. Furthermore recent observation of the latex dispersions by confocal laser scanning microscopy showed the presence of stable void structures in which practically no particles were found [29, 31, 78, 79]. The voids were sometimes as long as about 50 μm with a maximum cross-section of about 50 μm × 150 μm. Certainly such a large size cannot be detected by existing scattering techniques. Moreover, the size of the localized ordered structure of the ionic polymers, which could be estimated by recent neutron scattering measurements at low q's [65], was also so large that it could not be detected by conventional techniques which Benmouna et al. paid attention to;

(3) the correlation hole concept was derived by assuming a purely repulsive interaction. As was discussed in recent articles [49, 31], this type of interaction (for example, the DLVO potential or the Yukawa potential) assuredly provided an excellent agreement with the observed structure factor of latex particle dispersions. However, a satisfactory agreement with observations was obtained from the Sogami potential, which contains a short-range repulsion and a long-range attraction [50]. A similar agreement was also noticed by us for the shear modulus of latex dispersions [80]. Matsumoto and Kataoka [81] demonstrated that, when use is made of the Sogami potential, the order-disorder transition of latex systems could be accurately reproduced at a very low volume fraction as was observed by the experiments. These results indicate that the purely repulsive interaction cannot be regarded as the correct potential to describe various physico-chemical properties of colloidal systems, in contradiction to the majority opinion [82]. It is a dialectical error to consider the attractive contribution unnecessary, since the attraction has never been tested. Quite interestingly, this kind of error appears to be not infrequent. Readers should realize that the error involved is quite similar to that made by a zoology student in the cockroach experiment discussed by Eigen and Winkler [83].

We note that the repulsion-only assumption is often embodied in various ways. One of them is to introduce an effective radius of particles. For example the real particle is replaced by an "effective" particle having a hard-sphere radius plus the Debye length. It was claimed that such a mental excise was quite successful in explaining various colloidal phenomena [84]. However, it was announced recently that invoking such an effective radius entails an incredibly paradoxical situation, in which the Avogadro number cannot be one of the universal constants [85]. This is an example demonstrating the absurdity of the repulsion-only assumption.

For these reasons, it may be said that the use of the purely repulsive potential in the correlation hole theory is not justified. The correlation hole is a theoretical concept based on such a potential. Furthermore the theoretical outcome is inconsistent with observation. Therefore, such a statement by Halle, Wennerstrom and Piculell [86] based on the correlation hole concept that our interpretation (in terms of the spacial ordering) needs revision is clearly unwarranted. Similarly, the statements by Kaji et al. that the correlation hole concept was supported both by theory and experiments [87] and that the idea of identifying the observed interference peak as the Bragg peak may be negated [35] are spurious. It seems that these authors identify the gravy as the goose.

One more comment is necessary about the conformation of vinylic ionic polymers in solution. A lattice model was envisioned for ionic polymer solutions [88], in which extended polyion sections are arranged in parallel. Such an extended rod structure for ionic flexible polymers appears to be widely accepted. The rationale was the earlier observation that the exponent of the Mark–Houwink–Sakurada (MHS) equation was two when the intrinsic viscosity of salt-free solutions was determined by the so-called Fuoss plot [89]. Several authors, however, found the existence of a maximum in the viscosity-concentration curves, which was inconsistent with the Fuoss relation [89]. When the maximum was taken into consideration in the determination of the intrinsic viscosity, the exponent of the MHS equation was found to be 1.2 for high molecular weight fractions of sodium polystyrenesulfonates in water [90]. It is difficult to accept the interpretation that fully stretched, rodlike ionic polymers form a lattice structure. In short, micro-Brownian motion of the polymer chains is much more vigorous than previously thought. To rephrase, a much larger number of charges on a macroion chain appears to be required to keep the chain fully stretched.

7 Conclusion

After pointing out the shortcomings of some counterinterpretations, we draw attention to the following fact: the single, broad peak in SAXS and SANS has been observed not only for linear ionic polymers but also for globular proteins and colloidal particles. This means that any correct interpretation of the scattering peak should be commonly valid to these solutes. In this respect, it seems reasonable to take the single, broad peak as implying the special (though highly distorted) ordering and the peak position (or $2D_{exp}$) as the spacing between the centers of mass of the solutes. As mentioned above, the recent viscosity measurements showed that the flexible ionic polymers may be not fully stretched out like rods but may take a more or less coiled-up conformation. This new information on the chain conformation is also consistent with the interpretation of the scattering peaks mentioned above. Figure 7-1 sketches a simplified picture of the ordered structure for flexible ionic polymers. It should be re-emphasized that such an ordered distribution of ionic polymers in solutions and colloidal particles in dispersions is formed at low salt conditions, and that the ordered structure exists as a "cluster" of a limited size when the concentrations of polymers and particles are relatively low. The USAXS measurements showed that structures in latex dispersions are certainly formed in the interior part of the dispersion. Its lattice system could be estimated from the Bragg peak positions. The 3D paracrystal theory indicated that the structure is highly distorted, justifying the appearance of the single, broad peak. The SANS measurements showed a sharp upturn of the scattering intensity at low angles, indicating that

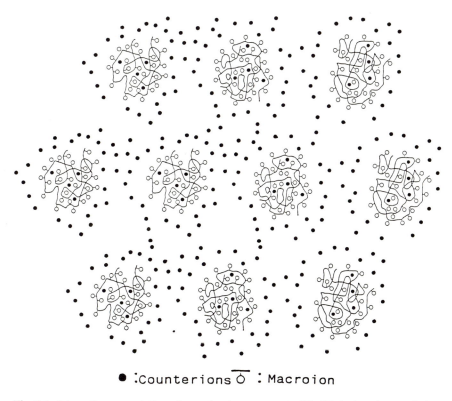

● :Counterions ○ : Macroion

Fig. 7-1. Schematic representation of an ordered arrangement of flexible ionic polymer solutions

solutions are not homogeneous in solute distribution. The size of the ordered cluster in the solutions was estimated from the upturn.

Acknowledgments: The authors' sincere thanks go to Helmut Ringsdorf, Mainz, for his kind encouragement to write the present article, to Prof. K. Dušek for his SAXS measurements of a NaPSS sample (Fig. 2-3), and also to Dr. Dietmar Schwahn, Jülich, for his providing us with the opportunity to carry out the neutron scattering measurements at low angles. Last but not least, the authors gratefully acknowledge the cooperation of their coworkers, whose names are referred to in the references.

8 List of Abbreviations

a	distance between neighboring lattice points in the ideal lattice (Average nearest neighbor interparticle distance)
Δa	standard deviation of the probability function of lattice points
bcc	body-centered cubic
d	density of polymer
$2D_{exp}$	Bragg spacing between macroions or ionic particles
$2D_0$	average spacing between macroions or ionic particles to be obtained from overall concentration
$d\Sigma(q)/d\Omega$	absolute scattering intensity of neutron
$[F(q)]^2$	form factor
$f_{\pm}$	mean activity coefficient of electrolyte
fcc	face-centered cubic
g	paracrystalline distortion factor
g(r)	radial distribution function
I	scattering intensity
k	a constant
m	electrolyte concentration $(mol\, l^{-1})$
M_w	weight-average molecular weight
N	number of macroions in one cluster
n	number of clusters in unit volume or number density of scatterer
N_A	Avogadro number
N_0	number of spherical scatterer in scattering volume
N_{tot}	total number of macroions or particles in the system
P(q)	intraparticle scattering function $(= kN_0[F(q)]^2$ in the present paper)
P_w	weight-average degree of polymerization
ρ_{dis} and $\rho_{cluster}$	average scattering length density in disordered regions and cluster, respectively
ρ_{polym} and ρ_{solv}	scattering length density of polymers and solvent, respectively

$\Delta\rho$	difference in scattering length density
q	scattering vector ($= 4\pi \sin\theta/\lambda$)
q_m	scattering vector at peak position
R	sphere radius
R_g	radius of gyration
sc	simple cubic
S(q)	interparticle structure factor
2θ	scattering angle
λ	wavelength
V	volume of cluster
Z(q)	lattice factor

9 References

1. Staudinger H (1960) Die Hochmolekularen Organischen Verbindungen, Kautschuk und Cellulose, Springer-Verlag, Berlin Heidelberg New York p 333
2. Bernal JD, Fankuchen I (1941) J Gen Physiol 25: 111
3. Riley DP, Oster G (1951) Discuss Faraday Soc 11: 107
4. Schulman JH, Riley DP (1948) J Coll Sci 3: 383
5. Brady CW (1951) J Chem Phys 19: 1547
6. Guinand S, Boyer-Kawenoki P, Dobry A, Tonnelat J (1949) C R Hebd Seances Acad Sci 229: 143
7. Doty P, Steiner RF (1949) J Chem Phys 17: 743; (1952) J Chem Phys 20: 85
8. Fuoss RM, Edelson D (1951) J Polymer Sci 6: 767
9. Veis A, Eggenberger DN (1953) J Am Chem Soc 76: 1560
10. Lin SC, Schurr JM (1978) Biopolymers 17: 425
11. Lin SC, Lee WI, Schurr JM (1978) Biopolymers 17: 1041
12. Schmitz KS (1990) An introduction to dynamic light scattering by macromolecules, Academic, San Diego, pp 377–396
13. Schmitz KS, Parthasarathy N (1981) In: Chen SH, Chu B, Nassal R (eds) Scattering techniques applied to supramolecular and nonequilibrium systems. Plenum, New York
14. Cotton JP, Moan M (1976) J Phys (Paris) 37: L-75
15. Frank HS, Thompson PT (1959) J Chem Phys 31: 1086
16. Debye PJW, Hückel E (1923) Physik Z 24: 185
17. Robinson, RA, Stokes RH (1959) Electrolyte solutions. Butterworth and Co Ltd, London, Chap 9
18. Desnoyers JE, Conway BE (1964) J Phys Chem 68: 2305
19. Bahe LW (1972) J Phys Chem 76: 1062
20. Bahe LW, Parker D (1975) J Am Chem Soc 97: 5664
21. To save space, only a review article is given here: Ise N (1971) Adv Polymer Sci 7: 536
22. Ise N, Okubo T, Hiragi Y, Kawai H, Hashimoto T, Fujimura M, Nakajima A, Hayashi II (1979) J Am Chem Soc 101: 5836
23. Plestil J, Mikes J, Dusek K (1979) Acta Polym 30: 29
24. Ise N, Okubo T, Yamamoto K, Kawai H, Hashimoto T, Fujimura M, Hiragi Y (1980) J Am Chem Soc 102: 7901
25. Ise N, Okubo T, Kunugi S, Matsuoka H, Yamamoto K, Ishii Y (1984) J Chem Phys 81: 3294
26. Yoshikawa Y, Matsuoka H, Ise N (1986) British Polym J 18: 242
27. Williams C, private communication
28. Ise N (1986) Angew Chem 25: 323
29. Ito K, Yoshida H, Ise N (1992) Chem Lett 1992: 2081
30. Ise N, Okubo T (1980) Acc Chem Res 13: 303

31. Dosho S, Ise N, Ito K, Iwai S, Kitano H, Matsuoka H, Nakamura H, Okumura H, Ono T, Sogami IS, Ueno Y, Yoshida H, Yoshiyama T (1993) Langmuir 9: 394
32. Matsuoka H, Ise N, Okubo T, Kunugi S, Tomiyama H, Yoshikawa Y (1985) J Chem Phys 83: 378
33. Patrowski A, Gulari E, Chu B (1980) J Chem Phys 78: 4187
34. Ishii Y, Matsuoka H, Ise N (1986) Ber Bunsenges Phys Chem 90: 50
35. Kaji K, Urakawa H, Kanaya T, Kitamaru R (1984) Macromolecules 17: 1835
36. Glatter O (1982) In: Kratky O, Glatter O (eds) Small-angle X-ray scattering. Academic Press, NY
37. Matsuoka H, Murai H, Ise N (1988) Phys Rev B 37: 1368
38. Sakka, S (1988) Science of the sol-gel method. Shoufu-Sha Publishing Co, Tokyo, p 31
39. In our original paper, d was assumed to be 1.576 and it was found that $2D_{exp}$ values were very close to $2D_0$ values, indicating the one-state structure. This seemed to be reasonable in light of the low charge density of the particles. Further study is certainly necessary to reach the final conclusion as to which of the two-state structure or the one-state structure is real for this system
40. Hayter JB, Penfold J (1981) Mol Phys 42: 109
41. Hansen JP, Hayter JB (1982) Mol Phys 46: 651
42. Hosemann R, Baguchi SN (1962) Direct analysis of diffraction by matter. North Holland, Amsterdam
43. Yarruso DJ, Cooper SL (1981) Macromolecules 16: 1871
44. Matsuoka H, Tanaka H, Hashimoto T, Ise N (1987) Phys Rev B 36: 1754
45. Matsuoka H, Tanaka H, Iizuka N, Hashimoto T, Ise N (1990) Phys Rev B 41: 3854
46. Härtel W, Versmold H, Wittig U (1984) Ber Bunsenges Phys Chem 88: 1063
47. Ottewill RH (1980) Progr Coll Polym Sci 67: 71
48. Cebula DJ, Goodwin JW, Jeffrey GC, Ottewill RH, Parentich A, Richardson RA (1983) Faraday Discus Chem Soc 76: 37
49. Sood, AK (1991) In: Ehrenreich H, Turnbull D (eds) Solid state physics. Academic Press, San Diego CA p 2
50. Sogami IS, Ise N (1984) J Chem Phys 81: 6320
51. Salgi P, Guerin JF, Rajagopalan R (1992) Colloid Polym Sci 270: 785
52. Overbeek, JThG (1987) J Chem Phys 87: 4406
53. Matsuoka H, Kakigami K, Ise N, Kobayashi Y, Machitani Y, Kikuchi T, Kato T (1991) Proc Natl Acad Sci USA 88: 6618
54. Matsuoka H, Kakigami K, Ise N (1991) The Rigaku Journal 8: 21
55. Matsuoka H, Kakigami K, Ise N (1991) Proc Japan Academy 67B: 170
56. Matsuoka H et al (1991) Photon Factory Activity Report 9: 331
57. Bonse U, Hart M (1966) Z Phys 189: 151
58. Alefeld B, Schwahn D, Springer T (1985) Nucl Inst Methods Phys Res A239: 229
59. Kose A, Ozaki M, Takano Y, Kobayashi Y, Hachisu S (1973) J Coll Int Sci 44: 330
60. Ito K, Nakamura H, Ise N (1986) J Chem Phys 85: 6136
61. Lambard J, Zemb Th (1991) J Appl Cryst 24: 255
62. Chu B, Li Y, Gao T (1992) Rev Sci Instrum 63: 4128
63. Pahl P, Bonse U (1991) J Appl Cryst 24: 771
64. North AN, Rigden JS, Mackie AR (1992) Rev Sci Instrum 63: 1741
65. Matsuoka H, Schwahn D, Ise N (1991) Macromolecules 24: 4227
66. Schmitz KS, Lu M, Singh N, Ramsay DJ (1984) Biopolymers 23: 1637
67. Drifford M, Dabliez JP (1985) Biopolymers 24: 1501
68. Sedlak M, Konak C, Stepanek P, Jakes J (1987) Polymer 28: 873
69. Förster S, Schmidt M, Antonietti M (1990) Polymer 31: 781
70. Sedlak M, Amis EJ (1992) J Chem Phys 96: 817, 826
71. Matsuoka H, Hattori N, Ise N (in preparation)
72. Alefeld B, Schwahn D, Springer T (1989) Nucl Inst Methods Phys Res A274: 210
73. Guinier A, Fournet G (1955) Small-angle scattering of X-rays. Wiley, NY
74. Nierlich M, Williams CE, Boue F, Cotton JP, Daoud M, Farnoux B, Jannink G, Picot C, Moan M, Wolff C, Rinaudo M, deGennes PG (1979) J Phys (Paris) 40: 701
75. Boue F, Daoud M, Nierlich M, Jannink G, Benoit H, Dupplessix R, Picot C (1977) Proc Symp Neutron Inelastic Scattering 1: 563
76. Ito K, Ise N (1987) J Chem Phys 86: 6502
77. Benmouna M, Weill G, Benoit H, Akcasu Z (1982) J Phys 43: 1679
78. Ito K, Yoshida H, Ise N (1992) Chem Lett 2081

79. Ito K, Yoshida H, Ise N (1993) Science (in press)
80. Ito K, Sumaru K, Ise N (1992) Phys Rev B 46: 3105
81. Matsumoto M, Kataoka Y (1988) In: Ise N, Sogami I (eds) Ordering and organization in ionic solutions. World Scientific, Singapore, p 574
82. Overbeek, JThG (1990) Faraday Discuss 90: 183
83. Eigen M, Winkler R (1976) Das Spiel: Naturgesetze steuern den Zufall. Piper, München, Chapter 17 (English version (1981) Law of the game: how the principles of nature govern chance. Penguin Books, New York). The story goes as follows: A zoology student intended to prove the fact that cockroaches have auditory organs on their legs. He trained and trained cockroaches and finally succeeded: when he ordered to cockroaches, "Forward march", they moved forward. Then he took all legs from one cockroach and again ordered, "Forward march". The cockroaches moved forward except for the one without legs. Based on this observation, the student proudly claimed that the experiment has conclusively proven that the cockroaches hear with legs
84. Brenner SL (1976) J Phys Chem 80: 1473; Barnes CJ, Chan DYC, Everett H, Yates DE (1978) J Chem Soc Faraday Trans 2 74: 136; Hachisu S, Takano K (1982) Adv Colloid Interface Sci 16: 233; Furusawa K, Yamashita S (1982) J Colloid Interface Sci 89: 574; Okubo T (1987) J Chem Phys 86: 2394, 5182, 5528; (1988) 88: 2083, 6581; (1987) Colloid Polym Sci 265: 522, 597; (1988) 266: 1042, 1049
85. Ito K, Ieki T, Ise N (1992) Langmuir 8: 2952
86. Halle B, Wennerstrom H, Piculell L (1984) J Phys Chem 88: 2482
87. Kaji K, Kurita K (1984) Fijikkusu. Maruzen Publishing Co, Tokyo 2: 103 (in Japanese)
88. Urakawa H, Kanaya T, Kaji K, Kitamaru R (1982) Polymer Reprints, Japan 31: 2245
89. Fuoss RM (1952) Faraday Discuss 11: 125
90. Yamanaka J, Matsuoka H, Kitano H, Hasegawa M, Ise N (1990) J Am Chem Soc 112: 587
91. van der Maarel JRC, Groot LCA, Hollander JG, Jesse W, Kuil ME, Leyte JC, Leyte-Zuiderweg LH, Mandel M, Cotton JP, Jannink G, Lapp A, Farago B (1993) Macromolecules 26: 7295

Editor: Prof. H. Ringsdorf
Received: June 1993

Note added during print: We are pleased to read a recent article by Maarel et al. [91], who observed the upturn at low angles ($q < 0.02$ Å^{-1}) in the scattering profile of polystyrenesulfonate solutions, declining the widely believed view advanced by Boue et al. [75] that the upturn is due to impurities.

Static and Dynamic Light Scattering on Moderately Concentrated Solutions: Isotropic Solutions of Flexible and Rodlike Chains and Nematic Solutions of Rodlike Chains

G.C. Berry

Department of Chemistry, Carnegie Mellon University, Pittsburgh, PA, USA

The static and dynamic light scattering from moderately concentrated polymer solutions is discussed, including the light scattering from both isotropic and nematic solutions. Polarized light scattering is discussed for flexible and semiflexible chain polymers to elucidate the thermodynamic and dynamic information accessible from the light scattering data in the range of concentrations between the dilute and concentrated ranges. The behavior under both Flory Theta conditions (vanishing second virial coefficient) and in good solvents is considered. Depolarized light scattering is discussed for semiflexible and rodlike chains, including both ordered and nematic solutions. In the latter case, the depolarized scattering of fully aligned nematic solutions is discussed in terms of the Frank curvature elasticities and the Ericksen–Leslie anisotropic viscosity coefficients.

List of Symbols

$a(\rho/L, z)$	The ratio $A_2/A_2^{(R)}$; see Eq. 5
$a_1(\rho/L)$	A function in the excluded volume; see Eq. 6
A_2, A_3	The second virial coefficient, the third virial coefficient, etc.; see Eq. 5
$A_2^{(R)}$	The second virial coefficient for a rodlike chain; see Eq. 5
$\hat{A}_i(S)$	The dimensionless function $\hat{A}_i/(1 - S^2)$; see Eqs. 128–133
$\hat{A}_i(A_2 Mc)$	The dimensionless ratio α_i/η^{ISO}; see Eq. 122
$b(q, c)$	A correlation length obtained from the dependence of $\mathbf{R}(q)$ on q; see Eqn. 30
$B(c)$	The function $\{F(0, c) - 1\}/c$; Eq. 8
c	The solute concentration (wt/vol)
$\hat{c}$	The reduced concentration $cN_A R_G^3/M$
c_{NI}	The concentration for the isotropic to nematic phase transition
$C^{(2)}$	The coefficient to u in an expansion of $M(u, 0)$; see Eq. 35
d_G	Geometric radius of a chain element
d_H	Hydrodynamic radius of a chain element; see Eq. 3
d_{Thermo}	Thermodynamic radius of a chain element; see Eq. 5
$D_\perp(c)$	The translational diffusion coefficient perpendicular to the chain axis, at concentration c; see Eq. 76
$D_{\perp; 0}$	The translational diffusion coefficient perpendicular to the chain axis, at infinite dilution; see Eq. 74
$D_\parallel(c)$	The translational diffusion coefficient parallel to the chain axis, at concentration c; see Eq. 76
$D_{\parallel; 0}$	The translational diffusion coefficient parallel to the chain axis, at infinite dilution; see Eq. 73
D_{gel}	The product fD_M; see Eq. 51
$D_{Hv}(c)$	The effective diffusion coefficient, given by $K_{1; Hv}(q, c, u_\pi)/6$ in the limit of small q
$D_M(c)$	The mutual diffusion coefficient, given by $K_1(q, c)/q^2$ in the limit of small q
$D_R(c)$	The rotational diffusion coefficient at concentration c; see Eq. 77
$D_{R; 0}$	The rotational diffusion coefficient at infinite dilution; see Eq. 72
$D_S(c)$	The self diffusion constant; see Eq. 23
D_T	Translational diffusion constant at infinite dilution
f	The function $1 + \Delta L/L(\infty)$; see Eq. 51
f_c	An optical coherence factor given by $g^{(2)}(0; q) - 1$; see Eq. 16
$F(q, c)$	The intermolecular structure factor $\mathbf{R}(q, c)/K_{op}cMP(q, c)$; see Eq. 7

$F_{Hv}(q, c)$ — The intermolecular structure factor
$R_{Hv}(q, c)/(3/5)K_{op}cM\delta^2 P_{Hv}((q, \vartheta)$; see Eq. 62

$F_{Vv}(q, c)$ — The intermolecular structure factor
$R_{Vv}(q, c)/K_{op}cMP_{Vv}((q, c)$; see Eq. 61

$F_\mu(A, i, n)$ — A geometric function appearing in an expression for $\Gamma_{Ai; n\mu}(q_\perp, q_\parallel)$; see Eqs. 95 and 96

$g(t/\tau_G)$ — The relaxation function for $G(t)$; see Eq. 45

$g^{(1)}(\tau; q, c)$ — The normalized electric field autocorrelation function, $g^{(1)}(\tau; q)$; see Eq. 16

$g^{(1)}_{Ai; n}(\tau; q_\perp, q_\parallel)$ — The normalized electric field autocorrelation function for a nematic medium; see Eq. 85

$g^{(2)}(\tau; q, c)$ — The normalized photon count autocorrelation function $G^{(2)}(\tau; q)/G^{(2)}(\infty; q)$

$g^{(2)}_{Ai; n}(\tau; q_\perp, q_\parallel)$ — The normalized photon count autocorrelation function for a nematic medium; see Eq. 85

$G(t)$ — The shear relaxation modulus; instantaneous value $G(0)$

G_e — The equilibrium shear modulus ($G_e = 0$ for a fluid)

G_N — The pseudo-network modulus of an entangled flexible chain polymer

$G^{(2)}(\tau; \mathbf{q}, c)$ — The unnormalized photon count autocorrelation function, also $G^{(2)}(\tau; \mathbf{q})$; see Eq. 1

$G^{(2)}_{Ai; n}(\tau; q_\perp, q_\parallel)$ — The unnormalized photon count autocorrelation function for a nematic medium

$h(\rho/L, d_H/L)$ — The ratio $3R_H/2R_G$; see Eq. 3

$H(q, c)$ — The function $\{F(q, c)^{-1} - 1\}/c\Gamma(c)P(q, c)$; see Eq. 9

k — Boltzmann's constant

$k(t/\tau_K)$ — The relaxation function for $K_{OS}(t)$; see Eq. 45

k_α — A constant in a representation of α_c; see Eq. 29

k_D — The coefficient of $[\eta]c$ in an expansion of $D_M(c)/D_T$; see Eq. 41

$k_\eta(\rho/L, d_H/L)$ — The ratio $M[\eta]/\pi N_A R_H R_G^2$; see Eq. 4

k_μ — The dimensionless ratio $\hat{k}_\mu/A_2 Mc$; see Eqs. 104–106

$\hat{k}_\mu$ — The dimensionless ratio $K_\mu/(kT/d_H)(A_2 Mc)^2$; see Eq. 103

$K_1(q, c)$ — The first-cumulant; see Eq. 18

$K_{1; Hv}(q, c, u_\pi)$ — The first-cumulant for the horizontally polarized scattering with vertically polarized light

K_{op} — An optical constant relating intensities to the Rayleigh ratio

K_{OS} — The equilibrium osmotic modulus

$K_{OS}(t)$ — The osmotic modulus; instantaneous value $K_{OS}(0)$

K_μ — For $\mu = S, T, B$, Frank curvature elasticities for splay, twist and bend distortions, respectively; see Eq. 90

$\hat{K}_\mu(q_\perp, q_\parallel)$ — A function appearing in an expression for $\Gamma_{Ai; n\mu}(q_\perp, q_\parallel)$; see Eqs. 95 and 96

$\ell(t)$ — The relaxation function for $L(t)$; see Eq. 45

L — Chain contour length

$L(t)$ The effective longitudinal modulus $K_{os}(t) + (4/3)G(t)$; instantaneous value $L(0)$

$L(\infty)$ The equilibrium value of $L(t)$; see Eq. 45

M Molecular weight

M_e The effective molecular weight between entanglement loci for undiluted polymer

M_L The mass per unit length, M/L

$M(u, c)$ The ratio $K_1(q, c)/q^2 D_M(c)$; see Eq. 22

$M_{Hv}(u, c, u_\pi)$ The ratio $K_{1;\,Hv}(q, c, u_\pi)/6D_{Hv}(c)$; see Eq. 81

$\mathbf{n}$ The director of a nematic phase

$n(t; \mathbf{q})$ Photon count at time t and wave vector $\mathbf{q}$

N_A Avagadro's constant

p An exponent in a power-law representation of $\Gamma(c)$; see Eq. 25

$p_i(X), m_i(X)$ For $i = 1, 2, 3$, functions appearing in the scattering function for rodlike chains

$P(q, c)$ The intramolecular structure factor; see Eq. 7

$P_{Hv}(q, \vartheta; \delta)$ The intramolecular structure factor for the vertically polarized component of the scattering with vertically polarized incident light; see Eq. 62

$P_{Vv}(q; \delta)$ The intramolecular structure factor for the vertically polarized component of the scattering with vertically polarized incident light; see Eq. 61

$\mathbf{q}$ The wave vector, with modulus $q = (4\pi/\lambda)\sin(\vartheta/2)$ for an isotropic medium

$q_\perp, q_\parallel$ Components of the scattering vector; see Eq. 85

$Q(q, c)$ The function $\{F(q, c) - 1\}/cB(c)P(q, c)$; see Eq. 8

r The weight factor r_1 for a case with two components in Eq. 17; see Eq. 21 Also, the weight factor $\Gamma_S/\sum \Gamma_\mu$ in Eq. 112

$r(\rho/L)$ The ratio $R_G^2/(L\rho/3)$ in the absence of excluded volume; see Eq. 2

$r_\mu(q, c)$ Weight factors in a representation of $g^{(1)}(\tau; q)$, also $r_\mu(q)$; see Eqs. 17 and 20

R The Gas Constant $N_A k$

$\mathbf{R}(q, c)$ Rayleigh ratio (Eq. 1); also $\mathbf{R}(q)$

$\mathbf{R}_{Hv}(q, c)$ The horizontally polarized component of the Rayleigh ratio for vertically polarized incident light

$\mathbf{R}_{Vv}(q, c)$ The vertically polarized component of the Rayleigh ratio for vertically polarized incident light

R_G Root-mean-square radius of gyration

R_H Hydrodynamic radius, defined as $kT/6\pi\eta_s D_T$

s Exponent in a power-law relation for G_N, $s = (1 - p)/3$

$s(z)$ A function in the excluded volume; see Eq. 6

$S(q, c)$ The total structure factor $\mathbf{R}(q, c)/\mathbf{K}_{op}cM$; see Eq. 7

$S_{Hv}(q, c; \delta)$ The total structure factor $\mathbf{R}_{Hv}(q, c)/(3/5)\mathbf{K}_{op}cM\delta^2$; see Eq. 62

$S_{Vv}(q, c; \delta)$ The total structure factor $\mathbf{R}_{Vv}(q, c)/\mathbf{K}_{op}cM$; see Eq. 61

t	Time, e.g., in a relaxation function
u	A unit vector along the axis of a rodlike chain
u	The function $(qR_G)^2$; u_π is the value of u for $\vartheta = \pi$
V(c)	The ratio $\Gamma(c)/3A_3Mc$; see Eq. 27
w	The solute weight fraction
$W(\rho/L)$	A function appearing in the optical anisotropy of a wormlike chain; see Eq. 59
X	The scattering parameter $(12\,u)^{1/2}$; see Eq. 64
z	A thermodynamic interaction parameter; see Eq. 5
$\Delta K_{OS}, \Delta G, \Delta L$	$K_{OS}(0) - K_{OS}$, $G(0) - G_e$, and $L(0) - L(\infty) = \Delta K_{OS} + (4/3)\Delta G$, respectively
α_c	The excludedvolume factor at concentration c; see Eq. 29
α_i	For i = 1–6, Ericksen–Leslie viscosity coefficients; see Eq. 91
α_0	The excluded volume ratio at infinite dilution; $\alpha_o^2 = R_G^2/(L\rho/3)$
$\alpha_\parallel, \alpha_\perp$	Polarizabilities parallel and perpendicular to the symmetry axis of the scattering element
$\gamma_\mu(q, c)$	Relaxation rates in a representation of $g^{(1)}(\tau; q)$, also $\gamma_\mu(q)$; see Eq. 17. Also, for nematic media, $\gamma_1 = \alpha_3 - \alpha_2$ and $\gamma_2 = \alpha_3 + \alpha_2$
$\hat{\gamma}_\mu(q, c)$	Reduced relaxation rates in a representation of $g^{(1)}(\tau; q)$, also $\hat{\gamma}_\mu(q)$; see Eq. 20
$\Gamma(c)$	The function $\{F(0, c)^{-1} - 1\}/c$; see Eq. 9
$\Gamma_{Ai;\,n\mu}(q_\perp, q_\parallel)$	Weight factors in a representation of $g_{Ai;\,n}^{(1)}(\tau; q_\perp, q_\parallel)$; see Eq. 86
δ	The optical anisotropy of a polymer chain; see Eqs. 57 and 62
δ_o	The optical anisotropy of a scattering element with molecular weight m_o
$\zeta(c)$	The segmental friction factor; see Eq. 42
η^{ISO}	The viscosity calculated as $\eta_s K_\eta N_A^2 M[\eta](cL/M_L)^3$; see Eq. 123
η_s	The solvent viscosity
η_μ	For μ = S, T, B, viscosity coefficients related to splay, twist and bend distortions, respectively
$\hat{\eta}_\mu(q_\perp, q_\parallel)$	A function in the numerator for $\hat{\tau}_\mu(q_\perp, q_\parallel)$; see Eq. 110
$[\eta]$	Intrinsic viscosity (vol/wt)
ϑ	The scattering angle
κ	The dimensionless parameter $K_1\tau_L$; see Eq. 51
λ	The wavelength of light in the scattering medium; λ_o the same *in vaccu*; also, with nematic media, $\lambda = -\gamma_2/\gamma_1$
μ	The exponent $\partial\ln[\eta]/\partial\ln M$; a subscript used in various places
$\xi_H(q, c)$	A correlation length in a representation of b(q, c); see Eq. 30
$\xi_P(q, c)$	A correlation length in a representation of b(q, c); see Eq. 30

$\xi_\Xi(c)$	A correlation length for hydrodynamic interactions; see Eq. 43
$\Xi(c)$	The molecular friction factor; $\Xi(0) = kT/D_T$
ρ	Chain persistence length
$\hat{\rho}$	The density of an undiluted polymer (vol/wt)
τ	The time in the photon count autocorrelation function
τ_K, τ_G, τ_L	Relaxation times; see Eq. 45
τ_μ	For $\mu = S, T, B$, relaxation times given by $\eta_\mu/K_\mu q^2$
$\hat{\tau}_\mu(q_\perp, q_\parallel)$	Functions in a representation of $g_{Ai;n}^{(1)}(\tau; q_\perp, q_\parallel)$; see Eqs. 86 and 110
φ	The solute volume fraction
ψ_j	Reduced virial coefficients $A_j M(M/N_A R_G^3)^{j-1}$ for integer i (e.g., $\psi_2 = A_2 M(M/N_A R_G^3)$, etc.); see Eq. 14
Ω_{23}	The ratio ψ_2^2/ψ_3
Ω_∞	The limiting value of Ω_{23} for flexible chains with strong excluded volume

1 Introduction

Electromagnetic scattering (light, X-ray and neutron) has long been used to characterize certain properties of polymers in dilute solutions [1–9]. In recent years scattering has also been applied to provide useful information on inter-molecular thermodynamics and dynamics on moderately concentrated solutions of macromolecules. Here, a moderately concentrated solution is one for which the concentration c (wt/vol) is much less than the density ρ of the undiluted polymer, but in the range of the reciprocal of the volume swept out by the chain in rotation about its center of mass; see below for a more precise definition. Static and dynamic light scattering from solutions in this concentration range will be discussed for two cases: isotropic solutions of flexible and rodlike chain polymers, and nematic solutions of rodlike chains. The former will emphasize polarized scattering, whereas the latter will emphasize depolarized scattering. (Here, polarized and depolarized scattering refer to scattering in the horizontal plane, with vertically polarized incident light and vertically or hori-zontally polarized scattered light, respectively.) The general principles of polar-ized and depolarized static and dynamic light scattering may be found in a number of monographs or reviews [1–9] – the nomenclature here will follow [6] for the most part. In particular, the principal function of interest here in either case is the (unnormalized) photon count autocorrelation function $G^{(2)}(\tau; \mathbf{q})$ computed from the photon count statistics for a given wave vector $\mathbf{q}$ [4, 6–9]:

$$G^{(2)}(\tau; \mathbf{q}) = \langle n(t; \mathbf{q}) n(\tau + t; \mathbf{q}) \rangle_t \qquad (1)$$

where $n(t)$ is the number of photons detected over the time interval t to $t + \Delta t$, and the average is over the time of the data acquisition, which must exceed the longest correlation times in the photon count statistics (see below); the effect of the interval Δt on the result is discussed in detail elsewhere [6, 8]. In the limit with τ longer than all such correlation times, $G^{(2)}(\tau; \mathbf{q})$ tends to an asymptotic limit $G^{(2)}(\infty; \mathbf{q})$, where $[G^{(2)}(\infty; \mathbf{q})]^{1/2}$ is proportional to the time-averaged Rayleigh ratio $\mathbf{R}(\mathbf{q})$ [4, 6] – notation to indicate the dependence of $G^{(2)}(\tau; \mathbf{q})$ and $\mathbf{R}(\mathbf{q})$ on solute concentration c is sometimes suppressed for convenience but, for example, $G^{(2)}(\tau; \mathbf{q}) = G^{(2)}(\tau; \mathbf{q}, c)$, $\mathbf{R}(\mathbf{q}) = \mathbf{R}(\mathbf{q}, c)$, etc.

1.1 Molecular Parameters

Several parameters determined in dilute solutions used in the following to characterize the chain conformation, hydrodynamics and thermodynamics are presented here for convenience – additional parameters are presented as needed. Further details on the functional forms summarized may be found elsewhere [1, 10–16]. For example, under Flory Theta conditions, given by the condition that the second virial coefficient A_2 is zero, the root mean square radius of gyration

R_G may be expressed in terms of the chain contour length L and the persistence length ρ [10, 12]:

$$R_G^2 = (L\rho/3) \, r(\rho/L) \tag{2a}$$

$$r(\rho/L) \approx \{1 + 4\rho/L\}^{-1} \tag{2b}$$

where Eq. (2b) is a useful approximation to the exact relation [15, 16]; $R_G^2 = L\rho/3$ for flexible chains ($\rho/L \ll 1$) and $R_G^2 = L^2/12$ for rodlike chains ($\rho/L \gg 1$). Similarly, under Flory Theta conditions, the hydrodynamic radius R_H may be expressed in the form [10, 12]:

$$R_H = (2/3)R_G h(\rho/L, d_H/L) \tag{3a}$$

$$h(\rho/L, d_H/L) \approx \left\{ \frac{1 + 4\rho/L}{1 + (16\rho/27L)\ln^2(3L/2d_H)} \right\}^{1/2} \tag{3b}$$

where d_H is the hydrodynamic diameter of the chain, often equated with the geometric diameter d_G of the chain, and Eq. (3b) is an approximation to the exact relation [15, 16]; R_H is calculated as $R_H = kT/6\pi\eta_s D_T$, with D_T the translational diffusion coefficient, and η_s the viscosity of the solvent. The intrinsic viscosity $[\eta]$ may be related to M, R_G and R_H by the expression [17]

$$M[\eta] = \pi N_A R_H R_G^2 k_\eta(\rho/L, d_H/L) \tag{4}$$

where $k_\eta(\rho/L, d_H/L)$ is unity for rodlike chains, and tends to a constant ($\approx 10/3$) for high molecular weight flexible chain polymers; N_A is Avogadro's number. The second virial coefficient may be expressed in the form [10, 12]

$$A_2 = A_2^{(R)} a(\rho/L, z) \tag{5a}$$

$$A_2^{(R)} = (\pi N_A/4M_L^2) d_{Thermo} \tag{5b}$$

$$z = (3d_{Thermo}/16\rho)(3L/\pi\rho)^{1/2} \tag{5c}$$

$$a(\rho/L, z) \approx \{1 + b_1(\rho/L)z/\alpha_0^3\}^{-1} \tag{5d}$$

where the thermodynamic diameter d_{Thermo} of the chain is a measure of the length scale of the segmental interactions, α_0 is the expansion factor for the radius of gyration (see below), $M_L = M/L$, and Eq. (5d) is an approximation to more complicated relations [15, 16]. The function $b_1(\rho/L)$ tends to a constant $b_1 \approx 2.865$ for flexible chains ($\rho/L \ll 1$), and to zero for rodlike chains ($\rho/L \gg 1$). The parameter $A_2^{(R)}$ is proportional to the binary cluster integral β_2 [10–14]. At the present time, it is beyond the scope of theory to provide estimates of $A_2^{(R)}$. However, d_{Thermo} reduces to zero at the Flory Theta temperature, for which A_2 is zero, is about equal to the geometrical chain diameter d_G for chains interacting through a hard-core repulsive potential, and can be much larger than d_G for polyelectrolyte chains; e.g., see the discussion in [15].

Finally, in solutions in good solvents, R_G is enhanced by an expansion parameter α_c for flexible chain polymers; α_c tends to a limit α_0 at infinite dilution, and to unity with increasing c. At infinite dilution, the effect of this chain expansion on R_G and R_H may be taken into account by multiplication of ρ by α_0^2 in the relations given above, e.g., at infinite dilution, $R_G^2 = (L\rho/3)\alpha_0^2$ for a flexible chain polymer. In general, α_0 may be expressed in the form [1, 6, 10–14]:

$$\alpha_0^2 = 1 + a_1(\rho/L)s(z) \tag{6}$$

where the function $a_1(\rho/L)$ tends to a constant $a_1 = 134/105$ for flexible chains $(\rho/L \ll 1)$, and to zero for rodlike chains $(\rho/L \gg 1)$. The function $s(z)$ is equal to z for small z, and tends to scale with $z^{2/5}$ for large z [1, 10–14]. Further, $s(z) \approx A_2 M^2 \alpha_0^3/4\pi^{3/2} N_A R_G^3$ for flexible chain polymers over a range of z [10].

1.2 Static Scattering from Isotropic Solutions

Certain general relations used with the polarized static scattering from isotropic solutions are gathered in this section. Relations for the depolarized scattering from isotropic solutions are given in the section on Isotropic Solutions of Semiflexible Chain Polymers, and the static scattering from ordered solutions is considered in the section on Nematic Solutions of Semiflexible Chain Polymers.

For isotropic solutions, the polarized scattering is a function of the modulus q of the wave vector $\mathbf{q}$, with $q = (4\pi/\lambda)\sin(\vartheta/2)$, where λ is the wavelength of the scattered light in the scattering medium, and ϑ is the scattering angle. As discussed below, for nematic solutions, the depolarized scattering depends on certain components of $\mathbf{q}$, selected by the orientation of the optical symmetry axis of the nematic fluid with respect to the polarizations of the incident and scattered beams.

In all cases of interest here, the scattering is in the Rayleigh–Gans limit [4, 6], so that in the absence of depolarized scattering (see below), the Rayleigh ratio may be expressed in the form [2–4, 6, 12]

$$\mathbf{R}(q, c) = K_{op} c M\, S(q, c) \tag{7a}$$

$$S(q, c) = P(q, c)F(q, c) \tag{7b}$$

where K_{op} is an optical constant, and $P(q, c)$ and $F(q, c)$ are the intramolecular and intermolecular structure factors, respectively; $P(0, c)$ is unity, but $F(0, c)$ will generally not be equal to unity except at infinite dilution. Two expressions are commonly employed to represent $F(q, c)$. Scattering theory leads naturally to the relation [3, 4, 6, 12]:

$$F(q, c) = 1 - cB(c)P(q, c)Q(q, c) \tag{8}$$

where $Q(q, c)$ depends on intermolecular interference, with $Q(0, c) = 1$ (Note:

the nomenclature for $Q(q, c)$ here differs from that in [6]). Alternatively, the inverse of $F(q, c)$ may be expressed in terms of an interference function $H(q, c)$:

$$F(q, c)^{-1} = 1 + c\Gamma(c)P(q, c)H(q, c) \tag{9}$$

which, together with the requirement $H(0, c) = 1$, may be considered to define $H(q, c)$ and $\Gamma(c)$ in terms of the more direct functions $B(c)$ and $Q(q, c)$, i.e.,

$$\Gamma(c) = B(c)/[1 - cB(c)] \tag{10a}$$

$$H(q, c) = Q(q, c) \frac{[1 - cB(c)]}{[1 - cB(c)P(q, c)Q(q, c)]} \tag{10b}$$

It is largely a matter of convenience as to which form is used for $F(q, c)$, e.g., one of the functions $H(q, c)$ or $Q(q, c)$ might be less dependent on q than the other, and therefore more convenient to use. With the use of Eq. (9),

$$\frac{K_{op}cM}{R(q, c)} = \frac{1}{S(q, c)} = \frac{1}{P(q, c)} + c\Gamma(c)H(q, c) \tag{11}$$

Some approximate theories lead directly to Eqs. (9) or (11) with $H(q, c)$ equal to unity for all q, see below. Note that if $H(q, c)$ is unity, then $Q(q, c)$, which is given by $H(q, c)\{1 + c\Gamma(c)\}/\{1 + c\Gamma(c)P(q, c)H(q, c)\}$, will increase to the limiting value $1 + c\Gamma(c)$ for large q.

With Eq. (9), thermodynamic information is found in the function $\Gamma(c)$, and thermodynamic and conformational information is represented in the functions $P(q, c)$ and $H(q, c)$. As is well known, in the limit of infinite dilution, both $H(q, 0)$ and $Q(q, 0)$ tend to unity for all q [3, 6, 12, 19] (the "single-contact" approximation), and R_G may be determined from $P(q, 0)$ [1–4, 6, 12]. In the limit $qR_G \gg 1$, $F(q, c)$ tends to unity, as then the scattering can only be sensitive to short-range correlations among the scattering elements, and will be dominated by intramolecular interference effects reflected in $P(q, 0)$, i.e., $c\Gamma(c)P(q, c)H(q, c) \ll 1$ (or $cB(c)P(q, c)Q(q, c) \ll 1$) for large q, and in consequence, $F(\infty, c) \approx 1$. This asymptotic behavior will require larger q with increasing $\Gamma(c)$, and may generally be beyond that available in light scattering experiments for moderately concentrated solutions, for which $c\Gamma(c)$ is large.

For a monodispersed solute, $F(0, c)$ is related to the equilibrium osmotic modulus K_{os} [1–4, 6, 12]:

$$F(0, c) = cRT/M \, K_{os} \tag{12}$$

$$K_{os} = c\partial\Pi/\partial c \tag{13}$$

where Π is the osmotic pressure (the relation between $F(0, c)$ and Π is more complex for a heterodisperse solute [3, 6, 12]). Of course, for dilute solutions, both $\Gamma(c)$ and $c\Gamma(c)H(q, c)$ may be expanded as power series in c [1–4, 6, 12]:

$$c\Gamma(c) = 2A_2 Mc + 3A_3 Mc^2 + \ldots = 2\psi_2\hat{c} + 3\psi_3\hat{c}^2 + \ldots \tag{14a}$$

$$c\Gamma(c)H(q, c) = 2\psi_2 W_2(q)\hat{c} + \{3\psi_3 W_3(q) + 4[P(q, 0)W_2(q)^2$$

$$- W_3(q)]\psi_2^2\}\hat{c}^2 + \ldots \tag{14b}$$

when A_3 is the third virial coefficient, the $\psi_j = A_j M(M/N_A R_G^3)^{j-1}$ are dimensionless virial coefficients, and $\hat{c} = cN_A R_G^3/M$ is a dimensionless concentration. These relations form the basis for the determination of M and A_2 (and sometimes A_3 as well) from scattering data on dilute solutions [1–4, 6, 12].

1.3 Dynamic Scattering from Isotropic Solutions

Certain general relations used with the polarized dynamic scattering from isotropic solutions are gathered in this section. Relations for the depolarized scattering from isotropic scattering are given in the section on Isotropic Solutions of Semiflexible Chain Polymers, and the dynamic scattering from ordered solutions is considered in the section on Nematic Solutions of Semiflexible Chain Polymers.

As mentioned above, in all cases of interest here, the scattering is in the Rayleigh–Gans limit [1–4, 6, 12], and it is assumed that the polarizability fluctuations may be treated as Gaussian stochastic variables [4, 6, 8], so that $G^{(2)}(\tau; q)$ may be expressed in the (normalized) form

$$g^{(2)}(\tau; q) = G^{(2)}(\tau; q)/G^{(2)}(\infty; q) \tag{15}$$

$$g^{(2)}(\tau; q) = 1 + f_C\{g^{(1)}(\tau; q)\}^2 \tag{16}$$

where $g^{(1)}(\tau; q)$ is unity for $\tau = 0$ and zero for very large τ, and the coherence factor f_C depends on the geometry of the scattering apparatus ($0 \leq f_C \leq 1$) [4, 6, 8]. The characterization parameters of interest are then represented in the thermodynamic functions related to $G^{(2)}(\infty; q)$, and the dynamic functions related to $g^{(1)}(\tau; q)$. In many cases, $g^{(1)}(\tau; q)$ may be represented as the sum of weighted exponentials, given by [4, 6]

$$g^{(1)}(\tau; q) = \sum r_\mu(q)\exp[-\tau\gamma_\mu(q)] \tag{17}$$

where $\sum r_\mu = 1$. With Eq. (17), $[G^{(2)}(\infty; q)]^{1/2}$ comprises the sum $[G^{(2)}(\infty; q)]^{1/2}$ $\sum r_\mu(q)$ of discrete contributions, with one term for each exponential function. This form will be utilized in discussion of the scattering from nematic solutions, and has been utilized in the discussion of dilute solutions in the presence of intermolecular association [16, 20], among other applications. In the discussion below, it will sometimes be convenient to scale the relaxation rates γ_μ by the first-cumulant K_1, where

$$K_1(q, c) = -\frac{1}{2}\lim_{\tau=0}\frac{\partial \ln[g^{(2)}(\tau; q, c) - 1]}{\partial \tau}. \tag{18}$$

With the summation in Eq. (17), the first-cumulant K_1 is given by

$$K_1(q, c) = \sum r_\mu(q, c)\gamma_\mu(q, c) \tag{19}$$

Thus, with $\hat{\gamma}_\mu = \gamma_\mu/K_1$, Eq. (17) becomes

$$g^{(1)}(\tau; q, c) = \sum r_\mu(q, c)\exp[-\tau K_1(q, c)\hat{\gamma}_\mu(q, c)] \tag{20}$$

with $\sum r_\mu \hat{\gamma}_\mu = 1$. In a number of cases discussed below, the sum is restricted to two terms. In such cases, we will use the convention $\hat{\gamma}_1 < \hat{\gamma}_2$, i.e., mode 1 will be the slow mode and mode 2 will be fast mode. For two terms, Eq. (20) becomes

$$g^{(1)}(\tau) = r\exp[-\tau K_1 \hat{\gamma}_1] + (1 - r)\exp[-\tau K_1 \hat{\gamma}_2] \tag{21a}$$

$$r = \frac{\hat{\gamma}_2 - 1}{\hat{\gamma}_2 - \hat{\gamma}_1} \tag{21b}$$

where the notation for the dependence on q and c is suppressed for convenience. In the application of expressions involving a sum of weighted exponentials to fit experimental data, it must be remembered that the inversion schemes used to compute the relaxation rates and their weight factors from data on $G^{(2)}(\tau; q, c)$ are limited in resolution, and that a "component" may actually represent an average over a range of closely spaced contributions, selected by q in the scattering experiment. As discussed below, nonexponential $g^{(1)}(\tau; q, c)$ for a monodisperse solute may have separate origins at infinite dilution and in concentrated solution.

In general, for polarized scattering from isotropic solutions, $K_1(q, c)$ may be expressed in the form [6, 14]

$$\frac{K_1(q, c)}{q^2 D_T} = \frac{D_M(c)}{D_T} M(u, c) \tag{22}$$

where $u = (qR_G)^2$, the mutual diffusion constant $D_M(c)$ reduces to the translational diffusion constant D_T at infinite dilution, $M(0, c) = 1$, and $M(u, 0)$ tends to unity for small u, see below. The mutual diffusion constant is conveniently represented in terms of the self-diffusion constant $D_S(c)$ by the expressions [6, 14]

$$\frac{D_M(c)}{D_T} = \frac{D_S(c)}{D_T} F(0, c)^{-1} \tag{23a}$$

$$\frac{D_S(c)}{D_T} = \frac{\Xi(0)}{\Xi(c)}(1 - \varphi) \tag{23b}$$

where $F(0, c)$ is discussed above, $\Xi(c)$ is the molecular friction factor, with $\Xi(0) = kT/D_T$, and φ is the solute volume fraction. These features are discussed in the following.

2 Isotropic Solutions of Flexible Chain Polymers

For very small concentration c, a polymer solution is microscopically inhomogeneous, with the average separation between polymer chains being much larger than their size as measured, for example, by R_G, and with intermolecular interactions essentially occurring pairwise; thus the scaling with $A_2 Mc$ in Eq. (14) at low c. The dimensionless virial coefficients $\psi_j = A_j M(M/N_A R_G^3)^{j-1}$ reach asymptotic limits for flexible chains with large M in good solvent systems [3, 10–13], so that in good solvents, $\Omega_{23} = (\psi_2)^2/\psi_3$ tends to a constant value Ω_∞ for large M for linear flexible chain polymers [6,10, 12, 21]. In the latter case, $c\Gamma(c)$ becomes a function of $A_2 Mc$ alone, leading to approximations useful in the extrapolation of data on dilute solutions $(A_2 Mc \ll 1)$ to the infinite dilution limit [1, 3, 6, 10, 12] and in the scaling of data with c in moderately concentrated solutions in good solvents [6, 14], see below. For example, for dilute solutions, the expression [6]

$$F(0, c)^{-1/2} = 1 + A_2 Mc - 1/2(1 - 3w_o/\Omega_\infty)(A_2 Mc)^2 + \ldots \qquad (24)$$

provides a useful approximation to $F(0, c)$ for flexible chains in good solvents, for which $w_o = \Omega_\infty/\Omega_{23}$ tends to unity. Thus, for dilute solutions in good solvents, plots of $(K_{op} cM/R(0, c))^{1/2}$ versus c tend to be more nearly linear than plots of $K_{op} cM/R(0, c)$ versus c.

In concentrated solutions, the solution approaches homogeneity on length scales much smaller than R_G, with radical change in behavior. For example, the well-known model of Flory and Huggins replaces the virial expansion in Eq. (14) in the representation of K_{OS} [1, 10, 14]. The intermediate concentration range of moderately concentrated solutions represents a concentration range for which the intermolecular interactions change between these two extremes. For flexible chain polymers, the concentration range of interest may be specified by the condition that the intramolecular repeat unity density $\rho_{INTRA} \propto N/R_G^3$ is comparable to the overall repeat unit density $\rho_{INTER} = cN_A/m_o$, where $N = M/m_o$ with M the molecular weight of a chain with root-mean-square radius of gyration R_G and repeat unit molecular weight m_o. This leads to the specification that the moderately concentrated solution regime corresponds to concentrations with $\hat{c} = cN_A R_G^3/M$ of order 1 to 10; an "overlap" concentration $c^* \propto M/N_A R_G^3$ is sometimes defined in this context [14, 22]. Since the intrinsic viscosity $[\eta]$ is essentially proportional to R_G^3/M for linear flexible chain polymers $([\eta] \approx 6.8 N_A R_G^3/M$ for high molecular weight flexible chain polymers [1, 12]), $\hat{c} \approx [\eta]c/6.8$, and the concentration range of interest can also be conveniently specified by the product $[\eta]c$, with moderately concentrated (or semi-dilute) solution behavior being anticipated in the range $1 < [\eta]c < 50$. Alternatively, for solutions in good solvents, the concentration can be scaled as $\hat{c} = A_2 Mc/\psi_{2,\infty}$ where $\psi_{2,\infty} \approx 6.7$ [10] is the limiting value of the penetration function $\psi_2 = A_2 M^2/N_A R_G^3$ for large M in good solvents [10–13, 23]; this scaling is not useful for solutions under Flory Theta conditions as A_2 is zero for that case. Two extremes of behavior may be considered, dependent on the value of the second virial coefficient A_2: behavior at the Flory Theta state, for which $A_2 = 0$, and behavior in "a very good solvent", for which $\psi_2 \approx \psi_{2,\infty}$. Of course, for many polymer/solvent compositions, these special conditions are not met, and, for example, ψ_2 increases with increasing M.

2.1 Thermodynamic Behavior

Thermodynamic and conformational characteristics are represented in the functions $P(q, c)$ and $F(q, c)$. Some theories of moderately concentrated solutions of flexible chains give $H(q, c)$ equal to unity for all q and c [14, 24, 25], but the exact behavior is not yet settled. The behavior of $H(q, c)$ is considered below for solutions at the Flory Theta temperature, and for solutions in good solvents, i.e., systems for which ψ_2 is large. Further, as discussed below, the principal effect on $P(q, c)$ appears to be that of the concentration dependence of R_G, rather than any appreciable change in the functional form of $P(q, c)$.

The series representation given by Eq. (14) is expected to fail as ĉ approaches and exceeds unity if $A_2 > 0$. For example, for systems with $A_2 > 0$, extension of the notion that $c\Gamma(c)$ may depend on $A_2 Mc$ alone in systems for which $\Omega_{23} = (\psi_2)^2/\psi_3$ is about equal to its asymptotic limit Ω_∞ for strong repulsive interchain interactions ("good solvents"), and the requirement that $\partial\Pi/\partial M = 0$ for a moderately concentrated solution, can be utilized to obtain a scaling-law for $c\Gamma(c)$ if the latter is expressed as a power-law relation [6, 14]. The result gives

$$c\Gamma(c) \propto (A_2 Mc)^{1+p} \gg 1 \tag{25}$$

with $p = 1/4$ for moderately concentrated solutions in a very good solvent [14]. Experimental data [6, 26–41] tend to link p to the exponent $v = \partial \ln R_G/\partial \ln M$ through the relation $p = (2 - 3v)/(3v - 1)$, which gives $p = 1/4$ if v adopts its asymptotic value $3/5$ for a good solvent [1, 6, 12–14], but will otherwise give slightly larger p as v decreases toward its value $1/2$ under Flory Theta conditions, or for small M, even under "good solvent" conditions. Of course, this scaling is not applicable as ψ_2 tends to zero, and some other form is needed to represent a wider range of c and solvent conditions for which v is smaller than its asymptotic value over the range of M of interest. The relation [6, 10, 42]

$$c\Gamma(c) = 2A_2 Mc\{1 + 2(w_o/p\Omega_\infty)A_2 Mc\}^p$$

$$+ 3A_3 Mc^2\{1 - \Omega_{23}/\Omega_\infty\}V(c) \tag{26}$$

provides a useful approximation to $c\Gamma(c)$ in both the dilute and moderately concentrated range, over a range of M, with the expectation that $w_o \approx 1$. The term explicit in A_3 is present to accommodate systems with $A_2 \approx 0$; it is negligible otherwise, and the function $V(c)$, discussed below, must be unity at low c, and is expected to be a constant at larger c. Examples of data compared with this relation with systems for which $A_2 > 0$ are given in Fig. 1. As discussed elsewhere, with the use of Eq. (26), plots of $\mathbf{K}_{op}cM/\mathbf{R}(0, c)$ versus $A_2 Mc$ are expected to be nearly coincident with plots of $\Pi M/RTc$ versus $A_2 Mc/2$ for a monodisperse solute [6]; data on both functions are included in Fig. 1. The expression provides a good fit to the data, with p given by the expression above, and for reasonable values of w_o.

Under the Flory Theta condition, since A_2 is zero, though A_3 (and higher virial coefficients) may be nonzero, Eqs. (14) and (26) reduce to the simple expression

$$\Gamma(c) = 3A_3 MV(c)c . \qquad (27)$$

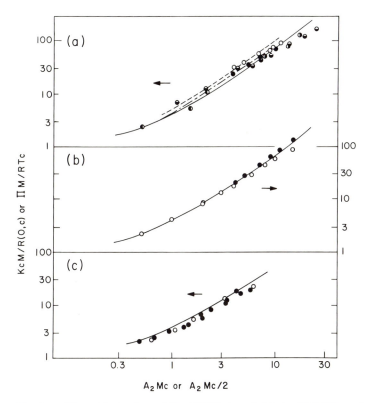

Fig. 1a–c. Bilogarithmic plots of $\mathbf{K}_{op}cM/\mathbf{R}(0, c)$ vs $A_2 Mc$ (or $\Pi M/RTc$ vs $A_2 Mc/2$) for several flexible chain polymers in good solvents. The *solid curves* represent the use of Eq. (26). The data sources are identified in [6, 10]: **a** $\mathbf{K}_{op}cM/\mathbf{R}(0, c)$ for polystyrene ($0.36 \leq 10^{-6} M_W \leq 7.6$) in benzene (15 °C); **b** $\mathbf{K}_{op}cM/\mathbf{R}(0, c)$ (*unfilled*) and $\Pi M/RTc$ (*filled*) for poly(α-methyl styrene) in toluene; **c** $\mathbf{K}_{op}cM/\mathbf{R}(0, c)$ for polystyrene ($10^{-6} M_W$ equal to 0.862 for *filled* and 1.50 for *unfilled*) in cyclopentane at T equal to the Flory Theta temperature plus 20 °C (no pips) and 35 °C (no pips) and 35 °C (pips). After Fig. 6 in Ref. [10]

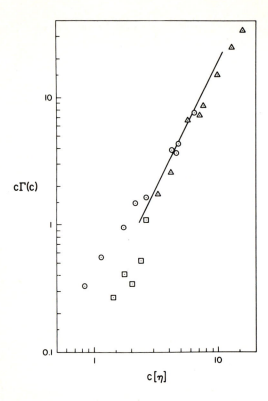

Fig. 2. Bilogarithmic plot of $c\Gamma(c)$ vs $[\eta]c$ for solutions of ploystyrene in cyclohexane at the Flory Theta temperature (34.8 °C): *circles* [18]; *squares* [30]; *triangles* [38]. The *line* represents Eq. (27) with $3A_3MV(c)/[\eta]^2 = 0.195$, i.e., $c\Gamma(c) = 0.195([\eta]c)^2$. After Fig. 6 in Ref. [10]

As shown in Fig. 2, $\Gamma(c)$ is found to be proportional to c for data on moderately concentrated solutions under the Flory Theta condition, implying that $V(c)$ has negligible concentration dependence in that range [30, 38, 39]. Indeed, according to a renormalization treatment, $V(c) \approx 1$ under the Flory Theta conditions [18, 40], permitting evaluation of A_3 from the observed data on $\Gamma(c)/c$ for the moderately concentrated solution. The finding that $A_3 > 0$ under Flory Theta conditions shows that the ternary cluster integral β_3 is not zero, even though the binary cluster integral $\beta_2 = 0$ [12, 13]. At the present time, it is beyond the scope of theory to provide estimates of either β_2 and β_3. It is believed, however, that it is reasonable for β_3 to be positive under conditions for which β_2 is zero [11, 13, 43, 44]. Since β_3 is very small, the ternary interaction parameter z_3 may be estimated from A_3 under Flory Theta conditions as $z_3 \approx \{3/(4\pi)^3\}\psi_3$. For example, the data in Fig. 2 give $z_3 \approx 0.0045$ for polystyrene in cyclohexane; $z_3 \approx 0.012$ for solutions of polystyrene in cyclopentane [40].

Additional thermodynamic and conformational information is available in the functions $P(q, c)$ and $H(q, c)$. The well-known Debye expression for $P(q, 0)$ is found to be a good representation for flexible chain polymers over a wide range in $u = (qR_G)^2$ [6, 45–47]:

$$P(q, 0) = (2/u^2)\{u - 1 + \exp(- u)\} \ . \tag{28}$$

Of course, this expression is inadequate for u large enough for the local structure of the chain to be important [45], but that is beyond the range accessible to light scattering (though important in X-ray and neutron scattering). The approximation $P(q, 0) \approx \{1 + (u/6)\}^{-2}$ is useful for $u < 2.5$, and provides the basis for an extrapolation method to determine R_G [6]. Although derived with Gaussian statistics suitable under the Flory Theta conditions, with $R_G^2 = \rho L/3 \propto M$, the effects of excluded volume interactions on $P(q, 0)$ appear to be satisfactorily taken into account by the use of Eq. (28) with the observed R_G, for which $R_G^2 \propto M\alpha_o^2$, where α_o is the excluded volume expansion factor at infinite dilution [6, 46, 47]. Here, we assume that for moderately concentrated solutions the variation of R_G with c provides the principal source of the dependence of $P(q, c)$ on c. Consequently, $P(q, c) \approx P(q, 0)$, provided the appropriate value of R_G is again used, with $R_G^2 \propto M\alpha_c^2$, where α_c is the ratio of R_G at concentration c to its value under Flory Theta conditions. The screening of excluded volume interactions causes α_c to decrease from α_o to unity with increasing c [1, 6, 10, 12, 14]. The assumptions that α_c scales with $A_2 Mc \propto \hat{c}$ in a good solvent, is independent of M, and a power-law in c for moderately concentrated solutions of very high molecular polymers in very good solvents leads to $\alpha_c^2 \propto \hat{c}^{-s}$ where $s = (1 - p)/3 = (2v - 1)/(3v - 1)$, with p and v defined above [6, 14, 28, 37] – $s = 1/4$ in the limit of full excluded volume, for which $v = 3/5$. One representation of this dependence useful from the dilute to concentrated range is given by [10]

$$\alpha_c = \alpha_o \{1 + (k_\alpha \hat{c})^2\}^{-s/4} ; \quad c < c_s \tag{29a}$$

$$\alpha_c = 1 ; \quad c > c_s \tag{29b}$$

where c_s is the concentration for which α_c is reduced to unity as c increases, k_α is a constant ($k_\alpha \approx 6.8$). Of course, the approximation that the functional form for $P(q, c)$ is invariant cannot be accurate under good solvent conditions, owing to the nongaussian chain character. Thus, for $qR_G \gg 1$, $P(q, c)$ reflects the thread-like nature of the chain, with $P(q, c) \propto q^{-\sigma'}$, where σ' depends on the short-range features of the chain conformation – e.g., σ' is 2 for a Gaussian chain, 5/3 for a flexible chain with full excluded volume, and unity for a rodlike chain [6]. The range of q for which this asymptotic behavior is observed will usually lie beyond that available in light scattering.

In addition to an estimate for $P(q, c)$, interpretation of data on $S(q, c)$ requires an evaluation of $H(q, c)$. Experience suggests that $H(q, c) \approx 1$ for dilute solutions (e.g., solutions with $[\eta]c < 0.1$), in which case the "single-contact approximation" applies, and data on $S(q, c)^{-1}$ vs $u = (qR_G)^2$ at different c form a series of parallel curves [1–6, 12, 19]. However, data are now available showing that in moderately concentrated solutions, $H(q, c)$ deviates significantly from unity under both Flory Theta conditions [18] and in good solvents [46, 48, 49]. The experimental evaluation of $H(q, c)$ is simpler under Flory Theta conditions as R_G may be considered to be independent of concentration in that case. The behavior for $H(q, c)$ under Flory Theta conditions calculated from

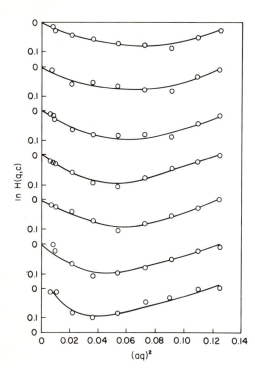

Fig. 3. $H(q, c)$ vs $(aq)^2 = (R_G q)^2/6$ for moderately concentrated solutions of polystyrene in cyclohexane at the Flory Theta temperature (34.8 °C), $M_w = 8.62 \times 10^5$, $[\eta] = 76$ ml/g, $R_G = 27$ nm: $[\eta]c$ equal to 0.60, 1.15, 1.73, 2.68, 4.64, 4.69, 6.54 from bottom to top. The curves are drawn merely to aid the eye. After Fig. 4 in Ref. [18]

experimental data on $S(q, c)$ and $\Gamma(c)$, and with $P(q, c) \approx P(q, 0)$ given by Eq. (28) using the experimental R_G are shown in Fig. 3 [18]. The shallow minimum in $H(q, c)$ corresponds to an observed maximum in $S(q, c)$ with increasing q. The situation is more complex in a good solvent, in that R_G is expected to decrease with increasing c until it is (about) equal to its value under Flory Theta conditions [6, 10, 14, 29, 37], but simpler in that interactions scaling with β_2 may be expected to dominate both intramolecular and intermolecular interactions. Data on dilute and moderately concentrated solutions of a high molecular weight poly(α-methyl styrene) in a good solvent ($\psi_2 \approx 5$) given in Fig. 4 exhibit the development of a minimum in $S(q, c)^{-1}$ with increasing q as $[\eta]c$ approaches the range for moderately concentrated solutions [46, 48, 49]. As may be seen, the curves for the dilute solutions tend to cross at intermediate u. This effect, which may be beyond the precision available in the data, would require $H(q, c)$ to vary with q and c. A clear minimum may be observed for solutions in the moderately concentrated range, such that the initial tangent to $S(q, c)^{-1}$ is negative, showing that $H(q, c)$ is clearly dependent on c in that range. The behavior under both Flory Theta conditions and in good solvents show that $H(q, c) \approx 1$ for dilute solutions, but that $H(q, c)$ exhibits a minimum with increasing qR_G as the concentration is increased. Thus $b(0, c)^2 = R_G^2/3$ in dilute solutions (for which $\Gamma(c)$ tends to zero and $H(q, c) \approx 1$), but $\partial S(q, c)/\partial q^2 > 0$ for

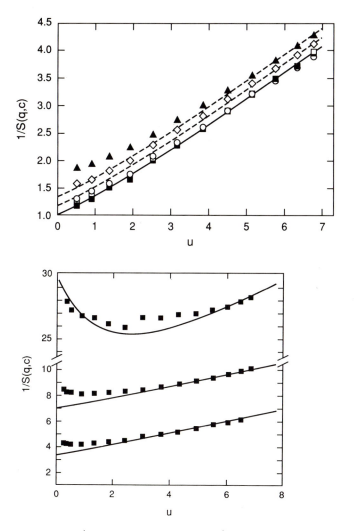

Fig. 4. $S(q, c)^{-1} = K_{op}cM/R(q, c)$ vs $u = (qR_G)^2$ for dilute to moderately concentrated solutions of poly(α-methyl styrene) in toluene (a good solvent) [46, 49]: $M_W = 2.96 \times 10^6$ and $R_G = 74$ nm. *Upper:* $[\eta]c$ equal to 0.023, 0.046, 0.095, 0.18, 0.30, and 0.53 for the *filled square, unfilled square, circle, unfilled diamond,* and *filled triangle,* respectively. *Lower:* $[\eta]c$ equal to 0.94, 1.94, and 5.30 from *bottom to top.* The *dashed curves* represent $P(q, c)^{-1} + cst$ using Eq. (28) with the value of R_G at infinite dilution: $cst = 0$ for the lowest curve, and otherwise is chosen to fit the data for the larger range of u; these curves are given merely to reveal the deviation of the data from the behavior at low c. The *solid curve* for the most concentrated solution is calculated using the approximation $H(q, c) \approx (P(2q, c))^2/P(q, c)$ as discussed in the text

small q for moderately concentrated solutions, with the result that $b(0, c)$ has no simple physical meaning.

The behavior for $H(q, c)$ complicates the interpretation of dependence of $S(q, c)$ on q for moderately concentrated solutions. A correlation length $b(q, c)$, and related lenghs ξ_P and ξ_H, may be defined from the observed Rayleigh ratio $R(q, c)$ by

$$b(q, c)^2 = R(0, c)\frac{\partial R(q, c)^{-1}(q, c)}{\partial q^2} \tag{30a}$$

$$b(q, c)^2 = \xi_P(q, c)^2 + \xi_H(q, c)^2 \tag{30b}$$

$$\xi_P(q, c)^2 = \frac{1}{1 + c\Gamma(c)}\frac{\partial P^{-1}(q, c)}{\partial q^2} \tag{30c}$$

$$\xi_H(q, c)^2 = \frac{c\Gamma(c)}{1 + c\Gamma(c)}\frac{\partial H(q, c)}{\partial q^2} \tag{30d}$$

Thus, in the limit of infinite dilution and zero q, $\xi_H(0, 0)$ is zero, and $b(0, 0)^2 = \xi_P(0, 0)^2 = R_G^2/3$. If $H(q, c)$ is independent of q, then in dilute or moderately concentrated solutions, $\xi_H = 0$, and $b(0, c)^2 = \xi_P(0, c)^2 = R_G^2/3\{1 + c\Gamma(c)\}$ reflects the variation of R_G and $c\Gamma(c)$ with c; this approximation is frequently assumed to be valid for moderately concentrated solutions, and is, for example, predicted in the so-called random phase approximation applied to flexible chain polymers [14, 24, 25]. Under these conditions, $b(0, c) = \xi_P(0, c) \propto \alpha_c[\rho L/c\Gamma(c)]^{1/2} \propto (\rho L)^{1/2}/(A_2 Mc)^{3/4}$ for moderately concentrated solutions of flexible chain polymers in good solvents, where the scaling laws for $\Gamma(c)$ and α_c given above are used. In general, however, $H(q, c)$ does not appear to be unity for all q, voiding this interpretation of b. Unlike $\xi_P(0, c)$, which must be positive, $\xi_H(0, c)^2$ may be either positive, negative, or essentially zero.

Since it appears that $H(q, c)$ returns to unity with moderately concentrated solutions under Flory Theta conditions for a range of q larger than q_{max}, it is possible to define and evaluate a correlation length $b(q > q_{max}, c)$ by the use of Eq. (30) for $q > q_{max}$. Data so obtained, shown in Fig. 5a, are seen to scale with $[\eta]c \propto \hat{c}$; the theoretical curve computed as $b(q > q_{max}, c) \approx \xi_P(0, c)$ using the data on $\Gamma(c)$ as a function of $[\eta]c$ provides a satisfactory fit to the data. This correlation length might, for example, be obtained in neutron scattering owing to the larger q-range of such data. Data on $S(q, c)$ for moderately concentrated solutions in good solvents are frequently interpreted to give a scaling length taken to be equivalent to $\xi_P(0, c)$, i.e., $b(0, c) \approx \xi_P(0, c) = \{(\rho L\alpha_c^2/3)/[1 + c\Gamma(c)]\}^{1/2} \propto (\rho L)^{1/2}/(A_2 Mc)^{3/4}$. In this interpretation, the variation of $b(0, c)$ with c reflects the variation of α_c and $c\Gamma(c)$ with c [6, 14, 29–41]. Examples of data reported for $b(0, c)$ for several good solvent systems are shown in Fig. 5b [39], using values of $R_{G,LS}$ and M_W determined by light scattering to compute an estimate $[\eta]_{CALC} = 2\pi N_A(R_{G,LS})^3/M_W$ of $[\eta]$. The theoretical expression for $\xi_P(0, c)/\{\rho L\alpha_0^2/3\}^{1/2}$ is also shown in Fig. 5b; the latter was computed with the relations given above for α_c/α_o and $\Gamma(c)$ and the approximations $A_2 M \approx [\eta] \approx \hat{c}/6.8$ for good solvent conditions. Although the experimental data for moderately concentrated solutions tend to the theoretical power-law

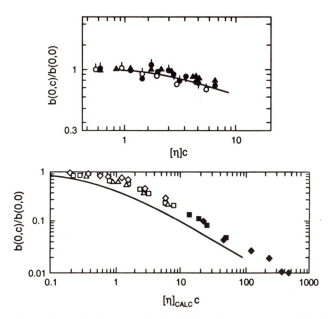

Fig. 5. Bilogarithmic plots of $b(0, c)/b(0, 0)$ (or $b(q > q_{max}, c)/b(0, 0)$) vs $[\eta]c$ for moderately concentrated solutions of flexible chain polymers under two thermodynamic conditions; $b(0, 0)^2 = R_G^2/3$: *Upper*: $b(q > q_{max}, c)/b(0, 0)$ determined under Flory Theta solvent conditions; the symbols are defined in Fig. 5 of [18]. The *solid curve* represents $\xi(0, c)/(R_G/3^{1/2})$, see Eq. (30): After Ref. [18]. *Lower*: $b(0, c)/b(0, 0)$ determined under good solvent conditions; the symbols are for data given in [39] and $b(0, c)$ determined by light scattering, *unfilled*, or neutron scattering, *filled*: polydimethyl siloxane, *diamonds*; polystyrene, *squares*; and poly(methyl methacrylate), *triangles*. The *solid curve* represents $\xi(0, c)/(R_G/3^{1/2})$, see Eq. (30)

behavior, the theoretical values of $\xi_P(0, c)/\{\rho L\alpha_0^2/3)\}^{1/2}$ are systematically smaller than the experimental data given for $b(0, c)/\{\rho L\alpha_0^2/3)\}^{1/2}$ in the moderately concentrated range; this discrepancy may be removed by arbitrary reduction of $[\eta]c$ in the experimental plot by a factor of 2.5. In part, the discrepancy may be due to the use of light scattering averages in the calculation of $[\eta]_{CALC}$, but that appears unlikely to be the sole source of the discrepancy. Thus, a contribution to the reported $b(0, c)$ due to some neglected variation in $H(q, c)$ with q may also be involved. Much of the data used in Fig. 5b were obtained from neutron scattering, carried out with q more than ten-fold larger than values in light scattering. Further, the experiments tend to be on polymers with a relatively large R_G, making the values of $qR_G > 1$ for most of the data. Consequently, the reported $b(0, c)$ may actually correspond to the correlation length $b(q > q_{max}, c)$ defined above. Indeed, data on $b(0, c)$ determined for solutions with $3 < [\eta]c < 10$, and for $qR_G < 1$ [40] show deviations from the scaling predicted for $\xi_P(0, c)$, suggesting that $H(q, c)$ was not independent of q in the range studied. By contrast, $b(q > q_{max}, c)$ was close to the predicted scaling behavior, suggesting that $H(q, c)$ was invariant with q for $q > q_{max}$.

Formally, $Q(q, c)$ may be expressed in terms of the pair correlation function $g(r, c)$ of the separation of molecular centers of gravity and certain intramolecular functions [2–4, 6, 10, 12, 27]. In this description, the vector between two scattering elements situated on two separate molecules may be replaced by the vector sum of three separate vectors: the vector between the two molecular centers of gravity, and the vectors between the scattering element and the center of mass on the molecule on which the element is found for each of two scattering elements. Then, if the relative orientations of these three vectors may be taken as a random variable [27]:

$$cB(c)Q(q, c) = ĉ\{\xi_g(c)/R_G\}^3 \frac{P_G(q, c)}{(P(q, c))^2}$$

$$\int_0^\infty 4\pi x^2 [1 - g(x, c)] \frac{\sin\{q\xi_g(c)x\}}{q\xi_g(c)x} \, dx \qquad (31a)$$

where $x = r/\xi_g(c)$, with $\xi_g(c)$ a concentration scaling length, and $P_G(q, c)$ is the scattering function for the scattering elements of a chain relative to the center of mass $(P_G(0, c) = 1)$. For $q = 0$, this gives $cB(c) \propto ĉ\{\xi_g(c)/R_G\}^3$, where the proportionality constant depends on $g(r, c)$. For a Gaussian chain [27]:

$$P_G(q, c) = \frac{\pi}{u}\exp(- u/6)\{\operatorname{erf}(u^{1/2}/2)\}^2 \qquad (31b)$$

In the evaluation of $P_G(q, c)$ for the nongaussian chain under good solvent conditions, it may be reasonable to use Eq. (31b) with u calculated using the value of α_c at the concentration of interest to evaluate R_G. In general, $g(r, c)$ may be represented in terms of the pair potential $V(r, c)$. Under good solvent conditions, it is reasonable to express $V(r, c)$ in the form $z\upsilon(r/\xi_g(c))$, where $\xi_g(c)$ is a scaling length for the intermolecular interactions, and z is the Fixman interaction parameter (see Eq. (5)). The situation may be more complex under Flory Theta conditions. Although $B(c)$ tends to zero with decreasing c under Flory Theta conditions, this does not imply that $g(r, c)$ tends to unity at low c under Flory Theta conditions, but rather that the integral in Eq. (31) tends to zero, whatever the nature of $g(r, c)$, and that the use of a single thermodynamic parameter (e.g., z) is not sufficient.

Prior work has often been based on an expansion of $g(r, c)$ about its dilute solution limit $g(r, 0)$ [1, 10, 12, 18, 27, 51]. Thus,

$$2\psi_2 W_2(q)P(q, c) = - R_G^{-3}\int G_{12} \frac{\sin(qS_{12})}{qS_{12}} \, dS_{12} \qquad (32a)$$

$$3\psi_3 W_3(q)P(q, c) = - R_G^{-6}\int\int\int\{[G_{12}G_{13}G_{23} + G_{13}G_{23}]$$

$$+ [1 + G_{12}][1 + G_{13}][1 + G_{23}]Y_{123}\}$$

$$\frac{\sin(qS_{12})}{qS_{12}} \, dS_{12}dS_{13}dS_{23} \qquad (32b)$$

where the integration in Eq. (32b) is over all separations S_{ij} between molecular centers that sum to zero, $G_{ij} = \exp[-V(S_{ij})/kT] - 1$ and $Y_{123} = \exp[-\Delta V(S_{12}, S_{13}, S_{23})/kT] - 1$, with $V(S_{ij})$ the potential of mean force for chains with centers separated by S_{ij}. Based on the observation reported above that $c\Gamma(c) \propto \hat{c}^2$ under Flory Theta conditions, it is reasonable to assume that $H(q, c)$ may be approximated by the term of order $\hat{c}^2$ in Eq. (14b): i.e., $H(q, c) \approx W_3(q)$. The behavior is attributed to the effects of a repulsive component in the intermolecular potential, reflecting the nonzero β_3 under conditions with zero β_2. As may be seen in Eq. (32a), A_2 is zero if the integral over G_{12} is zero, and does not require G_{12} itself to be zero. Nevertheless, it is usually assumed [12] that $V(S_{ij})$ is scaled by a single parameter β_2, and reduces to zero when A_2 vanishes under Flory Theta conditions, so that $\psi_2 = 0$, and the term in brackets in Eq. (32b) reduces to Y_{123}. The latter is often reasonably neglected, but in this formulation, ΔV should scale with β_3, making $\psi_3 > 0$. The alternative leaves G_{ij}, etc., nonzero and an unknown function of both β_2 and β_3 leads to an integration in Eq. (32b) that cannot be resolved at present. Unfortunately, no theoretical form is available for ΔV. It has been suggested that ΔV be approximated by an expression of the type found for binary interaction in the Flory–Krigbaum theory of A_2 [1, 10, 12], so that [18]

$$Y_{123} = -(Y\sigma_\varepsilon^3)^3 \exp[-\sigma_\varepsilon^2(S_{12}^2 + S_{13}^2 + S_{23}^2)/R_G^2] \qquad (33)$$

where both Y and σ_ε are expected to depend on β_3. Use of this relation (with $V(S_{ij}) = 0$) gives

$$\psi_3 = (\pi^3/3^{5/2})(Y\sigma_\varepsilon)^3 \qquad (34a)$$

$$W_3(q) = \exp[-\varepsilon u/3] \qquad (34b)$$

where $\varepsilon = (1/2\sigma_\varepsilon^2) - 1$. With this formulation, the maximum in $S(q, c)$ requires $\varepsilon > 0$; the experimental data in Fig. 3 give $\varepsilon \approx 10$ [18].

The preceding is certainly oversimplified, and does not, for example, provide insight to the origin of ε. A complete treatment would permit assessment of $g(r, c)$ over a range of concentration and thermodynamic conditions. Paradoxically, owing to dominance of interactions involving β_2 under good solvent conditions, it may be easier to develop a treatment of $g(r, c)$ in good solvents than under Flory Theta conditions. A recent use of Eq. (31) for good solvents utilized the Flory–Krigbaum potential for $\upsilon(r/\xi_g(c))$, with $\xi_g(c) = R_G$, and the Percus–Yevick method to compute $g(r, c)$ from $V(r, c)$ [48, 49]. In that case, the Percus–Yevick formalism provides an integral equation which may be solved numerically [52]. The Flory–Krigbaum potential does not give the observed dependence of A_2 on molecular weight, but the results are of interest for the predicted qualitative dependence of $F(q, c)$ on the two parameters qR_G and $A_2 Mc$. The results using the observed data on $A_2 M$ to scale $V(r, c)$ predict a shallow maximum in $S(q, c)^{-1}$, along with a marked dependence of $H(q, c)$ on qR_G over the entire range in Fig. 5, in qualitative disagreement with the observed behavior [49]. The discrepancy may be related in part to the neglect of a concentration dependence of $\xi_g(c)$.

It may be noted that for dilute solutions of spheres interacting through a hard–core potential, $H(q, c) = [P(2q, c)]^{1/2}/P(q, c)$ [6]. Use of this relation and $P(q, c)$ for the flexible chain, calculated with Eq. (28) and the value of R_G at the concentration of interest, gives $H(q, c)$ that first decreases, then increases with q, being close to unity for the range of u in Fig. 4. As may be seen in Fig. 4, this relation is in qualitative accord with the data, giving a slightly deeper minimum in $S(q, c)^{-1}$ than that observed for the moderately concentrated solutions. This qualitative correspondence is suggestive, and could lead to insights to develop an improved representation of $g(r, c)$, and hence $H(q, c)$ for moderately concentrated polymer solutions. The nature of the $Q(q, c)$ behavior corresponding to this approximation to $H(q, c)$ is shown in Fig. 6 for several $c\Gamma(c)$.

Anomalously intense scattering at small q is sometimes reported for moderately concentrated solutions of flexible chain polymers, especially for solutions in good solvents [53–58]. It has been noted that the effect is most pronounced with solutions that are in the entanglement regime [53], suggesting that the anomalous scattering might be associated with the large scale concentration fluctuations due to the entanglement pseudo-network. At the present time, however, it appears that with moderately concentrated solutions, this behavior may not reflect intrinsic properties of the moderately concentrated solution, but may rather be due to extrinsic impurities, despite heroic efforts to eliminate these by a variety of techniques.

2.2 Dynamic Behavior

Dynamic behavior is represented in the function $g^{(1)}(\tau; q, c)$, which is expected be exponential for a monodisperse flexible coil solute at infinite dilution for

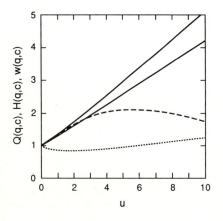

Fig. 6. The function $Q(q, c)$ corresponding to the approximation $H(q, c) \approx \{P(2q, c)\}^{1/2}/P(q, c)$, using Eq. (28) for $P(q, c)$, as discussed in the text. The *dotted curve* gives $H(q, c)$, the *dashed curve* gives $W(q, c) = P_G(q, c)/P(q, c)$ and the *solid curves* give $Q(q, c)$ for $c\Gamma(c)$ equal to 10 and 50 from bottom to top, respectively

scattering angles for which $qR_G \ll 1$ [4, 6]. Before turning to the principal topic of the scattering from moderately concentrated solutions, we remark briefly on behavior at infinite dilution as a basis for comparison. Although $g^{(1)}(\tau; q, 0)$ is exponential for $qR_G \ll 1$ (for a monodisperse solute), for $qR_G \gg 1$, internal modes owing to segmental motions in flexible chain polymers (or rotational modes in a rigid asymmetric molecule) intervene, resulting in nonexponential behavior, and a q-dependence to $K_1(q, 0)/q^2$, where $K_1(q, 0)$ is the limiting value of $K_1(q, c)$ at infinite dilution (similar notation will be used throughout). If qR_G is not too large, then Eq. (22) for $K_1(q, 0)$ may be expanded in u to give [4, 6]

$$\lim_{q=0} K_1(q, 0)/q^2 D_T = \lim_{q=0} M(u, 0) = 1 + C^{(2)}u + \ldots \tag{35}$$

where $C^{(2)}$ is a constant, depending on the chain conformation and the nature of the hydrodynamic interactions [4, 6, 59, 60]. Thus, for flexible coil chains, $(5\pi^2/4)C^{(2)}$ tends to 2.38 for large molecular weight M and 1.64 for small M [6, 60]. For larger u, the series expansion fails, and other methods are required to estimate $K_1(q, 0)$. For example, $K_1(q, 0)$ is expected to scale with q to a power other than two for very large q [61–73]. Renormalization group calculations have been employed to estimate $K_1(q, 0)$ over a wider range of qR_G to give the result [74–78]

$$K_1(q, 0)/q^2 D_T = M(u, 0) = N(\vartheta, \beta^*, \xi^*)/N(0, \beta^*, \xi^*) \tag{36a}$$

$$D_T = (kT/2^{1/2}\eta_e R_G)N(0, \beta^*, \xi^*) \tag{36b}$$

$$N(\vartheta, \beta^*, \xi^*) = \frac{1}{2\xi^*f(\vartheta)} \exp\left[\frac{\beta^*(I + L(\vartheta))}{4\pi^2} + \frac{3\xi^*V(\vartheta)}{8\pi^2}\right] \tag{36c}$$

where $\vartheta^{1/2} = qR_G \exp(-\beta^*I/4\pi^2)$, with $I \approx -1.350$, β^* and ξ^* are the fixed-points for excluded volume and hydrodynamic interactions, respectively, in the renormalization ($\beta^*/\pi^2 = 0$ and $1/2$, and $\xi^*/\pi^2 = 8/3$ and 2 for Flory Theta conditions and with excluded volume, respectively), $N(0, \beta^*, \xi^*)$ is the limiting value of $N(\vartheta, \beta^*, \xi^*)$ for small ϑ, and η_e is the effective viscosity, which is expected to be proportional to the solvent viscosity η_s, perhaps close to the limiting solvent viscosity at high frequency. The effective viscosity, η_e, appears explicitly in the relation for D_T, whereas $N(\vartheta, \beta^*, \xi^*)$, and hence $K_1(q, 0)/q^2 D_T$, depend only on qR_G and the fixed-point parameters. As anticipated, $R_H \propto R_G$ with this result. The functions $f(\vartheta)$, $L(\vartheta)$, and $V(\vartheta)$ are given in [77], but the latter two involve integrations that must be carried out numerically – in the limit of small ϑ, $L(\vartheta) \approx 0$, and $f(\vartheta) \approx -V(\vartheta) \approx 1/2$.

With these relations, $K_1(q, 0)/q^3$ approaches the limit $(kT/\eta_e)A_\infty$ for large qR_G, where A_∞ is 0.013 and 0.027 for Flory Theta conditions and with excluded volume, respectively. Experimental data have been interpreted to give η_e/η_s approximately 0.3–0.4, in order to obtain agreement of D_T with experimental data over a wide range of molecular weight, both under Flory Theta conditions, and in good solvents [77, 78]; alternative reasons for the departure of η_e/η_s

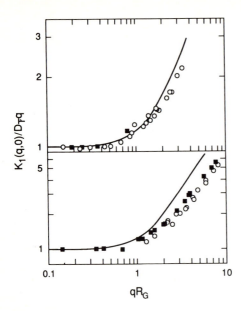

Fig. 7. Bilogarithmic plots of $K_1(q, 0)/q^2D_T$ vs $(qR_G)^2$ for moderately concentrated solutions of several flexible chain polymers in good solvents, lower, and under Flory Theta conditions, upper. The symbols are identified in Ref. [77]. The *solid curves* represent Eq. (36) calculated with a renormalization group treatment [76, 77]

from unity have been discussed [79]. The fit to $K_1(q, 0)/q^2D_T$ over a range of qR_G is shown in Fig. 7 for data discussed in [77]. The systematic deviations can be lessened by the inclusion of cross-over effects in the estimates of β^* and ξ^*, and by treating the proportionality between $9^{1/2}$ and qR_G as adjustable [77].

Treatments of the full behavior of $g^{(1)}(\tau; q)$ are less complete, even for infinite dilution. Approximate theories are available which predict that at infinite dilution, $g^{(1)}(\tau; q)$ can be represented by a sum of weighted exponentials (Eq. (17)), at least up to relatively large qR_G [4, 59] – with very large qR_G the behavior is more complex [61, 62]. At infinite dilution each of the terms contains a factor $\exp(-D_Tq^2\tau)$. Thus, at infinite dilution, $g^{(1)}(\tau; q)$ may be expressed in the form

$$g^{(1)}(\tau; q) = \exp(-D_Tq^2\tau) \sum_{\mu=1} r_\mu(q)\exp[-\tau\Delta\gamma_\mu(q)] \tag{37}$$

where $\Delta\gamma_\mu(q) = \gamma_\mu(q) - D_Tq^2$, and $\Delta\gamma_1(q) = 0$. At low q, theoretical treatments give $\Delta\gamma_2(q)$ equal to a constant $(2/\tau_1)$, where $\tau_1 \propto D_T/R_G^2 \propto M[\eta]$, represents the longest relaxation time for internal motion; higher order terms have negligible weight factors $r_\mu(q)$ at low q, but become increasingly important for larger q. In practice, a multitude of contributions are usually not seen, but usually, $g^{(1)}(\tau; q)$ may be represented with two weighted exponential terms, one (mode 1) corresponding to the center-of-mass diffusion, and scaling with q^2, and the other given by a term for which $\Delta\gamma_\mu(q)$ varies from scaling with q^0 to scaling with q^σ, where σ depends on the thermodynamic conditions, and the range of qR_G characterizing the experiment. For example, at intermediate qR_G, a mode with $\Delta\gamma_2(q) \propto q^2$ was observed in studies on solutions of a high molecular weight polymer under good solvent conditions, with $[\eta]c \approx 0.2$ [46]. The mode was attributed to longitudinal fluctuations in the intramolecular concentration, determined by the intramolecular longitudinal osmotic modulus L_{intra}, a kind of "breathing motion" of the chain observed on the intermediate range of qR_G studied [46, 80]. The diffusion character attributed to the mode is consistent with its larger rate insofar as L_{intra} exceeds the intramolecular elasticity kT, as would be expected. For larger qR_G, the scaling should change to reflect the behavior in $K_1(q, 0)$ discussed above. In a recent application of this expression, data were fitted using this expression with only two terms, rearranged to read

$$1 - \exp(D_T q^2 \tau)g^{(1)}(\tau; q) = r_2(q)\{1 - \exp[-\tau\Delta\gamma_2(q)]\} . \tag{38}$$

Data on dilute solutions in both a good solvent and at the Flory Theta temperature have been analyzed using this relation [73]. The function $\Delta\gamma_2(q)$, found to be independent of q for small q and to tend to proportionality with q^σ for large q, was fitted by the semi-empirical relation

$$\Delta\gamma_2(q) = \Delta\gamma_2(0) + k_\sigma q^\sigma \tag{39}$$

where k_σ and σ are constants. The term proportional to q^σ, which dominates the behavior at large q, represents the collective effects of all shorter relaxation times, too closely spaced to be resolved in the scattering experiment [73]. The data were fitted with $\sigma \approx 3$ at the Flory Theta temperature, and $\sigma \approx 3.85$ in the good solvent [73], both values being close to the predictions of certain theoretical calculations [61, 65]. In studies on solutions with $0.05 < [\eta]c < 0.2$ under Flory Theta conditions, $\Delta\gamma_2(0)$ was found to be independent of c [73]; this may reflect the result $D_M(c) \approx D_T$ for these conditions, see below. By contrast for solutions in a good solvent with $0.1 < [\eta]c < 0.6$, $\Delta\gamma_2(0)$ was found to increase linearly with c [73]; the increase appears to be larger than that in $D_M(c)/\alpha_c^2$, see below. Theoretical studies have predicted a certain dependence of the time constants on c in moderately concentrated solutions [81].

Returning to behavior for moderately concentrated solutions, $D_M(c)$ depends on both thermodynamics and hydrodynamics as expressed by Eq. (24). As discussed below, $D_M(c)$ may either increase or decrease with increasing c, depending on the balance of the thermodynamic and hydrodynamic factors [4, 6, 10, 14, 18, 29, 37, 47, 81–95]. For dilute solutions ($[\eta]c \ll 1$), $\Xi(c)$, like $F(0, c)$ may be expanded in a power-series to give [6, 10, 18]

$$\frac{\Xi(0)}{\Xi(c)} = 1 + k_1[\eta]c + k_2\Gamma(c)c + \ldots \approx [1 + c\Gamma(c)]^{k_2}\exp(k_1[\eta]c) \tag{40}$$

where the closed form is devised for use below. Combining the expressions for $\Xi(0)/\Xi(c)$ and $F(0, c)$,

$$D_M(c)/D_T = 1 + k_D[\eta]c + \ldots \tag{41}$$

where $k_D = (2 - k_2)(A_2 M/[\eta]) - k_1$, with both k_1 and k_2 dependent on the

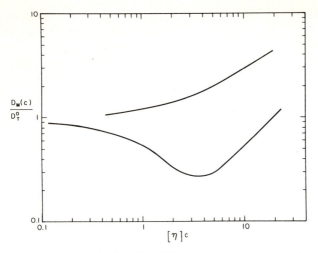

Fig. 8. Bilogarithmic plot of $D_M(c)/D_T$ vs $[\eta]c$ for solutions of several flexible chain polymers under good solvent conditions (e.g., $A_2M \approx [\eta]$), *upper curve* and under Flory Theta solvent conditions, *lower curve*, see Ref. [6, 10] for data sources. After Fig. 10 in Ref. [6]

solute: $k_1 = B(4\pi R_H^3/3M[\eta])$, with theoretical estimates $B \approx 1$ to 2.23 [6, 12, 82–85], and $k_2 \approx 1.2$ for flexible chain polymers [6, 12]. Consequently, for dilute solutions ($[\eta]c < 1$) $D_M(c)$ may increase or decrease with increasing c, dependent on $A_2M/[\eta]$, as shown in Fig. 8. For moderately concentrated solutions ($[\eta]c$ greater than about unity), the behavior is strikingly different in good solvents under the Flory Theta conditions. In either case, for moderately concentrated solutions [6, 10, 18],

$$\frac{\Xi(0)}{\Xi(c)} \approx F(0, c) \frac{R_H}{\xi_\Xi(c)} \frac{\zeta(0)}{\zeta(c)} \qquad (42)$$

where $R_H = \Xi(0)/6\pi\eta_s$ is the hydrodynamic radius (η_s being the solvent viscosity), $\zeta(c)$ is the segmental friction factor, and $\xi_\Xi(c)$ is a correlation length for hydrodynamic interactions – intermolecular correlations are nil for lengths exceeding $\xi_\Xi(c)$ [6, 14]. For moderately concentrated solutions [6, 14, 18],

$$\xi_\Xi(c) = R_H\{F(0, c)/K_\xi[\eta]c\}^{1/3} \qquad (43)$$

where K_ξ is a constant (for large M and in good solvents) equal to $N_A R_H^3/6M[\eta]$. Consequently, power-law behavior is expected for $\{R_H/\xi_\Xi(c)\}F(0, c)$ in moderately concentrated solutions. The concentration dependence of $\zeta(0)/\zeta(c)$ may usually be represented as $\exp\{C_1/(T - T_g + C_2)\}$, where C_1 and C_2 are essentially constants, whereas the glass temperature T_g may depend markedly on the polymer concentration [22, 96]. In good solvents, $D_M(c)/D_T$ is expected to increase continuously, approaching power-law behavior with $D_M(c)/D_T \propto ([\eta]c)^{(2+p)/3}$ for $[\eta]c > 1$, consistent with the behavior seen in Fig. 8.

Under Flory Theta conditions, the initial decrease in $D_M(c)/D_T$ with increasing c is expected to reverse, with $D_M(c)/D_T \propto [\eta]c$ in moderately concentrated solutions, again consistent with the behavior shown in Fig. 8, but not in quantitative agreement. The discrepancy is attributed to cross-over behavior for $\Xi(0)/\Xi(c)$. An approximate relation combining the behavior expected for both dilute and moderately concentrated solutions is given by [18]

$$\frac{\Xi(0)}{\Xi(c)F(0, c)} \approx \{[1 + c\Gamma(c)]^{1-k_2}\exp(-k_1[\eta]c)$$

$$+ (K_\xi c\Gamma(c)[\eta]c)^{1/3}\} \frac{\zeta(0)}{\zeta(c)} \tag{44}$$

The first term in the brackets, designed for the dilute solution behavior, becomes negligible for larger c, but remains important over the range of concentrations shown in Fig. 8. The data obtained under Flory Theta conditions are fitted satisfactorily using Eq. (44) with $k_1 = 0.95$ and $k_2 = 0.53$ (B = 1.7), which are in the ranges expected for flexible chain polymers [18, 82–86]. Neglect of the factor $\zeta(0)/\zeta(c)$ can cause apparent departure from the behavior described, especially for more viscous solvents, since this function does not scale with c in the same way as the remaining components of $\Xi(0)/\Xi(c)F(0, c)$.

As mentioned above, nonexponential $g^{(1)}(\tau; q)$ have been frequently reported for moderately concentrated solutions, especially under Flory Theta conditions [18, 97–107]. Although Monte Carlo simulations have been studied to simulate the scattering from moderately concentrated polymer solutions [108], as discussed below, an approximate, but straightforward, theoretical treatment for polymers in the chain entanglement regime appears to predict much of the observed behavior [18, 99, 109]. The latter, related to a model first developed to explain the scattering from gels [110], predicts two weighted exponential terms, with one of the modes related to the effects of the intermolecular entanglement pseudo-network (or the actual network if the material is a gel) on longitudinal fluctuations in the concentration, determined by the intramolecular longitudinal modulus L(t). In this case, L(t) comprises contributions from the osmotic modulus $K_{OS}(t)$ and the shear modulus G(t):

$$L(t) = L(\infty) + \Delta L\ell(t) \tag{45a}$$

$$L(\infty) = K_{OS} + (4/3)G_e \tag{45b}$$

$$\Delta L\ell(t) = \Delta K_{OS}k(t/\tau_K) + (4/3)\Delta Gg(t/\tau_G) \tag{45c}$$

where $\ell(t)$ is a function that decreases monotonically from unit for $t = 0$ to zero for very large t, $\Delta L = L(0) - L(\infty) = \Delta K_{OS} + (4/3)\Delta G$, with $\Delta G = G(0) - G_e$ and $\Delta K_{OS} = K_{OS}(0) - K_{OS}$, G(0) and $K_{OS}(0)$ the "instantaneous" moduli on the light scattering time scale, G_e the equilibrium modulus, and K_{OS} given by Eq. (13). The equilibrium modulus G_e is zero for a fluid and $\Delta G \gg \Delta K_{OS}$, but $G_e > 0$ for a gel, and ΔK_{OS} and ΔG may be comparable [96, 111–114].

In principle, the time dependence of $\ell(t)$ could be complicated. In the theory as developed for an entangled polymer, it is assumed that $\ell(t)$ relaxes from unity to zero with a certain relaxation time τ_L. Following the treatment in [18, 99, 109], the displacement $\mathbf{u}(\mathbf{r}, t)$ of a material point on a polymer chain at position $\mathbf{r}$ at ime t is assumed to obey the relation

$$\text{div } \sigma(\mathbf{r}, t) - \zeta_M \frac{\partial \mathbf{u}(\mathbf{r}, t)}{\partial t} + \mathbf{A}(\mathbf{r}, t) = 0 \tag{46}$$

where $\sigma(\mathbf{r}, t)$ is the stress tensor for a volume element containing the material point, $\zeta_M = N_A \Xi(c)c/M$ and $\mathbf{A}(\mathbf{r}, t)$ is randomly fluctuating force acting at $\mathbf{r}$ (i.e., the effects of Brownian motion among the solvent molecules on the material point). Following linear viscoelasticity, the stress tensor is formulated as [96, 114]

$$\sigma_{ij}(t) = 2 \int_{-\infty}^{t} G(t - s) \frac{\partial \varepsilon_{ij}(s)}{\partial s} ds$$

$$+ \delta_{ij} \int_{-\infty}^{t} [K_{os}(t - s) - \tfrac{2}{3} G(t - s)] \frac{\partial \varepsilon_{kk}(s)}{\partial s} ds \tag{47}$$

where the $\varepsilon_{ij}(t)$ are the elements of the strain tensor, with summation over repeated indices, and δ_{ij} is the Kronecker delta. Solving for the Fourier transform $\mathbf{u}^*(q, \omega)$ of the displacement (with $\mathbf{q} = (0, 0, q)$ and $\mathbf{u}^* = (u_1^*, u_2^*, u_3^*)$) gives [18, 99, 109]

$$\Omega^*(q, \omega) u_3^*(q, \omega) = 0 \tag{48a}$$

$$\Omega^*(q, \omega) = -i\omega\zeta_M - q^2\{L(\infty) + i\omega\Delta L \ell^*(\omega\tau_L)\} \tag{48b}$$

where $\ell^*(\omega\tau_L)$ is the Fourier transform of $\ell(t)$. With $\ell(t) = \exp(-t/\tau_L)$ and $\Omega^* = \Omega' + i\Omega''$,

$$\Omega'(q, \omega) = \frac{q^2}{1 + (\omega\tau_L)^2} \{L(\infty)[1 + (\omega\tau_L)^2] + \Delta L(\omega\tau_L)^2\} \tag{49a}$$

$$\Omega''(q, \omega) = \frac{q^2\omega\tau_L}{1 + (\omega\tau_L)^2} \left\{ \frac{\zeta_M}{q^2\tau_L} [1 + (\omega\tau_L)^2] + \Delta L \right\} \tag{49b}$$

Finally, with neglect of intramolecular dynamics, so that $M(u, c) = 1$ and K_1/q^2 is independent of q, $g^{(1)}(\tau; q, c)$ is calculated as $S(\tau; q, c)/S(0; q, c)$, where

$$S(\tau; q, c) = \int_{-\infty}^{\infty} \{\Omega''(q, \omega)/[\Omega^*(q, \omega)]^2\} \exp(i\tau\omega) d \ln\omega . \tag{50}$$

This calculation results in $g^{(1)}(\tau; q, c)$ as the sum of two weighted exponentials in the form given by Eq. (20) with

$$2\hat{\gamma}_\mu\kappa = (1 + f\kappa) + (-1)^\mu[(1 + f\kappa)^2 - 4\kappa]^{1/2} \tag{51}$$

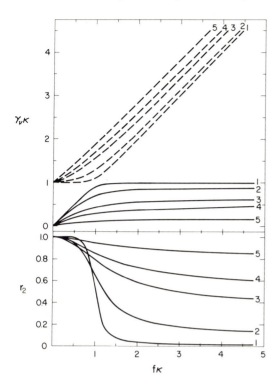

Fig. 9. Theoretical parameters in the two-exponent representation of $g^{(1)}(\tau; q)$ for an entangled flexible chain polymer fluid. *Upper:* $f\kappa = K_1 \tau_G = D_M \tau_L q^2$ for $f - 1 (= 4\Delta L/3L(\infty)) = 0.01$, 0.1, 0.5, 1.0, and 4.0 for *curves 1–5*, respectively, with $\hat{\gamma}_2$(---) and $\hat{\gamma}_1$(——). *Lower:* $r_1 = r$ for the same values of $f - 1$. After Fig. 1 in Ref. [18].

where $\mu = 1, 2$, $\kappa = K_1 \tau_L$, with $K_1 = q^2 L(\infty)/\zeta_M$, and $f - 1 = \Delta L/L(\infty)$. The parameters $\hat{\gamma}_1 \kappa$, $\hat{\gamma}_2 \kappa$, and r given by this theory are shown as functions of $f\kappa$ in Fig. 9. According to this result, the fast mode (corresponding to $\hat{\gamma}_2$) is diffusive in character, with a diffusion constant $D_{gel} = f D_M$, and with γ_2 scaling with q^2 for larger q. The slow mode (corresponding to $\hat{\gamma}_1$), is nondiffusive for large q, with γ_1 independent of q for larger q; it is only important for $f > 1$, and is entirely absent if f is unity, i.e., if $\Delta L/L(\infty) \ll 1$. Consequently, the slow mode is suppressed by conditions that decrease $\Delta L/L(\infty)$. For example, for $\kappa > 1$, $\gamma_1 = \hat{\gamma}_1 K_1 \approx (f \tau_L)^{-1}$, and $r_1 = r \approx (f - 1)/f$; in this regime, the weighting of mode 1 reflecting the relaxation l(t) increases in direct proportion to the relaxation weight ΔL, with corresponding decrease in mode 2 reflecting the diffusion, with rate $\gamma_2 = \hat{\gamma}_2 K_1 \approx q^2 D_{gel}$. This calculation is similar to that for the relaxing contributions to the longitudinal modulus in the scattering from undiluted polymers, with the scattering arising from density rather than concentration fluctuations [115]; in that case, of course, K_{OS} is replaced by the bulk modulus and the propagation of phonons gives rise to Brillouin scattering in addition to the scattering peaked around zero q. In that case, l(t) cannot be represented by a single exponential function, but can instead be represented as a weighted sum of exponentials [115]:

$$\ell(t) = \sum \ell_\mu \exp[-\lambda_\mu t/\tau_L] \tag{52}$$

where $\sum \ell_\mu = 1$. An approximate solution is then given for $g^{(1)}(\tau; q, c)$ for the limit with all the time constants $\tau_L/\lambda_\mu > 1/f D_M$, in the form of Eq. (17), with $r_\mu = \ell_\mu$ and $\gamma_\mu = \lambda_\mu/\tau_L$. More complex behavior can be expected for smaller τ_L/λ_μ. Analogous behavior may be obtained with the scattering from moderately concentrated solutions or gels, see below.

The neglect of internal dynamics in the preceding calculation appears to be reasonable for a moderately concentrated solution, but the treatment may be expected to fail for very large q, where the length scale probed in the scattering is no longer relevant to the length, important in the chain entanglement process.

For moderately concentrated solutions, it is assumed that $\Delta L \approx (4/3)G(0)$, i.e., relaxation of the osmotic modulus is presumed to be very fast on the time scale of the light scattering experiment. Then, $G(0)$ is taken to be the pseudo-network modulus G_N, given by the power-law relation [14, 114, 116]

$$G_N = \frac{cRT}{M_e} \left(\frac{c}{\hat{\rho}}\right)^{1+s} \tag{53}$$

where M_e is the molecular weight between entanglement loci (for undiluted polymer), $\hat{\rho}$ is the polymer density, and $s = (1 - p)/3$, so that $s = 0$ under Flory Theta conditions ($p = 1$), and increases to $1/4$ in a very good solvent ($p = 1/4$). The identification $G(0) \approx G_N$ assumes that relaxations occurring on a time scale faster than $\tau_N = \eta/G_N$ are not included in the $g^{(1)}(\tau; q)$ determined by light scattering, where η is the shear viscosity of the solution. Making use of this expression, and relations given above for K_{OS} [18],

$$f - 1 \approx \frac{4}{3} \frac{(\rho/c)^{1-4s}}{([\eta]_e \hat{\rho})^{1/\mu}} A([\eta]c) \tag{54}$$

where $[\eta]_e$ is the intrinsic viscosity for a chain with molecular weight M_e, $\mu = \partial \ln[\eta]/\partial \ln M$, and $A([\eta]c) = ([\eta]c)^{(1 - \mu)/\mu}/F(0, c)$ tends to be nearly independent of c [6, 18]. As expected from the nature of K_{OS}, f is larger under Flory Theta conditions than in a good solvent, making the nonexponential behavior more pronounced under Flory Theta conditions than in a good solvent. Results obtained in a study on solutions of polystyrene over a range of concentrations at the Flory Theta temperature are given in Figs. 10 and 11 [18]. The data were well fitted by Eq. (21) with two weighted exponentials, and as may be seen in Figs. 10 and 11, the data are in good agreement with the theory. Theoretical curves were calculated using adjustable parameters f and τ_L at each concentration, with the results that the measured $(f - 1)(c/\hat{\rho}) \approx 0.062$ is in good agreement with the value expected for the polymer used, and with $\tau_L \propto \tau_N = \eta/G_N$, as shown in Fig. 12 [18, 116].

The estimate $\ell(t) \approx g(t) \approx \exp(- t/\tau_L)$ used earlier is known to be inaccurate, as g(t) is not a simple exponential function [114]. On the other hand, if g(t) is expressed as a sum of exponentials, as would be expected [96, 111–114], then τ_L for a pseudo-exponential behavior might lie between τ_N and $\tau_c = \eta J_e^\circ$, where J_e° is the steady-state recoverable compliance; for linear flexible chain polymers

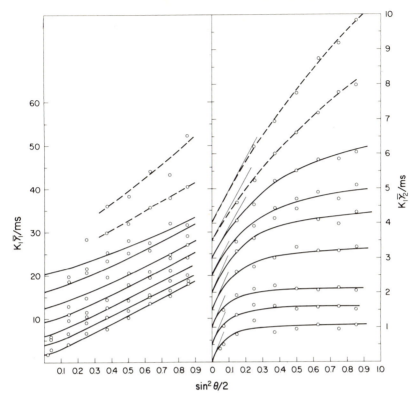

Fig. 10. Plots of $K_1\hat{\gamma}_\mu/ms^{-1}$ vs $sin^2(\vartheta/2)$ ($\mu = 1, 2$ for left and right, respectively) for moderately concentrated solutions of polystyrene in cyclohexane at the Flory Theta temperature (34.8 °C), $M_W = 8.62 \times 10^5$, $[\eta] = 76$ ml/g, $R_G = 27$ nm: $(\eta)c$ equal to 1.00, 2.12, 3.66, 4.25, 6.00, 6.54, 7.33, 7.91, 8.31 from bottom to top. The data for $K_1\hat{\gamma}^1$ and $K_1\hat{\gamma}^2$ are off set by 2.0 and 0.5 ms^{-1}, respectively, between data sets for clarity. The *solid curves* represent Eq. (51) using parameters discussed in the text; the *dashed lines* are merely to guide the eye. After Ref. [18]

with a narrow distribution of molecular weights, $J_e^\circ \approx 3J_N = 3/G_N$, but J_e° is increased markedly with increased molecular weight dispersion, whereas J_N is not [96, 114]. As seen in Fig. 12, the observed τ_L lie between τ_N and τ_c, being much closer to the former. Similar results have been reported on other systems under Flory Theta conditions [100]. This suggests that in the scattering experiments, the effective ℓ (t) on the time-scale of the scattering experiment was given by one or two exponential contributions to g(t) with time constants τ_L/λ_μ slightly larger than τ_N. Other investigators have reported a more complex behavior, with data fitted by Eq. (17) with several exponential terms [103–107]. On the basis of the known behavior of G(t) for $t > \tau_N$ [114], it seems unlikely that the reported behavior can represent the effects of nonexponential g(t) within the simplified analysis of $g^{(1)}(\tau; q)$ discussed above. The source of this discrepancy is not understood at present, but several possibilities may be

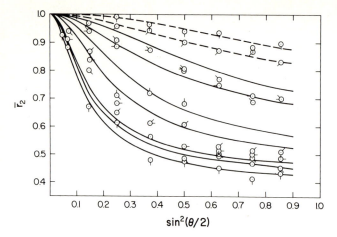

Fig. 11. Plot of r_1 vs $\sin^2(\vartheta/2)$ ($\mu = 1, 2$ for the moderately concentrated solutions of polystyrene in cyclohexane at the Flory Theta temperature (34.8 °C) identified in Fig. 10. The *solid curves* are calculated using Eq. (51) with parameters discussed in the text; the *dashed lines* are merely to guide the eye. After Ref. [18]

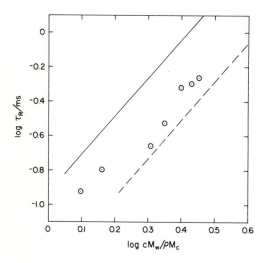

Fig. 12. Bilogarithmic plot of τ_T/ms vs $cM_w/\rho M_c$ for the moderately concentrated solutions of polystyrene in cyclohexane at the Flory Theta temperature (34.8 °C) identified in Fig. 10. The *solid* and *dashed lines* give $\tau_c = \eta J_e^0$ and $\tau_N = \tau_c/3 \approx \eta/G_N$, respectively, where J_e^0 is the steady-state recoverable compliance, and $G_N \approx e/J_e^0$. After Ref. [18]

suggested: owing to the effects of a higher solvent viscosity, the relaxation times may have been altered enough that the light scattering time scale encompassed a range for which both $k(t)$ and $g(t)$ contribute, with $L(0)$ larger than $(4/3)G_N$, on the time scale probed by the scattering experiment, giving $\ell(t)$ a multiple-exponential character, and deviation from the simple result given by Eqs. (21) and (51). Further, if the experimental time-scale encompasses times $\tau_L/\lambda_\mu < 1/D_M$, then $g^{(1)}(\tau; q)$ may not be simply interpreted in terms of a sum of weighted exponentials directly related to the sum of weighted exponentials in

ℓ (t). Rather, more complex mode coupling may have to be taken into account, leading to $g^{(1)}(\tau; q)$ in the form given by Eq. (17), but with a more complicated interpretation of the parameters r_μ and γ_μ than that given above.

3 Isotropic Solutions of Semiflexible Chain Polymers

Semiflexible chains are characterized by a conformation with a persistence length ρ that is an appreciable fraction of the contour length L [6, 7, 12, 117]. The chain adopts a flexible coil conformation as $\rho/L \ll 1$, and a rodlike conformation as $\rho/L \gg 1$. The study of semiflexible chains introduces some additional features, the principal one being the appearance of depolarized scattering. Unlike the case with flexible chain polymers, the variable A_2M and $[\eta]$ are not expected to scale in the same way with M in good solvent conditions. Thus, with $\rho/L \gg 1$ for rodlike chains, $a(\cdot) \approx 1$, $h(\cdot) \propto 3^{3/2}/2\ln(3L/2d_H)$, $k_\eta(\cdot) \approx 1$, $r(\cdot) \approx L/4\rho$ and $a_1(\cdot) \approx 0$. Therefore, for rodlike chains, $\hat{c} = cN_AR_G^3/M \propto L^2M_L^{-1}c \propto [\eta]c$ (neglecting the weak dependence of $\ln(3L/2d_H)$ on L), whereas $A_2Mc \propto Ld_{Thermo}M_L^{-1}c$ is not proportional to $\hat{c}$. Consequently, "moderately concentrated solution behavior" may be expected for different ranges of c for thermodynamic and hydrodynamic properties. For polymers with repeat unit geometric diameter d_G interacting through a hard-core potential, $d_{Thermo} \approx d_G$, but d_{Thermo} may exceed d_G with polyelectrolyte chains, or decrease to zero under Flory Theta conditions.

As discussed below, with rodlike polymers, moderately concentrated solutions may form an ordered mesophase if A_2Mc exceeds about 4 [7, 117–119]. On the other hand, deviations from simple power-series expansions in dynamic properties are expected for $[\eta]c$ of order unity [113]. It is not obvious that a concentration regime will exist between $[\eta]c \approx 1$ and $A_2Mc \approx 4$ for real molecules, see below.

3.1 Thermodynamic Behavior

Much of the formalism given in the *Introduction* applies with isotropic solutions of rodlike polymers, but owing to the anisotropy of the chain itself, the polarized scattering $R_{iso}(q, c)$ considered in the preceding is augmented by depolarized scattering $R_{aniso}(q, c)$ and a contribution $R_{cross}(q, c)$ that vanishes at zero q. Thus, for the vertically polarized component of the scattering with vertically polarized light [4, 6, 12]:

$$R_{Vv}(q, c) = R_{iso}(q, c) + (4/3)R_{aniso}(q, c) + R_{cross}(q, c) \tag{55}$$

whereas the depolarized scattering determined with the horizontally polarized component of the scattering with vertically polarized light is given by [4, 6, 12]:

$$\mathbf{R}_{Hv}(q, c) = \mathbf{R}_{aniso}(q, c) . \tag{56}$$

All of the general relations given in the preceding apply to $\mathbf{R}_{iso}(q, c)$, but $\mathbf{R}_{aniso}(q, c)$ requires the use of new relations. Thus, $\mathbf{R}_{aniso}(0, c) \propto m_o \delta_o^2 G_2(c)$, where δ_o is the optical anisotropy of the n scattering elements, each with molecular weight m_o, and $G_2(c)$ involves an ensemble average over the second Legendre polynomial $P_2(\cos \beta_{i_v, j_\mu})$, with β_{i_v, j_μ} the angle between cylindrically symmetric scattering elements i on chain v and j on chain μ [4, 6, 12, 120–124]:

$$\mathbf{R}_{aniso}(0, c) = (3/5)\mathbf{K}_{op}cm_o\delta_o^2 G_2(c) = (3/5)\mathbf{K}_{op}cM\delta^2 F_{Hv}(0, c) \tag{57a}$$

$$G_2(c) = (1/n) \left\langle \sum_{i_v} \sum_{j_\mu} P_2(\cos \beta_{i_v, j_\mu}) \right\rangle \tag{57b}$$

$$\delta^2 = \delta_o^2 G_2(0)/n . \tag{57c}$$

For scattering elements with cylindrical symmetry, δ_o is given by the polarizabilities $\alpha_\parallel$ and $\alpha_\perp$ parallel and perpendicular to the elemental symmetry axis, respectively:

$$\delta_o = \frac{\alpha_\parallel - \alpha_\perp}{\alpha_\parallel + 2\alpha_\perp} , \tag{58}$$

Note that $\mathbf{K}_{op}\delta_o^2 \propto (\alpha_\parallel - \alpha_\perp)^2$ may not be zero, even if $\mathbf{K}_{op}$ is zero. For rodlike chains at infinite dilution, $G_2(0) = n$, and $\mathbf{R}_{aniso}(0, c) = (3/5)\mathbf{K}_{op}cM\delta_o^2$.

For chains with persistence length ρ and contour length L, use of the wormlike chain model at infinite dilution gives [6, 12]

$$G_2(0)/n = W(\rho/L) \tag{59a}$$

$$W(\rho/L) = (2\rho/3L)\{1 - (\rho/3L)[1 - \exp(- 3L/\rho)]\} \tag{59b}$$

giving $\delta^2/\delta_o^2 = 2\rho/3L$ for flexible chains, and $\delta = \delta_o$ for rods.

The function $F_{Hv}(0, c)$ is expected to be much less dependent on c than $F(0, c)$ in dilute solutions, as is confirmed by experiment [6, 7, 120]. Thus, for dilute solutions of rodlike molecules [124],

$$F_{Hv}(0, c) = 1 + A_2 Mc/4 \approx \{F(0, c)\}^{-1/8} \tag{60}$$

where $F(0, c)$ is the function discussed in the Introduction, e.g., see Eqs. (12)–(14), etc. It is assumed that the higher order terms are negligible, even in moderately concentrated solutions of rodlike chains [7, 121, 123]. With this expression, the depolarized scattering is enhanced as the orientation fluctuations are enhanced by the approach to the nematic phase. A similar expression is probably a reasonable approximation for semiflexible chains, though it may overemphasize the dependence of $F_{Hv}(0, c)$ on c.

For $q \neq 0$, interference effects reduce the scattering, and contributions from $\mathbf{R}_{cross}(q, c)$ must be included. Thus, for diluted solutions ($A_2 Mc \ll 1$), the polarized scattering may be expressed in the form

$$\mathbf{R}_{Vv}(q, c) = \mathbf{K}_{op} c M\, S_{Vv}(q, c; \delta) \tag{61a}$$

$$S_{Vv}(q, c; \delta) = P_{Vv}(q; \delta) F_{Vv}(q, c; \delta) \tag{61b}$$

and the depolarized scattering may be expressed in the form

$$\mathbf{R}_{Hv}(q, c) = (3/5)\mathbf{K}_{op} c M S_{Hv}(q, \vartheta, c; \delta) \tag{62a}$$

$$S_{Hv}(q, \vartheta, c; \delta) = \delta^2 P_{Hv}(q, \vartheta) F_{Hv}(q, c) \tag{62b}$$

where the anisotropy δ at infinite dilution is given by Eq. (59) for wormlike chains. As indicated in Eq. (62), $P_{Hv}(q, \vartheta; \delta)$ depends explicitly on the scattering angle ϑ as well as on q. Since the chain conformation should not vary with c (even for a wormlike chain at low $A_2 Mc$), neither $P_{Vv}(q; \delta)$ nor $P_{Hv}(q, \vartheta; \delta)$ are expected to depend on c.

The function $F_{Vv}(q, c; \delta)$ includes contributions from $\mathbf{R}_{cross}(q, c)$ which vanish for zero q, so that $F_{Vv}(0, c; \delta)$ is given by

$$F_{Vv}(0, c; \delta) = F(0, c) + 4/5\delta^2 F_{Hv}(0, c). \tag{63}$$

It should be noted that A_3 and higher virial coefficients are expected to be small for rodlike polymers interacting through a hard-core potential, even if A_2 is large [7, 113, 120–123].

For rodlike chains the angular dependence may be expressed in the forms [4, 6, 125]:

$$(1 + 4\delta^2/5)P_{Vv}(q; \delta) = p_1(X) + (4/5)\delta^2 p_3(X)$$

$$+ (\delta - 2\delta^2)m_1(X) + (9/8)\delta^2 m_2(X)$$

$$+ \delta^2 m_3(X) \tag{64}$$

$$P_{Hv}(q, \vartheta) = p_3(X) + (5/16)(1 - \cos \vartheta)m_2(X) \tag{65}$$

where $X = (12\,u)^{1/2}$, and the p_μ and m_μ are given by

$$p_1(X) = (2/X^2)\{X\,Si(X) + \cos X - 1\} \approx 1 - 1/3u + \ldots \tag{66a}$$

$$p_2(X) = (6/X^3)\{X - \sin X\} \approx 1 - 3/5u + \ldots \tag{66b}$$

$$p_3(X) = (10/X^5)\{X^3 + 3X \cos X - 3 \sin X\} \approx 1 - 3/7u + \ldots \tag{66c}$$

$$m_1 = p_1 - p_2 \approx 4/15u + \ldots \tag{67a}$$

$$m_2 = 3p_1 - p_2 - 2p_3 \approx 16/35u + \ldots \tag{67b}$$

$$m_3 = p_3 - p_2 \approx 6/35u + \ldots \tag{67c}$$

As shown by the expansions, the p_μ are unity at zero u, whereas the m_μ, which include contributions from $\mathbf{R}_{cross}(q, 0)$, are zero at zero u; $Si(\cdot)$ is the sine integral. Note that the explicit dependence of $P_{Hv}(q, \vartheta)$ on the scattering angle is not important for small q. For small u, with neglect of the anisotropy,

$$P_{Vv}^{-1}(q; 0) = p_1^{-1}(u) = 1 + (1/3)u + (7/225)u^2 + \ldots \tag{68}$$

showing the expected initial tangent in common with all isotropic scatterers.

Including the effects of anisotropy, in forms useful for either rodlike or wormlike chains [6, 125],

$$P_{Hv}^{-1}(q; \delta) = 1 + (3/7)(f_3\delta/\delta_o)^2 u + \ldots \tag{69}$$

$$P_{Vv}^{-1}(q; \delta) = 1 + (1/3)J(\delta)u + \ldots \tag{70a}$$

$$J(\delta) = \frac{1 - (4/5)f_1\delta + (4/7)(f_2\delta)^2}{1 + (4/5)\delta^2} \tag{70b}$$

where f_1, f_2, and f_3 are functions of ρ/L, all tending to unity for ρ/L large enough to make the depolarized scattering significant.

As implied by these expressions, $\mathbf{R}_{Hv}(q, c, \vartheta)/c$ appears to be essentially independent of c for rodlike chains in dilute solutions, even for polyelectrolyte systems for which d_{Thermo}, and therefore A_2, is large owing to electrostatic effects [126]. Furthermore, the dependence on q of $\mathbf{R}_{Vv}(q, c)$ and $\mathbf{R}_{Hv}(q, c, \vartheta)$ leads to consistent estimates for R_G from the polarized and depolarized scattering. The ratio $\mathbf{R}_{Hv}(0, c)/\mathbf{R}_{Vv}(0, c)$ tends to $3\delta^2/(5 + 4\delta^2)$ at infinite dilution, providing a value of δ from which ρ may be determined if δ_o is known [4, 6]. Although δ_o can be nearly unity for rodlike chains since all the scattering elements share the same symmetry axis, for helical molecules with an overall rodlike symmetry, δ_o may be vanishingly small owing to the averaging over the elemental scattering elements arrayed along the helical path. For example, extensive studies have been reported on solutions of polyglutamates [7, 120, 127–129], which adopt a helical structure in certain solvents, with δ small enough that contributions of the anisotropy to the polarized scattering may be neglected in consideration of $\mathbf{R}_{Vv}(q, c)$ [120], that is, $\mathbf{R}_{Vv}(q, c) \approx \mathbf{R}_{iso}(q, c)$, with $\mathbf{R}_{iso}(q, c)$ represented by Eqs. (7)–(14).

Relatively little work has been done on the scattering of rodlike chains from moderately concentrated solutions. A major experimental difficulty is that of obtaining samples with a narrow distribution of chain lengths, and finding solvation conditions that eliminate intermolecular association, especially with rodlike polymers. The results on the scattering from the poly(benzyl glutamate) solutions [7, 120] showed that whereas $\mathbf{R}_{Hv}(0, c)/c$ was essentially independent of c for dilute solutions, $\mathbf{R}_{Hv}(0, c)/c$ increased slightly with c in the moderately concentrated regime, never, however, becoming large enough to void the approximation $\mathbf{R}_{Vv}(q, c) \approx \mathbf{R}_{iso}(q, c)$. The increase in $\mathbf{R}_{Hv}(0, c)/c$ at larger c suggests that enhanced orientational correlation among the chain elements results in

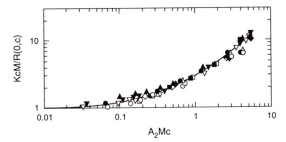

Fig. 13. Bilogarithmic plots of $K_{op}cM/R_{Vv}(0, c)$ vs A_2Mc for moderately concentrated solutions of poly(benzyl glutamate) in good solvent [120]. The curve represents the use of Eq. (63) and higher virial coefficients equal to zero (and $\delta \approx 0$). The data were collected in the temperature range 15 to 75 °C for samples with $10^{-3} M_w$ equal to 277, 179, 149, and 60; see Ref. [120] for a complete listing

increased $F_{Hv}(0, c)$ as A_2Mc approaches the value for the transition to the ordered state. The data on the moderately concentrated solutions of poly(benzyl glutamate) gave $F_{Hv}(0, c) > 1$, but smaller than that given by Eq. (60).

The scattering data on $K_{op}cM/R_{Vv}(0, c) = F(0, c)$ for moderately concentrated poly(benzyl glutamate) solutions are shown in Fig. 13. The results give $F(0, c)$ linear in A_2Mc for A_2Mc less than about 6 (see the curve in Fig. 13), as expected if $A_2 \neq 0$, but with A_3 and higher order virial coefficients negligibly small. With $A_2 = A_2^{(R)}$ the data give $d_{Thermo} \approx 1.6$ nm, in reasonable agreement with the expected geometric diameter d_G of the helical chain [7, 120]. Thus, the results are in reasonable accord with the expected behavior for rodlike chains interacting principally through a hard-core repulsive potential.

With A_3 and higher order virial coefficients negligibly small, $c\Gamma(c)H(q, c)$ should be represented by Eq. (14b) with retention of only the term proportional to $\hat{c}$, or $c\Gamma(c)H(q, c) \approx 2A_2^{(R)}Mc W_2(q)$ [121]. The random phase approximation has been used to estimate $S(q, c)$ for moderately concentrated solutions of rodlike chains [121, 123] to give a result with $P_{Vv}(q, c)^{-1}$ given by Eq. (64a), $\Gamma(c) = 2A_2^{(R)}M$, and [7, 117–119]

$$H(q, c) \approx 1 - \frac{3}{200} \frac{1}{1 - A_2Mc/4} u^2 + \dots \tag{71}$$

With these expressions, the correlation legnth $b \approx \xi_P$ to order q^2, and is expected to scale with A_2Mc, the behavior being more complicated for larger q. Data on isotropic polyglutamate solutions shown in Fig. 14 [120] exhibit the anticipated scaling, as shown by the curved line, with $6b/L$ tending to unity for small A_2Mc, and to scaling with $(A_2Mc)^{-1/2}$ for larger A_2Mc.

3.2 Dynamic Behavior

Owing to the molecular asymmetry of rodlike chains, dynamic light scattering is sensitive to both center-of-mass translation, and molecular rotation, with the

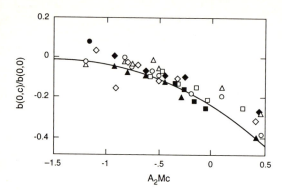

Fig. 14. Bilogarithmic plot of $b(0, c)/b(0, 0)$ vs $A_2 M_w c$ for moderately concentrated solutions of poly(benzyl glutamate) in good solvent for the polymers described in the caption to Fig. 13 [120]; $b(0, 0)^2 = R_G^2/3$. The *curve* represents $\xi_p(0, c)$ using the experimental data for $\Gamma(c)$ (see Eq. (30))

latter resulting in a relaxation rate that is independent of q [4, 6, 123, 128–130]. Thus, the behavior of rodlike chains introduces several concentration-dependent diffusion constants, in addition to $D_M(c)$: diffusion constants $D_\perp(c)$ and $D_\parallel(c)$ for translation perpendicular and parallel to the chain, respectively, and a rotational diffusion constant $D_R(c)$. These are discussed further below. At infinite dilution, $D_M(0) = D_T$, and $D_R(0) = D_{R,0}$, $D_\parallel(0) = D_{\parallel;0}$, and $D_\perp(0) = D_{\perp;0}$. For rodlike chains [12, 113, 128–130],

$$D_{R;0} = 2RT/15\eta_s M[\eta] \tag{72}$$

$$D_{\parallel;0} = D_{R,0} L^2/6 \tag{73}$$

$$D_{\perp;0} = D_{\parallel;0}/2 = D_{R,0} L^2/12 \tag{74}$$

With $D_T = (D_{\parallel;0} + 2D_{\perp;0})/3 = 2D_{\parallel;0}/3$, and $[\eta]$ given by Eq. (4) (with $k_\eta = 1$). With these expressions, $K_1(q, 0)/q^2 D_T$ calculated from the $g_{VV}^{(1)}(\tau; q, c)$ for the polarized scattering may be represented by Eq. (35) with $C^{(2)} = 1/30$ for rodlike chains [131]; $C^{(2)}$ has also been calculated for a weakly bending rodlike chain model [131].

For moderately concentrated solutions, the diffusivities depend markedly on concentration. A theoretical treatment for the behavior in moderately concentrated solutions gives $K_1(q, 0)/q^2 D_T$ as in Eq. (22), where $D_M(c)/D_T$ is given by Eq. (23a), with $F(0, c)^{-1} = 1 + 2A_2 Mc$ (since A_3 and higher virial coefficients are negligible), and with $M(u, c)$ given by [7, 121–123]:

$$M(u, c) = \left(1 - \frac{4}{45} u\, m(u, c)\right) R(u, c) \tag{75a}$$

$$m(u, c) = \frac{D_\parallel(c) - D_\perp(c)}{D_S(c)} f_T(4u/15) - \frac{D_R(c) L^2}{8\, D_S(c)} f_R(2u/15) \tag{75b}$$

$$R(u, c) = 1 - \{1 - P_{Vv}(u)\} \frac{2A_2 Mc}{1 + 2A_2 Mc} f_W(u, c) \tag{75c}$$

Here $D_R(c)$ is the rotational diffusion constant, $D_S(c) = \{D_\parallel(c) + 2D_\perp(c)\}/3$ (the factor $1 - \varphi$ may be neglected as φ is small for isotropic solutions of rodlike chains). The functions $f_R(x)$ and $f_T(x)$ are unity for small x, and the results of numerical calculations of these [123] can be approximated as $(1 + x^2)^{-1/2}$ within 15%; $f_W(u, c)$ is equal to unity within the same approximation. In dilute solutions, $m(u, c) \approx 3/8$, making $M(u, c) > 1$. In moderately concentrated solutions, the effects of intermolecular constraints on diffusion reduce the diffusivities. With an approximation expected to be reasonable when this entanglement effect is well developed [113]

$$D_\parallel(c) \approx D_\parallel(0) \gg D_\perp(c) \approx 0 \tag{76}$$

$$D_R(c) \approx D_R(0)\{1 + k_R([\eta]c/M_L)^{-2}\} \approx 0 \tag{77}$$

where k_R is of the order of 10^{-3} [117, 129–133]. Given the approximate behavior $[\eta] \propto L^2$ for rodlike chains, $D_R(c) \propto L^7 c^2$. With this limiting behavior,

$$M(u, c) \approx \left(1 - \frac{4}{15} uf_T(4u/15)\right) R(u, c) \tag{78}$$

and $M(u, c) < 1$, in contrast to the behavior in dilute solutions. Moreover, $D_M(c)/D_T \approx (1/2)(1 + 2A_2Mc)$ is predicted to exceed unity for $A_2Mc > 1/2$ with the use of these diffusivities.

Experimental data on moderately concentrated solutions of poly(benzyl glutamate) are in qualitative agreement with preceding work [7, 120]. Thus, $M(u, c) - 1$ calculated from the experimental data is found to change sign as predicted, and $D_M(c)$ increases with increasing c in the range studied. The data [120] on $D_M(c)/(1 + 2A_2Mc)D_T = \Xi(0)/\Xi(c) \approx D_S(c)/D_T$ are shown in Fig. 15, in comparison with Brownian dynamics simulations [132]; the latter were calculated for a chain with $L/d_H = 50$ [120].

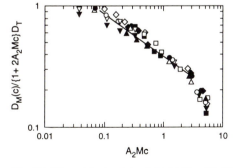

Fig. 15. Bilogarithmic plot of $D_M(c)/(1 + 2A_2Mc)D_T \approx D_S(c)/D_T$ for moderately concentrated solutions of poly(benzyl glutamate) in a good solvent for the polymers described in the caption to Fig. 13 [120]. The *curve* represents Brownian simulations in Ref. [133] for a rodlike chain with a length to diameter ratio of 50

An elegant matrix formulation has been developed [123] to give $g_{Vv}^{(1)}(\tau; q, c)$ for moderately concentrated isotropic solutions of rodlike chains, by numerical calculations, over a wide range of q and c. An approximate analytical form has also been developed [121, 122], limited to an undefinied range of small q. The latter treatment represents $g_{Vv}^{(1)}(\tau; q, c)$ as a weighted sum of two exponentials, as in Eq. (21). In this case, however, unlike the treatment discussed above for flexible chain polymers, both terms in $g_{Vv}^{(1)}(\tau; q, c)$ arise from diffusive effects, and both relaxation rates scale with q^2. One term is associated with fluctuations in concentration and the other with fluctuations in orientation, similar to the effects included in Eq. (75). In the treatment for the moderately concentrated isotropic solution, it is assumed that Eq. (76) applies, and that $D_R(c) \approx 0$ [113], with the results:

$$\hat{\gamma}_1 = 1 + 4x_1/15 \tag{79a}$$

$$\hat{\gamma}_2 = 1 + 4x_2/15 \tag{79b}$$

where x_1 and x_2 are the two roots of a quadratic equation in x, each functions of c/c_{NI}:

$$28(1 + 8c/c_{NI})x^2 - 15(4 - 67c/c_{NI})x - 315(1 - c/c_{NI}) = 0 . \tag{80}$$

The functions $\hat{\gamma}_1$ and $\hat{\gamma}_2$ are shown in Fig. 16 for the two reduced relaxation rates, along with the weight parameter for the faster mode (2), computed with Eq. (21b) to be $r_2 = x_1/(x_1 - x_2)$. As may be seen, the predicted deviation from single-exponential behavior is not strong if $8c/c_{NI} = 2A_2Mc$ is small, but becomes pronounced with increasing c/c_{NI}. This treatment does not include the nondiffusive relaxation discussed above for solutions of flexible chain polymers. There would appear to be no reason why such effects could not further complicate the behavior of $g_{Vv}^{(1)}(\tau; q, c)$ for moderately concentrated isotropic solutions of rodlike chains.

The depolarized scattering is related to orientation fluctuations, thereby being determined by rotational diffusion in dilute solutions, but dependent on translational modes as well in moderately concentrated solutions. In general,

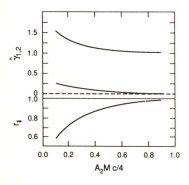

Fig. 16. Theoretically predicted values of $\hat{\gamma}_1$ and $\hat{\gamma}_2$ (upper) and r_2 (lower) vs A_2Mc for isotropic moderately concentrated solutions of rodlike chains [121, 122]

$g_{Hv}^{(1)}(\tau; q, c, \vartheta)$ determined from the depolarized scattering appears to be exponential, but it depends explicitly on the scattering angle ϑ as well as on q. Alternatively, $g_{Hv}^{(1)}(\tau; q, c, \vartheta)$ may be expressed in terms of the maximum value $u_\pi = (q_\pi R_G)^2$ of u available for a given R_G (i.e., $\vartheta = \pi$ and $q_\pi = 4\pi/\lambda$). In this case, the relations for the first-cumulant $K_{1;Hv}(q, c, u_\pi)$ may be expressed as [4, 7, 130]

$$\frac{K_{1;Hv}(q, c, u_\pi)}{6D_{R;0}} = \frac{D_{Hv}(c)}{D_{R;0}} M_{Hv}(u, c, u_\pi) \tag{81}$$

where $D_{Hv}(c)$ reduces to $D_{R;0}$ at infinite dilution, $M_{Hv}(0, c, u_\pi) = 1$, and $M_{Hv}(u, 0, u_\pi)$ tends to unity for small u. The function $D_{Hv}(c)$ is conveniently represented in terms of the rotational self-diffusion constant $D_R(c)$ by

$$\frac{D_{Hv}(c)}{D_{R;0}} = \frac{D_R(c)}{D_{R;0}} F_{Hv}(0, c)^{-1} \tag{82}$$

where $D_R(c)/D_{R;0}$ and $F_{Hv}(0, c)^{-1}$ are discussed above. Furthermore, calculations give [130]

$$M_{Hv}(u, c, u_\pi) = \{1 + \tfrac{1}{7} u m_{Hv}(u, c, u_\pi)\} R_{Hv}(u, c, u_\pi) \tag{83a}$$

$$m_{Hv}(u, c, u_\pi) = \frac{D_\|(c) - D_\perp(c)}{L^2 D_R(c)} f'_T(u, u_\pi) + 14 \frac{D_\perp(c)}{L^2 D_R(c)} + f'_R(u, u_\pi) \tag{83b}$$

$$R_{Hv}(u, c, u_\pi) = 1 + \{1 - P_{Hv}(u, u_\pi)\} \frac{A_2 Mc/4}{1 - A_2 Mc/4} f'_W(u, c). \tag{83c}$$

The functions $f'_T(u, u_\pi)$ and $f'_W(u, c)$ are each unity for small u, whereas $f'_R(0, u_\pi) = 0$; $f'_T(u, u_\pi)$ vanishes and $f'_R(u, u_\pi) \propto u$ for very large u [130]. Thus at infinite dilution, the modes involving translational motions are weak, and only affect $K_{1;Hv}(q, c, q_\pi)$ at large scattering angles, so that $M_{Hv}(u, c, u_\pi)$ is essentially unity for small u. For very large u, motions reflected in both $D_R(c)$ and $D_\perp(c)$ contribute to $K_{1;Hv}(q, c, q_\pi)$. In moderately concentrated solutions, the approximations given above give both $D_R(c)$ and $D_\perp(c)$ small owing to the constraints to motions lateral to the chain axis.

Relatively few data are available for comparison with these predictions, but it is reported [130] that the expected dependence of $D_R(c)$ on L is obtained from data [128] on $K_{1;Hv}(q, c, q_\pi)$ for moderately concentrated solutions of poly(benzyl glutamate), provided the analysis is made with the preceding relations, whereas the assumption that $R(u, c, u_\pi) F_{Hv}^{-1}(0, c) \approx 1$ results in a smaller dependence of $D_R(c)$ on L.

Data on $g_{Vv}^{(1)}(\tau; q, c)$ for moderately concentrated solutions of polyglutamates are found to be fitted by the weighted sum of as many as four exponential functions, as compared with the double-exponential behavior described in the preceding paragraph [120]. A very slowly decaying, angular dependent mode (nearly outside the parameters of the correlator used) was attributed to unidentified effects other than those discussed above, owing to the nature of its variation with scattering angle and concentration. The next two slowest modes appear to vary with concentration in the manner expected for modes 1 and 2 in the preceding, and scale with q^2, as expected. The origin of the very slow mode is obscure, and its characteristics only partially revealed. It may correspond to a nondiffusive relaxation, such as that discussed above in connection with Eq. (51) for flexible chain molecules.

4 Nematic Solutions of Semiflexible Chain Polymers

As mentioned in the preceding sections, if the solute polymer has a persistence length ρ that is an appreciable fraction of its contour length L, then it may form an ordered mesophase in moderately concentrated solutions, with symmetry between those of a liquid and a crystal. In the ordered mesophases of interest here, the material remains a fluid, but the solute exhibits orientational order while retaining translational disorder. The nematic mesophase will be assumed, for which the solute chains tend to be parallel to each other – more complex symmetries are known [134, 135]. Nematic phases may be expected with moderately concentrated solutions of rodlike chains, for which $\rho/L \gg 1$ for the isolated chain at infinite dilution, but may also form with moderately concentrated solutions of semiflexible chains for which ρ/L is smaller for the chain at infinite dilution [119]. In the latter case, the chain tends to adopt a rodlike conformation in the nematic phase, with increase in its persistence length on formation of the ordered phase [119, 136].

Nematic solutions present an anisotropic scattering medium, with fluctuations in molecular orientation resulting in appreciable depolarized scattering. Thus, in this case, it is the horizontally polarized component of the scattering with vertically polarized incident light that is of most interest [134, 135]. The scattering from concentration fluctuations is much smaller than that from orientation fluctuations. The anisotropy of a nematic mesophase of rodlike chains is measured by an order parameter S, given by [119, 134, 135]

$$S = (3\langle \mathbf{u} \cdot \mathbf{n} \rangle^2 - 1)/2 \qquad (84)$$

where $\mathbf{u}$ and $\mathbf{n}$ are unit vectors along the symmetry axes of the rodlike chain and the nematic mesophase, respectively, and the brackets indicate an ensemble average. Although $\mathbf{n}$ may exhibit spatial variation for a nematic phase not at equilibrium, or subject to external fields (e.g., magnetic, electric, surface, etc.), S is presumed to be invariant over small regions, and fixed by the solute length to diameter ratio L/d, the concentration c, and the nature of intermolecular interactions. To a first approximation, the nematic phase becomes stable when $A_2Mc > 4$ [119, 134, 135], and S is expected to increase with increasing A_2Mc.

The relations necessary to analyze the photon correlation (dynamic) light scattering from a nematic mesophase are summarized in the following. More detailed expositions may be found in the original literature [134, 135, 137, 138]. Nematic samples suitable for the scattering measurements discussed below must be fully aligned over the scattering volume, and that alignment must be known with respect to the geometric dimensions of the sample. This condition will not ordinarily be met by polymeric nematic fluids, but can be achieved by utilizing external fields to lift the degeneracy of the nematic alignment, including surface orientation effects and alignment in an external magnetic field [134, 135].

Owing to the anisotropy of the nematic mesophase, the notation must include not only the polarization directions of the incident and scattered light, but also information on the alignment of $\mathbf{n}$ with these and the scattering vector $\mathbf{q}$. The (unnormalized) photon count autocorrelation function will be denoted $G^{(2)}_{Ai;n}(\tau; q_\perp, q_\parallel)$, where $q_\parallel = |\mathbf{q} \cdot \mathbf{n}|$ and $q_\perp = |\mathbf{q} \times \mathbf{n}| = (q^2 - q_\parallel^2)^{1/2}$ are components of the scattering wave vector $\mathbf{q}$ ($q_\parallel = 0$ if the director is perpendicular to the scattering plane, and $q_\perp = 0$ if the director is parallel to the wave vector, see below). The notation {Ai; n} designates the orientations of the polarization of the detected scattered light (A), the incident polarization (i), and the director (n) with respect to the scattering plane, e.g., Vh; v will designate vertically (V) and horizontally (h) polarized scattered and incident light, respectively, and a vertically (v) oriented director. Geometrical arrangements of the polarizations of the scattered ray (A), the incident ray (i) and the director (n) such that one of the components $q_\parallel$ or $q_\perp$ of the scattering wave vector $\mathbf{q}$ is zero are especially useful, as discussed below.

With this notation, Eq. (16) takes the form

$$g^{(2)}_{Ai;n}(\tau; q_\perp, q_\parallel) = 1 + f[g^{(1)}_{Ai;n}(\tau; q_\perp, q_\parallel)]^2 \tag{85}$$

where $G^{(2)}_{Ai;n}(\infty; q_\perp, q_\parallel)$ is the limiting value of $G^{(2)}_{Ai;n}(\tau; q_\perp, q_\parallel)$ for large τ. As the scattering may involve contributions from two exponential terms to $g^{(1)}_{Ai;n}(\tau; q_\perp, q_\parallel)$, it is conveniently represented by Eq. (17) in the form

$$g^{(1)}_{Ai;n}(\tau; q_\perp \cdot q_\parallel) = \frac{\sum \Gamma_{Ai;n\mu}(q_\perp, q_\parallel)\exp[-\tau/\hat{\tau}_\mu(q_\perp, q_\parallel)]}{\sum \Gamma_{Ai;n\mu}(q_\perp, q_\parallel)} \tag{86}$$

$$[G^{(2)}_{Ai;n}(\infty; q_\perp, q_\parallel)]^{1/2} \propto \sum \Gamma_{Ai;n\mu}(q_\perp, q_\parallel) \tag{87}$$

for $\mu = S, T$. To proceed, theoretical expressions are needed for the functions $\hat{\tau}_\mu(q_\perp, q_\parallel)$ and $\Gamma_{Ai;n\mu}(q_\perp, q_\parallel)$ in terms of the relevant properties of the nematic mesophase. These are discussed below.

The scattering vector $\mathbf{q}$ is the difference between the wave vectors $\mathbf{k}_s$ and $\mathbf{k}_i$ of the scattered and incident beams, respectively. With a birefringent material, care must be taken to use the appropriate refractive index in calculating the variables $k_i = 2\pi n_i/\lambda_0$ and $k_s = 2\pi n_s/\lambda_0$, as well as i_μ and s_μ defined in the next section, where λ_0 is the wavelength of the incident light in vacuum, and n_i and n_s are the refractive indices for the propagation of the incident and scattered beams, respectively. Thus,

$$q^2 = (2\pi/\lambda_0)^2 \{(n_i - n_s)^2 + n_i n_s \sin^2(\vartheta/2)\} \tag{88}$$

where ϑ is the scattering angle. For a beam propagating with angle β between its direction and $\mathbf{n}$, and with polarization direction at angle φ from $\mathbf{n}$ (here, φ is 0 or $\pi/2$), the refractive index is given by

$$\frac{1}{n^2(\varphi, \beta)} = \left\{ \frac{\sin^2\varphi}{n_O^2} + (\cos^2\beta) \left[\frac{\sin^2\beta}{n_E^2} + \frac{\cos^2\beta}{n_O^2} \right] \right\} \tag{89}$$

where n_E and n_O are the extraordinary and ordinary refractive indices of the nematic medium, respectively.

4.1 Frank Elastic and Ericksen–Leslie Viscosity Coefficients

The calculation of the scattering from the nematic mesophase requires the use of an appropriate constitutive equation for the stress tensor. The static depolarized light scattering of a defect-free nematic mesophase reflects fluctuations in the curvature elasticity of the medium. The contribution $\sigma^{(e)}$ to the stress from the curvature elasticity involves a curvature free energy density W. For small distortions in the nematic field, W may be expressed in the form [134, 135, 139, 140]

$$2W = K_S(\operatorname{div} \mathbf{n})^2 + K_T(\mathbf{n} \cdot \operatorname{curl} \mathbf{n})^2 + K_B(\mathbf{n} \times \operatorname{curl} \mathbf{n})^2 \tag{90}$$

introducing Frank curvature elasticities K_S, K_T, and K_B for splay, twist and bend distortions, respectively. Molecular theories relating these to chain characteristics and the solute concentration are discussed below. The dynamic depolarized light scattering requires consideration of the viscous component $\sigma^{(v)}$ to the shear stress. The linear constitutive relation of Ericksen and Leslie is used for this purpose [134, 135, 139, 140]. Thus, in component form (in Cartesian coordinates) the viscous shear stress is given by

$$\sigma_{ji}^{(v)}(t) = \alpha_4 A_{ji} + (\alpha_5 + \lambda\alpha_2)(A_{jk}n_k n_i + A_{ki}n_k n_j)$$

$$+ [\alpha_1 - \lambda(\alpha_2 + \alpha_3)]n_j n_i n_k n_p A_{kp}$$

$$+ (\alpha_2/\gamma_1)n_j h_i + (\alpha_3/\gamma_2)h_j n_i \tag{91}$$

where the A_{kl} are the components of the symmetric rate of deformation tensor, $\gamma_1 = \alpha_3 - \alpha_2$, $\gamma_2 = \alpha_3 + \alpha_2$ and $\lambda = -\gamma_2/\gamma_1$. Although not needed here, the molecular field has components [134, 135, 139, 140]

$$h_i = \frac{\partial}{\partial x_j} \frac{\partial W}{\partial n_{i,j}} + \frac{\partial W}{\partial n_i} \tag{92}$$

in the absence of an external field. (The Parodi relation [134, 135, 139, 140] $\alpha_6 - \alpha_5 = \alpha_3 + \alpha_2$ is assumed.) The final two terms in $\sigma_{ji}^{(v)}(t)$ vanish in the absence of a molecular field (i.e., a uniform director field), making the stress tensor symmetric.

4.2 Thermodynamic Behavior

The static depolarized light scattering provides information on the fluctuations in the director field, which in turn are resisted by the molecular field. The fluctuations in the $\mathbf{n}(\mathbf{r})$ at position $\mathbf{r}$ may be separated into two components: $\delta n_1(\mathbf{r})$ in the plane formed by $\mathbf{q}$ and the undistorted director $\mathbf{n}$, and $\delta n_2(\mathbf{r})$ in the direction perpendicular to that plane. These are conveniently described in terms of an orthonormal coordinate system spanned by the unit basis vectors $\mathbf{e}_0$, $\mathbf{e}_S$, and $\mathbf{e}_T$, where $\mathbf{e}_0 = \mathbf{n}$, $\mathbf{e}_S = \mathbf{e}_T \times \mathbf{n}$, and $\mathbf{e}_T = (\mathbf{q} \times \mathbf{n})/|\mathbf{q} \times \mathbf{n}|$. Then, with $i_\mu = \mathbf{e}_\mu \cdot \mathbf{i}$ and $s_\mu = \mathbf{e}_\mu \cdot \mathbf{s}$, where $\mathbf{i}$ and $\mathbf{s}$ are unit vectors along the polarization of the incident and scattered beams, respectively, the fluctuations in the dielectric tensor may be expressed in the form [134, 135, 137]

$$\delta\varepsilon_{if} = \varepsilon_a \sum \delta n_\mu [i_\mu f_0 + s_\mu i_0] \tag{93}$$

where the sum is over the two components $\mu = S, T$ in the fluctuation, and $\varepsilon_a = \varepsilon_\parallel - \varepsilon_\perp$ is the dielectric anisotropoy. The strictly geometrical term in square brackets provides the means to select different modes of distortion in the depolarized light scattering. The scattered intensity, proportional to $\{G_{Ai;n}^{(2)}(\infty; q_\perp, q_\parallel)\}^{1/2}$, is calculated from the Fourier transform $\hat{W}(\mathbf{q})$ of $W(\mathbf{r})$, given by [134, 135, 137]

$$\hat{W}(\mathbf{q}) = \frac{1}{2} \sum \hat{K}_\mu(q_\perp, q_\parallel) |\delta n_\mu|^2 \tag{94}$$

where

$$\hat{K}_\mu(q_\perp, q_\parallel) = K_B q_\parallel^2 + K_\mu q_\perp^2 . \tag{95}$$

The equipartition theorem then gives $\langle |\delta n_\mu|^2 \rangle = kT/[K_B q_\parallel^2 + K_\mu q_\perp^2]$, and a light scattering intensity given by Eq. (87), with

$$\Gamma_{Ai;n\mu}(q_\perp, q_\parallel) = F_\mu(\mathbf{A}, \mathbf{i}, \mathbf{n})/\hat{K}_\mu(q_\perp, q_\parallel) \tag{96}$$

$$F_\mu(\mathbf{A}, \mathbf{i}, \mathbf{n}) = (i_\mu s_0 + s_\mu i_0)^2 \tag{97}$$

where $\mu = S, T$, respectively. Inspection of these relations shows that only the bend distortion is observed if $q_\perp = 0$, but that a mix of the twist and splay distortions will be observed if $q_\parallel = 0$, depending on the geometric functions $F_\mu(\mathbf{A}, \mathbf{i}, \mathbf{n})$.

Two geometrical arrangements are of particular interest, both described relative to a horizontal scattering plane, with vertically polarized incident light and horizontally polarized scattering, with **n** either vertical, Case 1, or horizontal, Case 2 [138, 141]. For Case 1, $q_{\parallel} = 0$ for any scattering angle, eliminating bend-modes from the scattering, but both F_S and F_T depend on the scattering angle. The scattering data give

$$[G^{(2)}_{Hv;v}(\infty; q_{\perp}, 0)]^{1/2} \propto \frac{\cos^2(\theta/2)}{K_S} + \frac{\sin^2(\theta/2)}{K_T} \qquad \text{Case 1} \qquad (98)$$

For Case 2, $F_S = 0$ for any scattering angle, thereby reducing Eq. (96) to a single term involving $F_T/\hat{K}_T(q_{\perp}, q_{\parallel})$, with elimination of splay modes. Two orientations of **n** in the scattering plane are of interest for Case 2: (a) an alignment with the normal to **n** at half the scattering angle θ makes $q_{\perp} = 0$, and (b) an alignment with **n** at half the scattering angle θ makes $q_{\parallel} = 0$ (both in the limit of small birefringence), so that the experimental data give, respectively,

$$[G^{(2)}_{Hv;h}(\infty; 0, q_{\parallel})]^{1/2} \propto \frac{\cos^2(\theta/2)}{K_B} \qquad \text{Case 2a} \qquad (99)$$

$$[G^{(2)}_{Hv;h}(\infty; q_{\perp}, 0)]^{1/2} \propto \frac{\sin^2(\theta/2)}{K_T}. \qquad \text{Case 2b} \qquad (100)$$

The latter arrangement ($q_{\parallel} = 0$) is most useful with homeotropically aligned samples, for which **n** is orthogonal to the sample plane, whereas the former ($q_{\perp} = 0$) is most useful with planar aligned samples, for which **n** is in the sample plane. It is convenient to eliminate the proportionality constant in these expressions by forming ratios of the measured photon count rates. Thus,

$$\left\{ \frac{G^{(2)}_{Hv;v}(\infty; q_{\perp}, 0)}{G^{(2)}_{Hv;h}(\infty; 0, q_{\parallel})} \right\}^{1/2} = \frac{K_B}{K_S} + \frac{K_B}{K_T} \tan^2(\theta/2) \qquad (101)$$

permitting assessment of the ratios of the Frank curvature elasticities [138, 141]. Although seldom used, the arrangement with horizontally polarized incident and scattered light, and horizontal **n** gives

$$[G^{(2)}_{Hh;h}(\infty; q_{\perp}, q_{\parallel})]^{1/2} \propto \frac{\sin^2(\theta)}{\hat{K}_S(q_{\perp}, q_{\parallel})} \qquad \text{Case 3} \qquad (102)$$

With solutions, concentrations fluctuations could contribute to $G^{(2)}_{Hh;h}(\infty; q_{\perp}, q_{\parallel})$, but this possibility may be evaluated by examination of $G^{(2)}_{Vv;v}(\infty; q_{\perp}, q_{\parallel})$, which should be nil if orientation fluctuations dominate the scattering.

Semi-empirical forms for the K_{μ} for rodlike chains are proportional to $(A_2Mc)^2$ and to coefficients $\hat{k}_{\mu}(\mu = S, T, B)$ that depend on the order parameter S [142–149]:

$$K_{\mu} = (kT/d_H)(A_2Mc)^2 \hat{k}_{\mu} \qquad (103)$$

Numerical computations [148, 149] of the K_{μ} for rodlike chains interacting through a hard-core repulsive potential give results that may be expressed in forms with $\hat{k}_\mu = (A_2Mc)k_\mu$, where [140, 141]:

$$k_S = (7/24\pi)(4S - 1) \tag{104}$$

$$k_T = k_S/3 \tag{105}$$

$$k_B = (12/7)k_S(3 - S)/(1 - S) \tag{106}$$

The order parameter S is a function of A_2Mc for equilibrium in the absence of external fields, with numerical computations [144, 148, 149] giving results that may be represented by the semi-empirical expression [141]

$$(1 - S)^{1/2}S^2/r_S \approx (3\pi)^{1/2}/2A_2Mc \tag{107}$$

$$r_S = 1 - (4/3)(1 - S)[1 - k_r(1 - S)] . \tag{108}$$

The form of these expressions is suggested by perturbation expansions [150, 151]. In a previously used version of these expressions [140], k_r was zero, but a slightly better numerical fit to the functions given by the perturbation expansions may be obtained with $k_r = 0.2315$. With the use of these relations, $1 - S^2 \propto (A_2Mc)^{-2}$ and power-law relations may be obtained. Thus, for large A_2Mc and the equilibrium S, $k_S \propto (A_2Mc)^0$, and $k_B \propto (A_2Mc)^2$. Plots of $\hat{k}_\mu$, each reduced by its asymptotic limit $\hat{k}_\mu(\infty)$ for large A_2Mc, are given in Fig. 17, showing the slow approach to the asymptotic limit.

In an alternative representation of this result, use is made of the relation $A_2Mc = \varphi L/d_G$ for rodlike chains interacting through a hard-core repulsive potential, with an effective interaction diameter equal to the geometric diameter d_G. Thus,

$$\hat{k}_\mu = b_\mu(\varphi L/d_G)^{c_\mu - 2} f_\mu(\varphi L/d_G) \tag{109}$$

where the $f_\mu(\cdot)$ increase slowly to unity with increasing $\varphi L/d_G$, and $b_\mu = (7/8\pi)$, $(7/24\pi)$, and $(4/3\pi^2)$, and $c_\mu = 1$, 1, and 3 for $\mu = S, T, B$, respectively [148, 149]. As may be seen, the asymptotic limits for the $\hat{k}_\mu$ are not reached until rather large

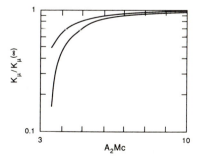

Fig. 17. Reduced plots of the Frank curvature elasticities K_μ vs $A_2Mc = \varphi L/d_G$ for rodlike chains interacting through a hard-core repulsive potential, where the asymptotic limit $K_\mu(\infty)$ for large $\varphi L/d_G$ is proportional to $\varphi L/d_G$ for $\mu = S, T$, and $(\varphi L/d_G)^3 \mu = B$ [148, 149]

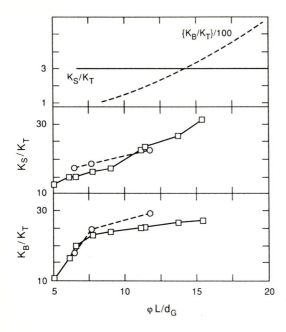

Fig. 18. Behavior of the Frank curvature elasticities. *Upper Panel*: The functions K_S/K_T and $(K_B/K_T)(\varphi L/d_G)^2$ vs $\log(\varphi L/d_G)$ for rodlike chains interacting through a hard-core repulsive potential [148, 149]. *Lower Panels*: Plots of K_B/K_T and K_S/K_T vs $A_2 Mc = \varphi L/d_G$ for nematic solutions of poly(benzyl glutamate) over a range of L for two concentrations ($\varphi = 0.16$ for the *squares*, and $\varphi = 0.20$ for the *circles* [148, 149]

$\varphi L/d_G$. Empirical fits to the numerical data give $f_\mu(x) \approx 1 - (a_\mu/x)^3$, where a_μ is 2.8 for $\mu = S, T$, and 3.3 $\mu = B$ – these are shown in comparison with the numerical data in Fig. 17. The calculated K_S/K_T and $(K_B/K_T)(\varphi L/d_G)^2$ are shown in Fig. 18 over a wide range of $\varphi L/d_G$ for rodlike chains interacting through a hard-core repulsive potential. Treatments with the wormlike chain model lead to similar results for the asymptotic limits, with $c_S = 1$ and a revised value of b_S, whereas for $\mu = T, B$, the chain length L is replaced by the persistence length ρ and $c_\mu = 1/3, 1$, respectively [148, 149, 152]. The change in K_B reflects the ability of the chain to bend, storing elastic energy intramolecularly instead of through the intermolecular process leading to K_B.

An extensive study of the curvature elasticity by static light scattering has been reported for nematic solutions of helical poly(benzyl glutamates) [148, 149], with less extensive studies on a rodlike poly(phenylene benzobisthiazole) [141, 153]. In the latter case, with solutions with $L_w/d_G \approx 100$ and $\varphi = 0.07$, the results gave $K_B/K_T \approx 8$ and $K_S/K_T \approx 14$. By comparison, with the predictions for rodlike chains interacting through a hard-core respulsive potential, $K_S/K_T = 3$, and $K_B/K_T \approx (32/7\pi)(\varphi L/d_G)^2$, or $K_B/K_T \approx 10$ for the polymer studied. The larger value reported for K_S/K_T may be related to osmotic contributions to K_S, similar to those discussed for wormlike chains [148].

The extensive data on the curvature elasticities determined on a series of poly(benzyl glutamates) at two concentrations are shown in Fig. 18, along with plots of K_S/K_T and K_B/K_T calculated for rodlike chains interacting through a hard-core repulsive potential [148, 149]. As may be seen, the behavior reported for the poly(benzyl glutamates) is not consistent with the predictions for rodlike chains. This behavior was attributed to departure from rigid rodlike chain conformation, modeled by wormlike chain statistics with persistence length ρ in the moderately concentrated solution. With ρ assumed to be independent of c and L, the relations discussed give $K_B/K_T \propto (\varphi\rho/d_G)^{2/3}$ and $K_S/K_T \propto (\varphi\rho/d_G)^{-1/3}(\varphi L_W/d_G)$ in the limit of large $\varphi L/d_G$. Consequently, the tendency of K_B/K_T to become independent of L/d_G for the sample at larger φ was attributed to a crossover to wormlike chain statistics for $L \approx \rho$ [148, 149]. It may be noted that $K_S/K_T > 3$, even for the shorter chains studied, presumably reflecting osmotic contributions to K_S [148].

4.3 Dynamic Behavior

The selection rules discussed above also apply to the dynamic scattering, permitting isolation of the contributions due to splay, twist and bend distortions. Thus, the $\Gamma_{Ai;n\mu}(q_\perp, q_\parallel)$ needed in Eq. (86) are given by Eq. (96), and the coherence times $\hat{\tau}_\mu(q_\perp, q_\parallel)$ are given by [134, 135, 137, 138]

$$\hat{\tau}_\mu(q_\perp, q_\parallel) = \hat{\eta}_\mu(q_\perp, q_\parallel)/\hat{K}_\mu(q_\perp, q_\parallel) \tag{110}$$

$$\hat{\eta}_\mu(q_\perp, q_\parallel) = \eta_B + (\eta_\mu - \eta_B)m_\mu(q_\perp/q_\parallel) \tag{111}$$

where $\hat{K}_\mu(q_\perp, q_\parallel)$ is given by Eq. (95), and the functions $m_\mu(q_\perp/q_\parallel)$ are discussed below (both increase from zero for small $q_\perp/q_\parallel$ to unity for large $q_\perp/q_\parallel$). The arrangements discussed above give [138, 141]

$$g_{Hv;v}^{(1)}(\tau; q_\perp, 0) = r\exp[-\tau/\tau_S] + (1-r)\exp[-\tau/\tau_T] \quad \text{Case 1} \tag{112}$$

$$g_{Hv;v}^{(1)}(\tau; q_\perp, q_\parallel) = \exp[-\tau/\hat{\tau}_T(q_\perp, q_\parallel)] \quad \text{Case 2} \tag{113}$$

$$g_{Hv;v}^{(1)}(\tau; 0, q_\parallel) = \exp[-\tau/\tau_B] \quad \text{Case 2a} \tag{114}$$

$$g_{Hv;v}^{(1)}(\tau; q_\perp, 0) = \exp[-\tau/\tau_T] \quad \text{Case 2b} \tag{115}$$

$$g_{Hv;v}^{(1)}(\tau; q_\perp, q_\parallel) = \exp[-\tau/\hat{\tau}_S(q_\perp, q_\parallel)] \quad \text{Case 3} \tag{116}$$

where $\tau_\mu = \eta_\mu/K_\mu q^2$, and $r = \Gamma_S/\sum\Gamma_\mu$ is given by

$$1/r = 1 + (K_S/K_T)\tan^2(\theta/2) . \tag{117}$$

Consequently, a two-exponential decay is expected in Case 1, and single exponential decays are expected in the remaining cases. If τ_S and τ_B differ sufficiently (as is the case), the two contributions to $g_{Hv;v}^{(1)}(\tau; q_\perp, 0)$ may be readily separated in Case 1. The functions $m_S(q_\perp/q_\parallel)$ and $m_T(q_\perp/q_\parallel)$ in $\hat{\tau}_S(q_\perp, q_\parallel)]$ and $\hat{\tau}_T(q_\perp, q_\parallel)$, respectively, are discussed below. Both $m_S(q_\perp/q_\parallel)$ and $m_T(q_\perp/q_\parallel)$ increase from zero for small $q_\perp/q_\parallel$ to unity for large $q_\perp/q_\parallel$. Consequently, the overall behavior does not scale simply with q^2. For these two cases,

$$q_\perp/q_\parallel = \frac{(n_\theta/n_E)(1 + g) - \cos\theta}{\sin\theta} \tag{118}$$

where n_θ is equal to $n(\varphi, \beta)$ for the appropriate geometry, and g is zero for Case 2, and equal to $\Delta n/n_o$ for Case 3.

The functions $m_\mu(q_\perp/q_\parallel)$ may be expressed as [137, 138, 141]

$$m_S(Q) =$$
$$\frac{(\eta_S - \eta_B)^{-1}\{(\eta_T - \eta_B)\eta_m + 2\eta_b[(\eta_T - \eta_S)(\eta_T - \eta_B)]\}^{1/2}Q^2 + \eta_b Q^4}{\eta_c + \eta_m Q^2 + \eta_b Q^4} \tag{119}$$

$$m_T(Q) = \eta_a Q^2/(\eta_c + \eta_a Q^2) \tag{120}$$

where $\eta_a, \eta_b,$ and η_c are the Miesowicz viscosities $(2\eta_a = -\alpha_4; 2\eta_b = \alpha_3 + \alpha_4 + \alpha_6;$ and $2\eta_c = -\alpha_2 + \alpha_4 + \alpha_5)$ [134, 135], and

$$\eta_m = \eta_c + \eta_b + \alpha_1 . \tag{121}$$

The viscometric functions may be obtained for rodlike chains interacting through a hard-core repulsive potential using expressions for the α_i given by Kuzuu and Doi [150, 151] as modified by Lee [154]. It is convenient to express α_i in the form [140]

$$\alpha_i = \eta^{ISO}\hat{A}_i(A_2Mc) \tag{122}$$

where η^{ISO} is the hypothetical viscosity for an isotropic solution of infinitely thin rodlike chains at the same φ and L [117, 140]:

$$\eta^{ISO} = \eta_S K_\eta N_A^2 M[\eta](cL/M_L)^3 \tag{123}$$

Similarly, the distortion viscosities may be represented in the form

$$\eta_\mu = \eta^{ISO}\hat{H}_\mu(A_2Mc) \tag{124}$$

for $\mu = $ S, T, B, where

$$\eta_T = \alpha_3 - \alpha_2 \tag{125}$$

$$\eta_S = \eta_T - \alpha_3^2/\eta_b \tag{126}$$

$$\eta_B = \eta_T - \alpha_2^2/\eta_c, \tag{127}$$

It is convenient to express the $\hat{A}_i$ as implicit functions of A_2Mc in terms of the order parameter S, in the form $\hat{A}_i = (1 - S^2)A_i(S)$, with [150, 151, 154]

$$A_1 = -\hat{p}^2[(2 - \varepsilon)/2]r_S S^2 \tag{128}$$

$$A_2 = -\hat{p}^2(1/2)(1 + \lambda^{-1})S \tag{129}$$

$$A_3 = \hat{p}^2(1/2)(1 - \lambda^{-1})S \tag{130}$$

$$A_4 = \hat{p}^2(1/105)[7(3 + \varepsilon) - 5(3 + 2\varepsilon)S - 3(2 - \varepsilon)r_S S^2] \tag{131}$$

$$A_5 = \hat{p}^2(1/7)[(5 + \varepsilon)S + (2 - \varepsilon)r_S S^2] \tag{132}$$

$$A_6 = -\hat{p}^2(1/7)(2 - \varepsilon)[S - r_S S^2] . \tag{133}$$

Here, ε is between zero and unity, and perturbation expansions are available giving S, λ, $\hat{p}$ and r_S as series in powers of $A_2 Mc$. The parameter ε, which is zero in the treatment of Kuzuu and Doi [150, 151], was added to the stress tensor by Lee in an *ad hoc* treatment [154] by analogy with work on related models. Semi-empirical representations of numerical calculations of S and r_S are given above, and the functions λ and p may be represented by the semi-empirical expressions:

$$\hat{p} = (8/3\pi)^{1/2} r_S^\delta \tag{134}$$

$$\lambda = r_S[1 - (1/3)(1 - S)r_S^\delta] . \tag{135}$$

In a previously used version of these relations [141], δ was zero, but a slightly better numerical fit to the functions given by the perturbation expansions may be obtained with $\delta = 0.075$. In these forms, the expressions emphasize the relation of the various reduced fucntions on the order parameter, which serves as a measure of the orientation in the system. For example, the expressions with k_r and δ equal to zero lead to simple analytical approximations to the A_i in terms of S [141]. For the model used, $1 - S^2 \propto (A_2 Mc)^{-2}$ and power-law relations may be obtained for the A_i, such that for large $A_2 Mc$, A_i scales as $(A_2 Mc)^0$ for i = 1, 2 and 5, and as $(A_2 Mc)^{-2}$ for i = 3, 4 and 6. Inspection of the relations for η_S and η_B shows that this behavior will tend to make $\eta_S \approx \eta_T$ with increasing $A_2 Mc$ owing to the decrease in α_3, whereas η_B may be appreciably smaller than η_T. Use of these A_i give power-law relations for large $A_2 Mc$ for $H_\mu = (1 - S^2)^{-1} \hat{H}_\mu$, such that H_μ scales as $(A_2 Mc)^0$ for $\mu = S$ and T, and as $(A_2 Mc)^{-4}$ for $\mu = B$ if $\varepsilon = 0$, or $(A_2 Mc)^{-2}$ if $\varepsilon > 0$ [150, 151, 154], exhibiting the anticipated behavior with $\eta_S \approx \eta_T$ and $\eta_S \ll \eta_T$ for large $A_2 Mc$.

The functions $m_S(q_\perp/q_\parallel)$ and $m_T(q_\perp/q_\parallel)$ calculated with the expressions for a rodlike chains interacting through a hard-core potential are given in Fig. 19, and ratios of the light scattering time constants for the same model are shown in Fig. 20, both calculated with $\varepsilon = 0$. Viscosities calculated for $\varepsilon = 0$ and 1 using the same model are given in Fig. 21. It may be seen that η_S/η_T is expected to approach unity with increasing $\varphi L/d_G$ for the rodlike model, whereas η_B/η_T is expected to tend to zero in the same limit. The latter behavior, coupled with that for K_B/K_T gives a ratio τ_B/τ_T of the light scattering time constants that decreases markedly with increasing $\varphi L/d_G$, as shown in Fig. 20. By contrast, τ_S/τ_T is not expected to vary much with $\varphi L/d_G$, and to be about equal to K_T/K_S.

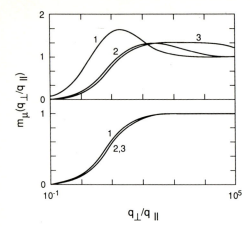

Fig. 19. The light scattering functions $m_\mu(q_\perp/q_\parallel)$ calculated for rodlike molecules interacting through a hard-core repulsive potential ($\varepsilon = 0$), for various values of $A_2Mc = (4/3)10^a$, with a equal to 0.5, 1, and 1.5, for curves 1, 2, and 3, respectively: *Upper*, $\mu = S$, and *lower*, $\mu = T$. After Fig. 2 in Ref. [141]

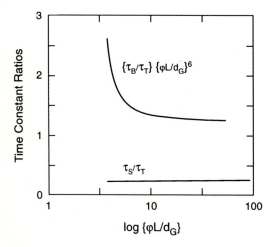

Fig. 20. Ratios of light scattering time constants for distortions of the director field for nematic solutions calculated for rodlike molecules interacting through a hard-core repulsive potential

An extensive study fo the curvature elasticity by dynamic light scattering has been reported for nematic solutions helical poly(benzyl glutamates) [148, 149], with less extensive studies on the rodlike poly(phenylene benzobisthiazole) [141, 153]. In the latter case, with solutions with $L_W/d_G \approx 100$ and φ from 0.035 to 0.07, the results gave $\tau_B/\tau_T \approx 3.5 \times 10^{-4}$. Although this ratio is in the range expected, neither time constant exhibited a strong dependence on φ, in contrast with the behavior expected, e.g., see Fig. 20. The extensive data on the distortion viscosities determined on a series of poly(benzyl glutamates) at two concentrations are shown in Fig. 18, along with plots of η_S/η_T and η_B/η_T calculated for rodlike chains interacting through a hard-core repulsive potential [148, 149]. As may be seen, η_S/η_T was found to be nearly independent of L, and about equal to unity, as expected, and η_B/η_T was small and decreased with increasing L.

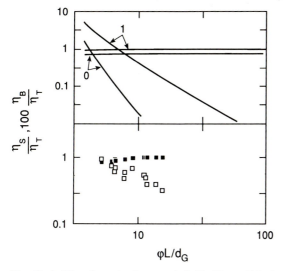

Fig. 21a,b. Viscosity ratios for nematic fluids. *Upper*: Calculated ratios for rodlike chains interacting through a hard-core repulsive potential for two choices (0 or 1) of ε as indicated. In each case, the nearly horizontal curves give η_S/η_T, and the other curves give $100 \times \eta_B/\eta_T$. *Lower*: Data on nematic solutions of poly(benzyl glutamate) [148, 149]; the *filled* and *unfilled symbols* give η_S/η_T and $100 \times \eta_B/\eta_T$, respectively

Although qualitatively similar, the behavior reported for the poly(benzyl gluta-mates) is not quantitatively consistent with the predictions for rodlike chains interacting through a hard-core potential. For example, η_B/η_T did not decrease as rapidly with increasing L as expected, and η_B tended to become independent of L with increasing L, at about the same L for which K_B was observed to become independent of L. As with the behavior discussed above for the Frank curvature elasticities, this behavior was attributed to departure from rigid rodlike chain conformation, modeled by wormlike chain statistics with persistence length ρ in the moderately concentrated solution [148].

Acknowledgment. It is a pleasure to acknowledge partial support by a grant from the National Science Foundation, Division of Materials Research, Polymers Program for the preparation of this manuscript. Discussions with Professor G.D. Patterson, and permission to make use of unpublished data of Dr. Seong Kim in his laboratory are also acknowledged.

5 References

1. Flory PJ (1953) Principles of polymer chemistry. Cornell University Press, Ithaca
2. Huglin MB (ed) (1972) Light scattering from polymer solutions. Academic, New York
3. Casassa EF, Berry GC (1975) In: Slade Jr PE (ed) Techniques and methods of polymer evaluation. Part 1. Polymer molecular weights. Marcel Dekker, New York, Chapt 5

4. Berne BJ, Pecora R (1976) Dynamic light scattering. Wiley-Interscience, New York
5. Burchard W (1983) Adv Polym Sci 48: 1
6. Berry GC (1987). In: Mark H, Overberger CG et al. (eds) Encyclopedia of polymer science and engineering. Wiley, New York, 8: 721
7. Russo PS (1993) In: Brown W (ed) Dynamic light scattering, the method and some applications. Oxford, New York
8. Pike ER, Jakeman E (1974) Adv Quantum Electron 2: 1
9. Schmitz KS (1990) An introduction to dynamic light scattering by macromolecules. Academic, San Diego
10. Casassa EF, Berry GC (1988) In: Allen G (ed) Comprehensive polymer science, vol 2. Pergamon, New York
11. Oono Y (1985) Adv Chem Phys 61: 301
12. Yamakawa H (1971) Modern theory of polymer solutions. Harper & Row, New York, Chs 3 and 6
13. Freed KF (1987) Renormalization group theory of macromolecules, Wiley, New York, Ch 11
14. deGennes PG (1979) Scaling concepts in polymer physics. Cornell University Press, Ithaca, New York, Chs 3, 7 and 8
15. Wei-Berk C, Berry GC (1990) J Polym Sci (B) Polym Phys 28: 1873
16. Furukawa R, Berry GC (1985) Pure Appl Chem 57: 913
17. Berry GC (1988) J Polym Sci (B) Polym Phys 26: 1137
18. Chen SJ, Berry GC (1990) Polymer 31: 793
19. Zimm BH (1948) J Chem Phys 16: 1099
20. Wei-Berk C, Berry GC (1990) J Appl Polym Sci 45: 261
21. Stockmayer WH, Casassa EF (1952) J Chem Phys 20: 1560
22. Berry GC, Fox TG (1968) Adv Polym Sci 5: 261
23. Fujita H (1990) Polymer solutions. Elsevier, Amsterdam
24. Edwards SF (1966) Proc Phys Soc 88: 265
25. Benoit H, Benmouna M (1984) Polymer 25: 1059
26. Benoit H (1963). In: Kerker M (ed) Electromagnetic Scattering; ICES I. Pergamon, Elmworth, NY, p 285
27. Benoit H, Picot C (1966) Pure Appl Chem 12: 545
28. Patterson GD, Flory PJ (1972) J Chem Soc Faraday Trans 68: 111
29. Daoud M, Cotton JP, Farnoux B, Jannink G, Sarma G, Benoit H, Duplessix R, Picot C, deGennes PG (1975) Macromolecules 8: 804
30. Candau F, Strazielle C, Benoit H (1976) Eur Polym J 12: 95
31. Nose T, Chu B (1979) Macromolecules 12: 590, 1123
32. Chu B, Nose T (1980) Macromolecules 13: 122
33. Roots J, Nyström B (1980) Macromolecules 13: 1595
34. Noda I, Kato N, Kitano T, Nagasawa M (1981) Macromolecules 14: 668
35. Amerzadeh T, McDonnell ME (1982) Macromolecules 15: 927
36. Wiltzius P, Haller HR, Cannell DS, Schaefer DW (1983) Phys Rev Lett 51: 1183
37. Schaefer DW (1984) Polymer 25: 387
38. Štěpánek P, Perzynski R, Delsanti M, Adam M (1984) Macromolecules 17: 2340
39. Lapp A, Picot C, Strazielle C (1985) J Physique Lett 46: L-1031
40. Hager BL, Berry GC, Tsai HH (1987) J Polym Sci (B) Polym Phys 25: 387
41. Ricker M, Schmidt M (1991) Makromol Chem 192: 693
42. Muthukumar M (1986) J Chem Phys 85: 4722
43. Oyama T, Oono Y (1976) J Phys Soc Japan 41: 228
44. Oyama T, Oono Y (1977) J Phys Soc Japan 42: 1348
45. Flory PJ (1979) Statistical Mechanics of chain molecules. Wiley-Interscience, New York
46. Kim SH, Ramsay DJ, Patterson GD, Selser JC (1990) J Polym Sci (B) Polym Phys 28: 2023
47. Patterson GD, Berry GC (1988) In: Nagasawa M (ed) Molecular conformation and dynamics of macromolecules in condensed systems. Elsevier, New York, p 73
48. Kim SH (1988) Static and dynamic light scattering of poly(α-methylstyrene) in solution, PhD Dissertation. Carnegie Mellon University, Pittsburgh PA
49. Patterson GD, Francis R (1991) private communication
50. Benoit H, Goldstein M (1953) J Chem Phys 21: 947
51. Flory PJ, Bueche AM (1958) J Polym Sci 27: 219
52. Root LJ, Stillinger FH, Washington GE (1988) J Chem Phys 88: 7791
53. Hager BL, Berry GC, Tsai HH (1978) Polym Prep Am Chem Soc Div Polym Chem 19(2): 719

54. Guenet JM, Nigel FFW, Ellsmore PA (1983) Polymer Comm 24: 230
55. Gan JYS, Francois J, Guenet JM (1986) Macromolecules 19: 173
56. Wendt E, Springer J (1988) Polymer 29: 1301
57. Medjahdi G, Sarazin D, François J (1990) Eur Polym J 26: 823
58. Medjahdi G, Sarazin D, François J (1991) Macromolecules 24: 4138
59. Perico A, Piaggio P, Cunnieberti C (1975) J Chem Phys 62: 2690
60. Akcasu AZ, Benmouna M, Han CC (1980) Polymer 21: 866
61. deGennes PG (1967) Physics 3: 37
62. Dubois-Violette E, deGennes PG (1967) Physics 3: 181
63. Silbey R, Deutch JM (1972) J Chem Phys 57: 5010
64. Akcasu Z, Gurol H (1976) J Polym Sci, Polym Phys Ed 14: 1
65. Freire JJ, de la Torre JG (1980) Chem Phys 49: 139
66. Tsunashima Y, Nemoto N, Kurata M (1983) Macromolecules 16: 1184
67. Nemoto N, Makita Y, Tsunashima Y, Kurata M (1984) Macromolecules 17: 425
68. Wiltzius P, Cannell DS (1986) Phys Rev Lett 56: 61
69. Tsunashima Y, Hirata M, Nemoto N, Kurata M (1987) Macromolecules 20: 1992
70. Tsunashima Y, Hirata M, Nemoto N, Kajiwara K, Kurata M (1987) Macromolecules 20: 2862
71. Martin JE (1986) Macromolecules 19: 1278
72. Bhatt M, Jamieson AM, Petschek RG (1989) Macromolecules 22: 1374
73. Ellis AR, Schaller JK, McKiernan ML, Selser JC (1990) J Chem Phys 92: 5731
74. Oono Y, Freed KF (1981) J Chem Phys 75: 1009
75. Ohta T, Oono Y (1982) Phys Lett A 89: 460
76. Shiwa Y (1987) Phys Rev Lett 58: 2102
77. Shiwa Y (1991) J Phys A: Math Gen 248: L579
78. Berry GC (1992) Chemtracts-Macromol Chem 3: 237
79. Fixman M (1990) J Chem Phys 92: 6858
80. Patterson GDC, Berry GC (1991) J Noncrystalline Solids 131–133: 799
81. Yamakawa H (1962) J Chem Phys 36: 2295
82. Imai S (1969) J Chem Phys 50: 2116
83. Pyun CW, Fixman M (1964) J Chem Phys 41: 937
84. Mulderije JJH (1980) Macromolecules 13: 1526
85. Muthukumar M (1984) Macromolecules 17: 971 (and references therein)
86. King TA, Knox A, Lee WI, McAdam JDG (1973) Polymer 14: 151
87. Mandema W, Zeldenust H (1977) Polymer 18: 835
88. Han CC, McCrackin FL (1979) Polymer 20: 427
89. Schmidt M, Burchard W (1981) Macromolecules 14: 210
90. Pritchard MJ, Caroline D (1981) Macromolecules 14: 424
91. Novotny VJ (1983) J Chem Phys 78: 183
92. Munch JP, Hild G, Candau S (1983) Macromolecules 16: 71
93. Tsunashima Y, Nemoto N (1983) Macromolecules 16: 1941
94. Noda I, Higo Y, Ueno N, Fujimoto T (1984) Macromolecules 7: 1055
95. Brown W, Johnsen RM (1985) Macromolecules 18: 379
96. Ferry JD (1980) Viscoelastic properties of polymers, 3rd Edn. Wiley, New York
97. Chu B, Nose T (1979) Macromolecules 12: 599
98. Amis EJ, Janmey PA, Ferry JD, Yu H (1983) 16: 441
99. Adam M, Delsanti M (1985) Macromolecules 18: 1760
100. Štěpánek P, Koňák Č, Jakeš J (1985) Polym Bull 16: 67
101. Brown W (1986) Macromolecules 19: 386, 3006
102. Wenzel M, Burchard W, Schätzel K (1986) Polymer 27: 195
103. Brown W, Johnsen RM, Štěpánek P, Jakeš J (1988) Macromolecules 21: 2859
104. Nicolai T, Brown W (1990) Macromolecules 23: 3150
105. Nicolai T, Brown W, Hvidt S, Heller K (1990) Macromolecules 23: 8088
106. Nicolai T, Brown W, Johnsen RM, Štěpánek P (1990) Macromolecules 231: 2859
107. Fang L, Brown W (1990) Macromolecules 23: 3284
108. Rodriquez AL, Rey A, Freire JJ (1992) Macromolecules 25: 3266
109. Brochard F (1983) J Phys (Paris) 44: 39
110. Tanaka T, Hocker LO, Benedek GB (1973) J Chem Phys 59: 5151
111. Graessley WW (1975) Adv Polym Sci 16: 1
112. Graessley WW (1982) Adv Polym Sci 47: 68
113. Doi M, Edwards SF (1986) The theory of polymer dynamics, Clarendon, Oxford

114. Berry GC, Plazek DJ (1986) In: Uhlmann DR, Kreidl NJ (eds) Glass: science and technology, Vol 3. Academic, New York, Ch 6
115. Patterson GD (1983) Adv Polym Sci 48: 125
116. Park JO, Berry GC (1989) Macromolecules 22: 3022
117. Berry GC (1989) In: Adams WW, Eby R, McLemore (eds) The materials science and engineering of rigid rod polymers. Mat Res Soc Pittsburgh PA, p 181
118. Onsager L (1949) Ann NY Acad Sci 59: 519
119. Flory PJ (1984) Adv Polym Sci 59: 1
120. DeLong LM, Russo PS (1991) Macromolecules 24: 6139
121. Shimada T, Doi M, Okano K (1988) J Chem Phys 88: 2815, 7181
122. Doi M, Shimada T, Okano K (1988) J Chem Phys 88: 4070
123. Maeda T (1989) Macromolecules 22: 1881
124. Benoit H, Stockmayer W (1956) J Phys Radiumn 17: 21; english translation in: Light Scattering from Dilute Polymer Solutions, McIntyre D, Gornick F (eds) Gordon and Breach, New York, Ch 9
125. Berry GC (1978) J Polym Sci Symp Ed 65: 143
126. Cotts PM, Berry GC (1983) J Polym Sci: Polym Phys Ed 21: 1255
127. Russo PS, Karasz FE, Langley KH (1984) J Chem Phys 80: 5312
128. Zero KM, Pecora R (1982) Macromolecules 15: 87
129. Keep GT, Pecora R (1988) Macromolecules 21: 817
130. Maeda T (1990) Macromolecules 23: 1464
131. Schmidt M, Stockmayer WH (1984) Macromolecules 17: 509
132. Teraoka I, Ookubo N, Hayakawa R (1985) Phys Rev Lett 55: 2712
133. Bitsania I, Davis HT, Tirrell M (1990) Macromolecules 23: 1157
134. de Gennes PG (1974) The physics of liquid crystals. Clarendon, Oxford
135. Chandrasekhar S (1977) Liquid Crystals. Cambridge University Press, Cambridge
136. Yoon DR, Flory PJ (1989) In: Adams WW, Eby R, McLemore (eds) The materials science and engineering of rigid rod polymers. Mat Res Soc Pittsburgh PA, p 11
137. van der Meulen JP, Zÿlstra RJJ (1984) J Physique 45: 1627
138. Taratuta V, Hurd AJ, Meyer RB (1985) Phys Rev Lett 55: 246
139. Leslie FM (1979) Adv Liq Cryst 4: 1
140. Berry GC (1988) Mol Cryst Liq Cryst 165: 333
141. Desvignes N, Suresh K, Berry GC (1993) J Appl Polym Sci, Symp Ed (in press)
142. Straley JP (1973) Phys Rev A 8: 2181
143. Poniewierski A, Stecki J (1979) Mol Phys 38: 1931
144. Lee S-D (1987) J Chem Phys 87: 4972
145. Lee S-D, Meyer RB (1986) J Chem Phys 84: 3443
146. Odijk T (1986) Liq Cryst 1: 553
147. Vroege GJ, Odijk T (1987) J Chem Phys 87: 4223
148. Lee S-D, Meyer RB (1990) Liq Cryst 7: 15
149. Lee S-D, Meyer RB (1991) In: Ciferri A (ed) Liquid crystallinity in polymers. VCH Publications, New York, p 343
150. Kuzuu N, Doi M (1983) J Phys Soc Jpn 52: 3486; (1984) 53: 1031
151. Kuzuu N, Doi M (1983) J Phys Soc Jpn 52: 3486; (1984) 53: 1031
152. deGennes P (1977) Mol Cryst Liq Cryst 34: 177
153. Se K, Berry GC (1987) Mol Cryst Liq Cryst 153: 133
154. Lee S-D (1988) J Chem Phys 88: 5196

Editor: Prof. Glöckner
Received May 1993

Author Index Volumes 101-114

Author Index Vols. 1-100 see Vol. 100

Améduri, B. and *Boutevin, B.*: Synthesis and Properties of Fluorinated Telechelic Monodispersed Compounds. Vol. 102, pp. 133-170.

Amselem, S. see Domb, A. J.: Vol. 107, pp. 93-142.

Arshady, R.: Polymer Synthesis via Activated Esters: A New Dimension of Creativity in Macromolecular Chemistry. Vol. 111, pp. 1-42.

Baltá-Calleja, F. J., González Arche, A., Ezquerra, T. A., Santa Cruz, C., Batallón, F., Frick, B. and *López Cabarcos, E.*: Structure and Properties of Ferroelectric Copolymers of Poly(vinylidene) Fluoride. Vol. 108, pp. 1-48.

Barshtein, G. R. and *Sabsai, O. Y.*: Compositions with Mineralorganic Fillers. Vol. 101, pp.1-28.

Batallán, F. see Baltá-Calleja, F. J.: Vol. 108, pp. 1-48.

Barton, J. see Hunkeler, D.: Vol. 112, pp. 115-134.

Berry, G.C.: Static and Dynamic Light Scattering on Moderately Concentraded Solutions: Isotropic Solutions of Flexible and Rodlike Chains and Nematic Solutions of Rodlike Chains. Vol. 114, pp. 233-290.

Bershtein, V.A. and *Ryzhov, V.A.*: Far Infrared Spectroscopy of Polymers. Vol. 114, pp. 43-122.

Binder, K.: Phase Transitions in Polymer Blends and Block Copolymer Melts: Some Recent Developments. Vol. 112, pp. 115-134.

Boutevin, B. and *Robin, J. J.*: Synthesis and Properties of Fluorinated Diols. Vol. 102. pp. 105-132.

Boutevin, B. see Amédouri, B.: Vol. 102, pp. 133-170.

Burban, J. H. see Cussler, E. L.: Vol. 110, pp. 67-80.

Candau, F. see Hunkeler, D.: Vol. 112, pp. 115-134.

Chow, T. S.: Glassy State Relaxation and Deformation in Polymers. Vol. 103, pp. 149-190.

Cussler, E. L., Wang, K. L. and *Burban, J. H.*: Hydrogels as Separation Agents. Vol. 110, pp. 67-80.

Dimonie, M. V. see Hunkeler, D.: Vol. 112, pp. 115-134.

Doelker, E.: Cellulose Derivatives. Vol. 107, pp. 199-266.

Domb, A. J., Amselem, S., Shah, J. and *Maniar, M.*: Polyanhydrides: Synthesis and Characterization. Vol.107, pp. 93-142.

Dubrovskii, S. A. see Kazanskii, K. S.: Vol. 104, pp. 97-134.

Ezquerra, T. A. see Baltá-Calleja, F. J.: Vol. 108, pp. 1-48.

Subject Index

Springer-Verlag
and the Environment

We at Springer-Verlag firmly believe that an international science publisher has a special obligation to the environment, and our corporate policies consistently reflect this conviction.

We also expect our business partners – paper mills, printers, packaging manufacturers, etc. – to commit themselves to using environmentally friendly materials and production processes.

The paper in this book is made from low- or no-chlorine pulp and is acid free, in conformance with international standards for paper permanency.